# The Urban Book Series

**Aims and Scope**

The Urban Book Series is a resource for urban studies and geography research worldwide. It provides a unique and innovative resource for the latest developments in the field, nurturing a comprehensive and encompassing publication venue for urban studies, urban geography, planning and regional development.

The series publishes peer-reviewed volumes related to urbanization, sustainability, urban environments, sustainable urbanism, governance, globalization, urban and sustainable development, spatial and area studies, urban management, urban infrastructure, urban dynamics, green cities and urban landscapes. It also invites research which documents urbanization processes and urban dynamics on a national, regional and local level, welcoming case studies, as well as comparative and applied research.

The series will appeal to urbanists, geographers, planners, engineers, architects, policy makers, and to all of those interested in a wide-ranging overview of contemporary urban studies and innovations in the field. It accepts monographs, edited volumes and textbooks.

More information about this series at http://www.springer.com/series/14773

Simone Sandholz

# Urban Centres in Asia and Latin America

## Heritage and Identities in Changing Urban Landscapes

Springer

Simone Sandholz
Institute of Geography
University of Innsbruck
Innsbruck
Austria

ISSN 2365-757X ISSN 2365-7588 (electronic)
The Urban Book Series
ISBN 978-3-319-43734-7 ISBN 978-3-319-43735-4 (eBook)
DOI 10.1007/978-3-319-43735-4

Library of Congress Control Number: 2016947388



Printed on acid-free paper

This Springer imprint is published by Springer Nature
The registered company is Springer International Publishing AG Switzerland

# Foreword

Inner cities are of crucial importance for all inhabitants of cities. Collective memory continues to live in their built environment and public spaces. In inner cities, the histories and fates of cities inscribe themselves in particularly powerful ways and strongly influence the identification patterns of inhabitants with their city. It seems that this fact has been forgotten in the Global North and especially in the Global South. The desire to modernize the urban at any cost, as well as the consequences of an ever-increasing displacement of urban growth and urban functions towards suburban areas has had profound repercussions for inner cities. In countries of the Global North and particularly in Europe—at least since the 1970s—the far-reaching consequences of insensitive managing of inner cities have been acknowledged. Protection of historical monuments, restructuring measures, and cautious city renewals have been the answer. With some years of delay this awareness of the urbanistic, socio-economic and sociocultural importance of inner cities can now also be observed in the countries of the Global South. In the face of undamped urban growth, mostly driven by economic interests leading to ever more social and socio-spatial fragmentation with profound implications for inner cities, international organizations are particularly concerned with inner city maintenance and renewal. The question of how to value inner cities from an intercultural perspective arises. Are concepts like protection of historical monuments, city as cultural heritage, but also inner city maintenance, restructuring, and renewal universally comparable, or should we rather conceptualize them as 'transfers' from a 'hegemonic' Eurocentric vision? How do politicians and urban planners deal with inner cities in the face of rapid urban growth; in the face of the dominance of entrepreneurial value realization; in the face of global role models such as waterfront projects, etc.; but also in the face of ever more fragmented city structures?

Those are some backgrounds and starting points for Simone Sandholz's book, a dissertation in urban geography. As an architect by training with a special expertise in protection of historical monuments and resource management she wrote her dissertation with strong links to urban sociology and urban policy. As such the book is a truly interdisciplinary project and at the same time an intercultural one, as the author has undertaken an intercultural and comparative research, building on her long-standing expertise in different regions of the Global South. Three case studies stand at the forefront of the analysis: Recife in Brazil, Kathmandu in Nepal and Yogyakarta in Indonesia; three cities that at first glance seem to have little in common. Nevertheless, all three have put strong emphasis on their respective inner cities in the last years; particularly in what concerns their problems, potentials and future perspectives. Simone Sandholz provides a multifacetted picture of the changes in the inner cities as well as of the respective policies and planning measures for inner city renewal. Against this backdrop she pays special attention to the perceptions and valuations of stakeholders, particularly by future urban planners who will decide the fate of the inner cities. We are in the century of cities. Inner cities merit our special attention. Simone Sandholz's book does not only represent an important contribution to the study of three emblematic cases of inner city development in their respective individuality and in their comparative dimension. What is more, she provides a detailed account of how these inner city developments are embedded in global discourses and policy approaches.

Prof. Martin Coy
Institute of Geography
University of Innsbruck
Austria

# Preface

Many cities around the globe, (still) comprise of historic fabric, particularly in the urban core areas. Very often intangible values, such as beliefs, events, habits or rituals are intrinsically linked with it.

At the same time, ongoing urbanization and changing preferences or lifestyles potentially alter the urban outline. Particularly economic development has the potential to fundamentally change the urban appearance of a quarter or even the city as a whole. While an economic decline very often results in decay of the historic fabric, an economic upturn might have even bigger impacts, like verticalization or gentrification. In parallel, changing urban paradigms and wishes for 'modernity' are triggering supposedly fashionable urban appearance, leading to an increasing grade of uniformity. In this way, historic fabric often is sacrificed for the sake of presumed modernity, without consideration of the intangible added values. Particularly cities in developing countries are struggling somewhere within a transformation between a 'Third-World City' on the way to become a 'global' one, leaving the 'old' and presumably outdated behind.

Processes to preserve urban heritage—tangible and increasingly intangible ones—do exist. However, often legal policies are insufficient or outdated in their approach, and their execution may be deficient. Furthermore, global paradigms on heritage and preservation mostly emerged from a Euro-American background, and are not always adequate for different cultural or regional backgrounds. While the global reference frame, with actors like UNESCO or different regional networks, is slowly changing and moving towards more holistic concepts of heritage considering intangible values as well as the global variety of cultural backgrounds and traditions, the national and urban are often lacking behind—particularly intangible values are hardly addressed. This is again especially true for the Global South where traditional understanding of conservation as a process, and not something static, and related practices may even contradict international paradigms and even relevant legislation. In addition, these cities are confronted with comparably higher challenges: namely urbanization, lack of available budgets, high levels of informality and insufficient means of protection. As a consequence, historic fabric is destroyed; intangible assets like traditional customs and beliefs are not cherished and may vanish easily.

Therefore, this research work aims at gaining a better understanding of the challenges that cities in the Global South are facing, regarding the preservation of their tangible and intangible heritage. It argues that urban heritage has a value going beyond the mere object value; constituting a crucial source of identity for urban inhabitants. The same is true for the urban intangible values and practices which often are associated with places or buildings. The empirical research is based on case studies of Kathmandu in Nepal, Yogyakarta in Indonesia and Recife in Brazil; three cities that still comprise of core areas with a high percentage of historic fabric and distinct cultural expressions.

The comparative study of the three areas reveals the similarities and differences of urban conservation policies, past and present upgrading strategies in the core areas and the importance of tangible and intangible heritage. All three cities share that urban heritage, habits and beliefs are still of importance to the population. While there are significant differences in the kind and level of protection the legal system provides, partly uncontrolled urban dynamics pose a threat to all of them.

After analysing the importance of heritage in shaping urban identities, central conclusions are drawn on the meaningfulness of global heritage paradigms in local non-Western contexts, and the need for integrated approaches considering the different facets of heritage as a whole.

Innsbruck, Austria Simone Sandholz

# Contents

# List of Abbreviations

| | |
|---|---|
| ADB | Asian Development Bank |
| AD Diper | *Agência de Desenvolvimento Econômico de Pernambuco* (Economic Development Agency Pernambuco), Brazil |
| ASEAN | Association of Southeast Asian Nations |
| BPPI | *Badan Pelestarian Pusaka Indonesia* (Indonesian Heritage Trust) |
| CBD | Central Business District |
| CIAM | Congrès Internationaux d'Architecture Moderne |
| CIAV | International Committee for Vernacular Architecture |
| CoE | Council of Europe |
| Condepe-Fidem | *Agência Estadual de Planejamento e Pesquisa de Pernambuco* (State Agency for Planning and Research of Pernambuco), Brazil |
| CPS | *Instituto da Cidade do Recife Engenheiro Pelópidas Silveira* |
| DIY | *Daerah Istimewa Yogyakarta* (Yogyakarta Special Province) |
| DoA | Department of Archaeology, Nepal |
| DPPC | *Diretoria de Preservação do Patrimônio Cultural* (Directorate for the Preservation of Cultural Heritage), Recife Municipality, Brazil |
| EU | European Union |
| FIDEM | *Fundação de Desenvolvimento Metropolitano* (Urban Development Foundation, Pernambuco, later Condepe-Fidem), Pernambuco, Brazil |
| FIFA | *Fédération Internationale de Football Association* (International Federation of Association Football) |
| Fundarpe | *Fundação do Patrimônio Histórico e Artístico de Pernambuco* (Foundation for the Historic and Artistic Heritage of Pernambuco), Brazil |
| GIZ | *Deutsche Gesellschaft für Internationale Zusammenarbeit* |
| HUL | Historic Urban Landscape |

| | |
|---|---|
| ICCROM | International Centre for the Study of the Preservation and Restoration of Cultural Property |
| ICOMOS | International Council on Monuments and Sites |
| IEP | *Imóveis Especiais de Preservação* (Special Properties of Preservation), Brazil |
| IFLA | International Federation of Library Associations and Institutions |
| IPHAN | *Instituto do Patrimônio Histórico e Artístico Nacional* (National Historic and Artistic Heritage Institute), Brazil |
| IUCN | International Union for Conservation of Nature |
| JICA | Japan International Cooperation Agency |
| KIP | Kampung Improvement Program |
| LDC | Least Developed Country |
| Mata Atlântica | Atlantic Rainforest |
| MDGs | Millennium Development Goals |
| NGO | Non-Governmental Organization |
| NSET | National Society for Earthquake Technology, Nepal |
| OAS | Organization of American States |
| OUV | Outstanding Universal Value |
| OWHC | Organization of World Heritage Cities |
| PCH | *Programa Integrado de Reconstruçao das Cidades Historicas* (Integrated Programme for the Reconstruction of Historic Cities), Brazil |
| PPP | Public–private partnership |
| PPSH | *Plano de Preservação dos Sítios Históricos* (Preservation Plan for Historic Sites of the Metropolitan Region of Recife) |
| PT | *Partido dos Trabalhadores* (Workers' Party) |
| RM | *Região Metropolitana* (Metropolitan Area), Brazil |
| RMR | *Região Metropolitana do Recife* (Metropolitan Area of Recife) |
| SDGs | Sustainable Development Goals |
| SPA | Setor de *Preservação Ambiental* (Environmental Protection Sector), Recife, Brazil |
| SPHAN | *Serviço do Patrimônio Histórico e Artístico Nacional* (National Historic and Artistic Heritage Service), Brazil |
| SPR | Setor de *Preservaçã Rigorosa* (Rigorous Protection Sector), Recife, Brazil |
| TU | Tribhuvan University, Kathmandu, Nepal |
| UFPE | *Universidade Federal de Pernambuco* (Federal University of Pernambuco) |
| UGM | Universitas Gadjah Mada, Yogyakarta, Indonesia |
| UN | United Nations |
| UNDP | United Nations Development Programme |
| UNEP | United Nations Environment Programme |
| UNESCO | United Nations Educational, Scientific and Cultural Organization |
| UN-HABITAT | United Nations Human Settlements Programme |

| | |
|---|---|
| UNICAP | *Universidade Católica de Pernambuco* (Catholic University of Pernambuco) |
| URB | *Empresa de Urbanização do Recife* (Urbanisation Enterprise of Recife) |
| USA | United States of America |
| VOC | *Vereenigde Oost-Indische Compagnie* (Dutch East India Company) |
| WHC | World Heritage Committee |
| ZEIS | *Zona Especiais de Interesse Social (*area of particular social interest), Brazil |
| ZEPH | *Zonas Especiais de Preservação do Patrimônio Histórico-Cultural* (Special Zones for the Preservation of historic-cultural Heritage), Recife, Brazil |

# List of Figures

# List of Tables

# Chapter 1
# Introduction

**Abstract** Worldwide cities are undergoing fundamental transformations. A large part of these shifts is taking place in the Global South, where urban change is comparably more dynamic, but often following global trends and resulting in increasing uniform urban layouts. At the same time trends towards more regionalism can be observed, backed by identity discourses. Historic city centres in particular became focal points of this debate, suffering continuous pressure of transformation while being portrayed as the holder of urban uniqueness. Cities worldwide made a turn towards the appreciation of their tangible and intangible heritage. This is particularly challenging in the Global South where heritage was somehow regarded as luxury or touristic feature for a long time, while focal areas of intervention were related to provision of adequate housing and infrastructure for growing cities. This research analyses the importance of urban heritage and its potential for a sustainable urban development in the case of three selected cities in Asia and Latin America. The overall goal is to gain a deeper understanding on the sociocultural construction of identity in historic city centres in the Global South, its differences and similarities against the background of the global heritage debate.

**Keywords** Historic cities · Urban heritage · Cultural context · Research approach

"*What is so special about old buildings?*" This reply somehow depicts the reaction of many people who came to know about the research topic. Why should someone take care about old things, especially about old urban things when we can have new things and modern cities instead? Who should have an interest in preserving such remnants from the past that apparently do not touch the individual person's life anyway? These supposedly simple questions aggregate many aspects that are worth being thought about, and trying to answer depicts some current global societal discourses behind.

The recent decades have witnessed some enormous transformations worldwide. Cities are now home to more than half of the world's population, sparking the debate on liveable cities and urban development, especially in cities of the Global South where most of the urban growth and transformation is taking place (cf. Coy



S. Sandholz, *Urban Centres in Asia and Latin America*,
The Urban Book Series, DOI 10.1007/978-3-319-43735-4_1

and Kraas 2003; Kraas 2010; Simone 2010; Angel et al. 2011; Qadeer 2012; Mucke 2014; United Nations 2014). Globalization has resulted in increasingly uniform urban layouts, while at the same time a trend back to regionalisms can be observed, linked to discourses of a local, urban, regional or national identity (Aronczyk 2013). Global concepts like the 'World Heritage' foster the identification and promotion of unique spots, always linked to the 'spirit' or identity of a place. Historic city centres in particular became focal points of this debate, suffering continuous pressure of transformation while being portrayed as the holder of the urban uniqueness. Cities in the Global South are particularly threatened by the loss of their urban heritage as they are witnessing the highest rate of urban change, formal and informal, while often lacking the economic means and adequate planning tools to face these challenges.

This is why it is worth caring about old buildings—in other words: the construction of an urban identity in historic city centres of the Global South is a relevant subject to do research on.

## 1.1 Problem Statement

It seems to be widely agreed that historic cities (manifested in their nucleus, the historic centre) are 'of value'. If so, why are then such spaces under constant compulsion of replacement? And what does this 'value' mean? Who has defined these tangible or intangible values and how far are they considered in the cities of the Global South? Simple questions—but complex answers.

Throughout the past decades also a shift in urban planning and renewal paradigms can be observed, from demolition and single house renovation measures with hardly any inclusion of social aspects to today's flagship projects like waterfront developments, often focusing on urban marketing aspects and middle to upper class. In parallel, also more holistic approaches including participatory planning processes and public–private partnerships came up.

At the same time the international 'homogenization of cities' (Löw 2008) is of threat for local culture and identity. Euro-centric planning concepts are implemented to achieve pretended locational advantages in a globalized world. This turns to be a downward spiral, not only because planning concepts are or were implemented that are already outdated and replaced in the places they originated from; they can even cause negative impacts for local place attachment as being designed for different cultural contexts.

Urbanisation in the Global South was and is taking place comparably faster than in the Euro-American area (United Nations 2014). Velocity of urban change has increased tremendously during twentieth century, in inner cities very often large-scale demolitions and new planning can be found. In close interaction with the—easily and immediately visible—loss of urban patterns and building fabric one also finds loss of use and place attachment (Ashworth and Tunbridge 2000; Gaebe 2004). Massive demolition of historic urban areas took place in the 1950s and

1960s in Europe and North America, followed by Arab and Latin American Countries in the 1960s and 1970s. In Asian cities, this process is still ongoing (Bandarin and van Oers 2012). Preservation of (urban) heritage in the Global South was somehow regarded as luxury, while focal areas of intervention were related to provision of adequate housing and infrastructure, or coping with the growing number of urban habitants.

Taking a look at the discourses on urban heritage, historic cities, or upgrading of historic parts of the city, one common feature becomes clear: despite the fact that the focal point of interest—the historic built environment in urban cores—is supposed to be everlasting and permanent, the discussion about it and its perception is not at all.

Discourses have changed over time in two ways, in terms of the content and scale. While the discourse content has emerged from local buildings to ensembles, whole city centres and from tangible to both, tangible and intangible values (Ahmad 2006; Falser 2010), also the involved actors and role models have changed from local to global, as can be seen in, e.g., the importance of UNESCO or ICOMOS Charters and Recommendations for the actual global discourses on historic cities (see Chap. 3).

Subsequently, the local or urban identity manifested in the historic city centre is not something automatically existent, but constructed and based on the communication about (urban) heritage. Apparently, cities worldwide made a turn towards the appreciation of their tangible past. Increasingly, intangible values are recognised, and included in the creation of a kind of brand or identity around heritage. Surprisingly, this seems to be true for any (historic) city independent of its location in the Global North or South, or its cultural background, although the understanding of heritage has been determined by European and/or Northern American paradigms until quite recently (Sullivan 1993; Ashworth and Tunbridge 2000; Albert 2013). This does not exclude the specific location/city from the list of drivers but attaches importance to other non-tangible drivers of importance for the construction of an urban identity and its manifestation in the historic centre.

The main focus of this research is to

> Gain a deeper understanding on the sociocultural construction of identity in historic city centres in the Global South, its differences and similarities against the background of the global heritage debate.

As shown in Fig. 1.1, the differentiation between the historic centre and the city as a whole is presumed to be based on values ascribed to the centre. In this context, a systemic approach seems suitable to clearly elaborate the construction of a local urban identity based on values ascribed to the centre or parts of it. Visualised in Fig. 1.1, the differentiation between the historic centre and the rest of the city is assumed to be based on the centre's value as build manifestation of the past (cf. Ashworth and Tunbridge 2000; Rautenberg 2011), showcasing urban history against comparably modern and international parts of the city and permitting a feeling of continuity and 'spatial sacralisation' (cf. Wöhler 2008). To incorporate

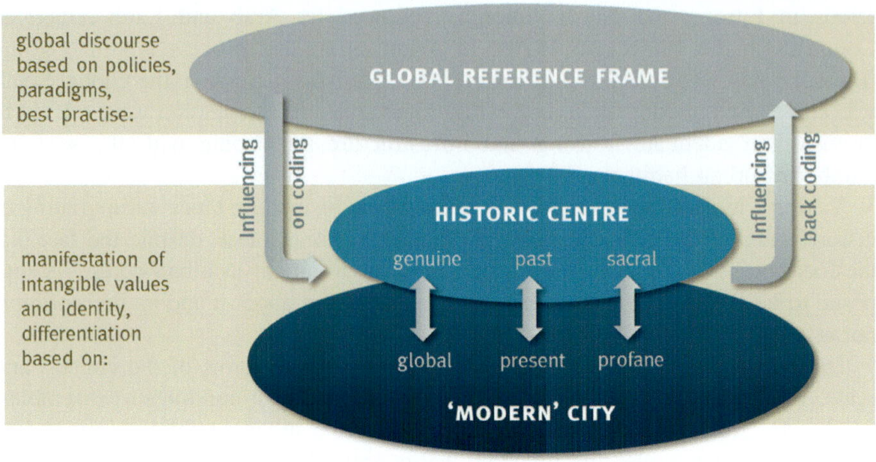


**Fig. 1.1** Levels of investigation and interlinkages

the single case into the global context an analysis of the global reference frame (e.g., flagship projects or policies, and charters determining global discourses) is essential.

## 1.2 Main Hypotheses and Deduction of Key Research Questions

This research analyses the importance of global heritage discourses in the case of selected cities in the Global South. The focal point is the question how far originally Euro-American approaches of heritage conservation have been transferred to other countries and how far they still persist, particularly in planning and legislation. It also stresses the regeneration of historic urban cores and how far they consider urban heritage in its different forms of expression, namely tangible and intangible heritage.

The main hypothesis is therefore: The conservation and regeneration of historic city centres is influenced by global paradigms with a predominant Euro-American background which still persists. Locally suitable inner-city conservation and regeneration in developing or emerging countries makes demands going beyond European concepts. Cultural assets (intangible heritage) are more difficult to preserve than physical elements (tangible heritage), but at the same time essential for the construction of a local identity. Current urban planning is not always adapted to the local context, while the preservation of historic centres, their form, functions and local identity can contribute to a more inclusive and sustainable city.

Based on this hypothesis, the following four central research questions can be derived:

**1** What is or are the current paradigms related to tangible and intangible heritage, on a global, national and local scale? Is regeneration carried out in the historic centres of case study cities, how important is it, and are concepts of place attachment considered in planning?

**2** Which objects are of value for local people, what kind of values do they attach to certain places/locations/buildings and how far are they considered in urban planning and regeneration? Are there differences between the case study cities and/or the cultural contexts, and how are they considered in planning processes?

**3** What kinds of modifications are needed in current urban renewal planning paradigms to adapt them to sociocultural conditions in South and South-East Asian cities as well as in Latin America?

**4** Are there "general" findings in terms of place attachment and planning policies that are valid in both cultural contexts, what kind of results could be fed back into global debates on historic urban cores, their values, maintenance and regeneration?

To answer the questions, three cities in Asia and Latin America are analysed (cf. Fig. 1.2): the city of Kathmandu in Nepal, Yogyakarta in Indonesia and Recife in Brazil. In all three cities ongoing inner-city conservation and re-use projects are analysed—examining their successes, results, and interdependencies with the totality of the city. The goal is to draw conclusions on future needs and necessities towards sustainable revitalization projects for the three cities and beyond.

**Fig. 1.2** Location of the case study cities

## 1.3 Research Outline

After the problem statement and the deduced research questions this research will shed a light on global discourses of the construction of heritage and identities. Figure 1.3 gives an overview of the structure.

Chapter 2 will give an overview on the theoretical background of this research. Starting with an overview on discourses on place attachment and identity, the debate on urban heritage focusing on the global level will be tackled, to shed a light on global reference frame influencing single cities. Thus, despite being centred in human geography, the research also involves aspects from other disciplines such as urban planning, conservation and social sciences.

Subsequently, the context concerning the preservation of cultural heritage and urban renewal will be assessed. Therefore, the global context is described, followed by an analysis of different world regions, including a critical assessment on how the Euro-American roots of the global heritage system are still prevalent. After focussing on urban heritage discourses and paradigms in particular, processes of urban renewal will be highlighted. To do so, the inclusion of urban heritage in such renewal processes will be described and evaluated in the context of the Global North and South.

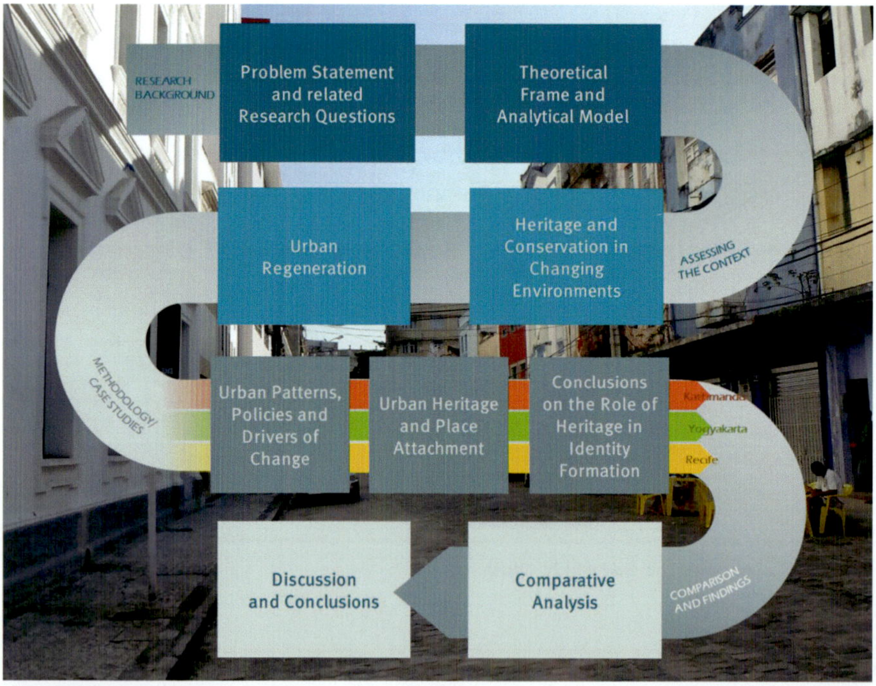


**Fig. 1.3** Outline of the research

After providing such background information, the empirical results will be presented in Chap. 5. To do so, the outcomes of all three case studies will be illustrated. In each case, a comprehensive overview on urban history and recent developments will be given first, before the main urban and heritage policies are assessed. Based on this, urban dynamics and phases of urban renewal are illustrated. Inner-city patterns and how such tangible and attached intangible assets are perceived are described next.

Each case study ends with conclusions on the role of urban heritage in shaping the urban identity of the respective city, before comparing the outcomes of the different case studies. Chapter 5 concludes with a discussion based on the case studies' results with reference to the theoretical background. General discussion and conclusions are given in Chap. 6 followed by final considerations in Chap. 7.

## References

Ahmad Y (2006) The scope and definitions of heritage: from tangible to intangible. Int J Herit Stud 12(3):292–300

Albert M-T (2013) Heritage studies—paradigmatic reflections. In: Albert M-T, Bernecker R, Rudolff B (eds) Understanding heritage: perspectives in heritage studies. de Gruyter, Berlin/Boston, pp 9–17

Angel S, Parent J, Civco DL, Blei AM (2011) Making room for a planet of cities. Lincoln Institute of Land Policy, Cambridge

Aronczyk M (2013) Branding the nation: the global business of national identity. Oxford University Press, New York

Ashworth GJ, Tunbridge JE (2000) The tourist-historic city. Retrospect and prospect of managing the heritage city. Advances in tourism research series. Elsevier, Oxford

Bandarin F, van Oers R (2012) The historic urban landscape: managing heritage in an urban century. Wiley-Blackwell, Sussex

Coy M, Kraas F (2003) Probleme der Urbanisierung in den Entwicklungsländern. Petermanns Geogr Mitt 147:32–41

Falser MS (2010) From the Venice Charter (1964) to the Nara Conference (1994)—changing concepts of authenticity? In: Falser MS, Lipp W, Tomaszewski A (eds) Theory and practice in conservation and preservation—interactions between theory and practice. Proceedings of the international conference of the ICOMOS international scientific committee for the theory and philosophy of conservation and restoration, pp 23–27 April 2008 (Vienna). Edizioni Polistampa, Florenz, pp 115–132

Gaebe W (2004) Urbane Räume. Eugen Ulmer GmbH & co., Stuttgart

Kraas F (2010) Urbanisierung als weltweite Herausforderung. In: Debiel T, Messner D, Nuscheler F, Roth M, Ulbert C (eds) Globale Trends 2010. Frieden, Entwicklung, Umwelt. Stiftung Entwicklung und Frieden, Frankfurt, pp 181–199

Löw M (2008) Soziologie der Städte. Suhrkamp, Frankfurt am Main

Mucke P (2014) Urbanisation—trends and risk assessment. In: Bündnis Entwicklung Hilft (Alliance Development Works) and United Nations University—Institute for Environment and Human Security (UNU-EHS) (ed) World Risk Report 2014—Focus: The city as a risk area. Bündnis Entwicklung Hilft (Alliance Development Works) and United Nations University—Institute for Environment and Human Security (UNU-EHS), Berlin/Bonn, pp 5–10

Qadeer MA (2012) Urban Development. In: Sanyal B, Vale LJ, Rosan CD (eds) Planning ideas that matter: livability, territoriality, governance, and reflective practice. MIT Press, Cambridge, pp 207–232

Rautenberg M (2011) Industrial heritage, regeneration of cities and public policies in the 1990s: elements of a French/British comparison. Int J Herit Stud 18(5):513–525. doi:10.1080/13527258.2011.637945

Simone A (2010) City life from Jakarta to Dakar. Movements at the Crossroads. Routledge, New York

Sullivan S (1993) Conservation policy delivery. In: Mac Lean MGH (ed) Cultural heritage in Asia and the Pacific: conservation and policy. Proceedings of a symposium held in Honolulu, Hawaii, September 8–13, 1991. The Getty Conservation Institute, Los Angeles, pp 15–26

United Nations (2014) World urbanisation prospects—highlights. United Nations, New York

Wöhler K (2008) Heritagefication: Zur Vergegenwärtigung des Kulturerbes. In: Luger K, Wöhler K (eds) Welterbe und Tourismus: Schützen und Nützen aus einer Perspektive der Nachhaltigkeit. StudienVerlag, Innsbruck, pp 43–58

# Chapter 2
# Theoretical Background and Research Approach

**Abstract** Different concepts and definitions from a range of scientific disciplines are related to memory, identity and attachment. Increasingly, such concepts are introduced in urban areas, based on the premise that people link to certain objects which are part of the urban identity and make a city unique. Such assets are manifold and combine built and non-built spaces, in other words tangible and intangible, places and place attachment. Since around two decades, the contribution of heritage to the formation of urban identity is as well considered in international heritage studies. Urban heritage and its manifold expressions *are addressed in different studies from the Euro-American or Western* context, and a growing number of research and case studies on cities and areas in the Global South. In this research, a mix of quantitative and qualitative methods is applied, including field surveys, questionnaires and expert interviews. A subsequent comparative analysis was done to assess potential similarities and differences among the three case studies, namely Kathmandu in Nepal, Yogyakarta in Indonesia and Recife in Brazil, three cities which still comprise of a historic core area.

**Keywords** Urban identity · Place attachment · Cultural context

Many different types of disciplines are dealing with cities. Architects are concerned with buildings, urban planners and engineers with its urban patterns and infrastructure. Ecologists may deal with green areas, parks and urban flora and fauna, while conservation scientists will care about cultural and natural heritage assets. This may already be closer to sociologists, (environmental) psychologists or anthropologists that deal with urban societal and social matters, with the same being true for human geographers. In fact, there is a whole variety of disciplines that are dealing with 'urban' issues, on different scales and in different perspectives, and this list is by far not complete. In the end a cross-disciplinary approach, combining elements and inputs from various disciplines, seems to be necessary to access the 'urban DNA' (cf. Mueller-Haagen et al. 2014) and assets that make a city unique. Such assets are manifold and combine built and non-built ones, in other words tangible and intangible, places and place attachment, from different times and actors.

 
S. Sandholz, *Urban Centres in Asia and Latin America*,
The Urban Book Series, DOI 10.1007/978-3-319-43735-4_2

This research builds on two main aspects. The first question is on people's links to certain objects, people, or groups whether it is called collective memory, identity or attachment. Physical space on different scales is one object of research in this context, often on a neighbourhood or urban scale. Research on this part is carried out in different disciplines, mostly within social sciences. The second aspect refers to urban heritage, how it is defined, what kind of values are associated with it and by whom. Most literature is concerned with the 'formal' aspects of heritage, e.g. its fabric, heritage policies or its values in a more science-based view. Classically, this is topic in planning or heritage studies, to a lesser extent in tourism-related disciplines like tourism geography, which is considering more the process and consequences of turning heritage to tourism destinations. Both aspects overlap when it comes to the question of how far historic urban fabric forms a part in the construction of an urban identity of a city's inhabitants, or how far today's urban dwellers feel attached to the urban past and the urban heritage. Surprisingly, there is not much research in this specific field. Mostly, either heritage values are predefined, based on national or global definitions, or attachment is analysed to other physical spaces, e.g. neighbourhoods or quarters the peer group inhabits, not to historic core areas. When it comes to questions of urban identity and how far the historic centre is part of that, research often seems tourism focused.

To bridge the described gap this chapter will first give an overview of the main concepts and definitions related to memory, identity and attachment, their authors and disciplinary as well as regional backgrounds, before linking urban heritage with (urban) identity. Subsequently, the relevance of the research presented will be illustrated against the background of previous studies in different countries and contexts before deducing the research methodology.

## 2.1 Urban Concepts of Collective Memory, Identity and Place Attachment

The ground-breaking work (Jacobs 1961) on 'collective memory' is Halbwachs' book (1980 [1950]) of the same title. In this publication he argues that the individual memory of the past is composed of two kinds of elements, a social memory induced by external or common sources, and a personal one from the individual itself

> While the collective memory endures and draws strength from its base in a coherent body of people, it is individuals as group members who remember. While these remembrances are mutually supportive of each other and common to all, individual members still vary in the intensity with which they experience them. I would readily acknowledge that each memory is a viewpoint on the collective memory, that this viewpoint changes as my relationships to other milieus change (Halbwachs 1980 [1950]: 48).

He is making a clear distinction between history and memory, with memory being dynamic and changing. History seeks to be objective while memory is emotional (Halbwachs 1980 [1950]; François and Schulze 2005; Fenster 2010;

Petermann 2014). To French sociologist Rautenberg (2010: 133) societies need to invent collective imaginaries of themselves '*in order to know what they are*'. As a consequence there are as many memories as there are groups (Halbwachs 1980 [1950])—a concept which found entry in different recent documents and charters, in particular the 1994 Nara Document of Authenticity (see Sect. 3.4), pointing out the relativity, plurality and variability of values over time and in different cultural surroundings (Rautenberg 1998, 2010, 2011; Scazzosi 2011). Already in the early 1960s famous Canadian author Jacobs (1961) said that:

> Cities need old buildings so badly it is probably impossible for vigorous streets and districts to grow without them [...] for really new ideas of any kind—no matter how ultimately profitable or otherwise successful some of them might prove to be—there is no leeway for such chancy trial, error and experimentation in the high-overhead economy of new construction. Old ideas can sometimes use new buildings. New ideas must use old buildings (Jacobs 1961: 187ff).

At that time she was commenting on her discontent on ongoing urban renewal practises that did not pay attention to the needs of different actor groups, criticizing modernist planning. Although her book on 'The Death and Life of Great American Cities' is now more than 50 years old her concepts still seem to be up-to-date, as proven by a number of journal papers, newspaper articles, blogs, etc. (cf. Greenwald 2013; Donnelly 2014; Schubert 2014) that are still—or again—referring to Jacobs. For example, Sharifi and Murayama (2013) refer to Jacobs's ideas as they investigate how traditional urban patterns can inspire planners to come up with more socially sustainable urban patterns for the case of Iran.

Alongside the work of Jacobs, the US-American urban planner and author Kevin Lynch wrote about place 'legibility' and 'imageability'. In his most influential book "The Image of the City" (1960) he emphasizes on the presence of time and history in any urban environment, and how these factors affect people. This urban environment or environmental image is composed of structure (spatial relations of an object to the observer or other objects), identity (the composition of individual elements to a determined and separable entity in the urban context) and a distinct meaning, in other words the emotional or practical signification for the user or observer (Lynch 1960; Seifert 2011). Lynch states about the image and aesthetics of a city that

> There seems to be a public image of any given city which is the overlap of many individual images. Or perhaps there is a series of public images, each held by some significant number of citizens. Such group images are necessary if an individual is to operate successfully within his environment and to cooperate with his fellows. [...] Each individual picture is unique, with some content that is rarely or never communicated, yet it approximates the public image, which, in different environments, is more or less compelling, more or less embracing (Lynch 1960: 46).

One thing Lynch and Jacobs share is an actor-centred perspective which focusses on the inhabitants, visitor or user of any urban area, rather than dealing with the planners. Both of them pay attention to urban spaces and their specific patterns which have the potential to be of certain—and specific—value for different actors. People feel different in different (urban) surroundings and spaces. In this context one very often speaks about the 'aura' of a distinct place or its 'genius loci'—a phrase derived from ancient Roman mythology where it signified spirits protecting a certain temple, area place, etc. (Castello 2010). Mostly 'genius loci' is used to describe a certain atmosphere or spirit of a place (Knirsch 2004). Norwegian architect and architectural historian-theoretician Norberg-Schulz (1982, 2013) then used the phrase 'genius loci' for his phenomenological analysis of cities. To him (2013: 273) "*The concrete things which constitute or given world are interrelated in complex and perhaps contradictory ways*", composed of built assets, natural assets and intangible assets like feelings. He promotes traditional urban and building forms, which he sees as the "*basis for bringing about a deeper symbolic understanding of places*" (Jive'n and Larkham 2003: 70). Who wants to experience the 'genius loci' of a city, has to enter into a dialogue with people and things (Greverus 2008; Brakman 2011). Jive'n and Larkham (2003) find aspects of genius loci apparent in many design-led considerations of traditional settlements, as e.g., Sentosa (2001: 255) explored for traditional cosmological beliefs, societal structures and traditional measurements that shaped what he terms the "*genius loci within Balinese dwellings environments*".

Another milestone in this discourse is Pierre Nora's article "Between memory and history: *les lieux de mémoire*" (1989). Since then numerous publications on '*lieux de mémoire*', '*Erinnerungsorte*', places of remembrance, etc., followed, in different languages and contexts. To him, a French historian, a '*lieu de mémoire*' is a place 'where memory crystallizes', e.g. also in '*French Marseillaise*', the revolutionary calendar or the Tricolore. Nora (1989) defines memory as 'something life' which was and is generated by living societies, and in a permanent evolution. Memories can be forgotten, deformed, manipulated and revived. The concept of '*lieux de mémoire*' has been transferred and analysed in various national contexts, e.g. as '*Erinnerungsorte*' in Germany, listing material and immaterial items like the Reichstag, Berlin Wall, Auschwitz or the national anthem (François and Schulze 2005; Saretzki 2008).

The ongoing occupation with the memory topic indicates changes taking place in its interpretation. Particularly globalization in its different facets is triggering this shift, resulting in a growing need for particular sites of memory, a 'spatialization' of memory, as it is no longer part of daily life and rituals (Nora 1989; Fenster 2010; Werlen 2014). British sociologist Anthony Giddens noticed a dislocation of space from place in modernity, a contrast of modernity and tradition

> In traditional cultures, the past is honoured and symbols are valued because they contain and perpetuate the experience of generations. Tradition is a mode of integrating the reflexive monitoring of action with the time-space organisation of the community. It is a means of handling time and space, which inserts any particular activity or experience within the continuity of past, present, and future, these in turn being structured by recurrent social practices. Tradition is not wholly static, because it has to be reinvented by each new generation as it takes over its cultural inheritance from those preceding it. Tradition does not so much resist change as pertain to a context in which there are few separated temporal and spatial markers in terms of which change can have any meaningful form (Giddens 1990: 37).

Spatially based identity has turned to be a research topic in social sciences in the 1970s (Weichhart 2004). In cultural and social geography, Werlen (1997) has pointed out the need to question essentialist views of spatial and cultural concepts. In this sense, space is not understood as something objectizable or measurable but as a rational category, defined and perceived by individuals rooted in their personal cultural and societal backgrounds. Spatial patterns are culturally encoded and reflect societal structures (Werlen 1997; Dürr 2005; Werlen and Lippuner 2007; Senil 2011). Spaces can be charged with a sense, they can be interpreted. Identity, as understood after the cultural turn, is relational. To establish a certain identity (nation, class, race, space, etc.) a distinction from other identities is needed, manifested in events, images and imaginations, often linked to certain spaces (Lossau 2014; Sen 2007). In recent human geography concepts such constructivist thoughts are increasingly adopted in spatial settings that are considered to adopt different and new meanings (Werlen 1997; Lossau 2014). British geographer Doreen Massey has argued for the importance of place. Places to her are dynamic, even conflicting, and with multiple identities:

> We need, therefore, to think through what might be an adequately progressive sense of place, one which would fit in with the current global-local times and the feelings and relations they give rise to, and which would be useful in what are, after all, political struggles often inevitably based on place. The question is how to hold on to that notion of geographical difference, of uniqueness, even of rootedness if people want that, without it being reactionary (Massey 1991: 26).

Therefore she is claiming that a sense of place can only be understood and constructed by not looking at a single space but rather by linking it to places beyond, by considering "*a global sense of the local, a global sense of place*" (Massey 1991: 29). Space and identity are interdependent and interrelated. From a social scientist perspective, tangible things become only of value if some imaginations are attached. When adopting this perspective it is no longer possible to analyse space 'per se' but how space is constructed and part of identification processes (Sörensson 2008; Weichhart 2010; Lossau 2014).

Authors like Nora have contributed to the reinvention of the remembrance topic within the past 50 years and in particular since the turn of the millennium. Subsequently, aspects of remembrance and identity—that Nora states to concentrate in particular places and things—have become important research topics. This is particularly true for cultural and social geography dealing with human construction of their environments (Petermann 2014).

Concepts of territoriality, on how people relate to space in built environments, have been developed in the 1980s (Hillier and Hanson 1984). Since then the local scale is of growing importance in geographical research, e.g. investigating on urban quarters instead of the city as a whole (Reuber 1993, 2014; Paasi 2004). This importance of scale and the differentiation between areas and territories shows how much space and identity are constructed (Wagner 2008). French sociologist Henri Lefebvre is one of the pioneers in this research field, in particular his publication on 'The Production of Space' (1991), first published in 1974. There he distinguishes different kinds of space, mental ones and real or social and physical space we live in. He argues that space is a social construction based on values and meanings, thus going far beyond a natural scientific understanding of space. In a later publication he states "*monumental buildings mask the power and the arbitrariness of power beneath signs and surfaces which claim to express collective will and collective thought*" (Lefebvre 1991: 143). In this sense, a monument is a potential and pivotal mediator between groups who claim some kind of—even diverging—ownership (in physical, historical or cultural terms) over a certain site and those who cannot (Di Giovine 2011). Lefebvre's achievements in conceptualizing space and its production are still influencing on urban research, e.g. 'Local versus Global' trends on an urban scale (cf. Greverus 2008, refers to Lefebvre in her paper on aesthetics of urban diversity; Roy 2009 and her analysis on the relevance of Lefebvre's concepts for cities in the Global South; or Frehse 2013, and her paper on the potential to use Lefebvre's methods for Latin American urban research).

According to French sociologist Pierre Bourdieu, space is hierarchical in a hierarchical society, even if that is not immediately apparent (Bourdieu 1989; Jung 2010). His spatial concept was later pursued and expanded by various researchers, among others by Löw. To her, space is produced; it is composed of a material base and cognitive efforts of perception, remembrance and imagination. Assessing this 'inherent logic' is indispensable to understand a city, which can be compared to an organism with a distinct character (Löw 2008, 2011).

Feelings towards places are embedded in cultural milieus and will therefore differ (Low 1992). People are linked to place by means of beliefs and practices. Urban anthropologist Low (1992) therefore distinguishes six different types of symbolic linkages between people and land

- Genealogic linkage through (family) history;
- Linkage through loss of land or destruction of a community;
- Economic linkage through ownership, inheritance, or politics;
- Cosmological linkage through religious, spiritual, or mythological relationships;
- Linkage through religious or secular pilgrimage and celebratory, cultural events; and
- Narrative linkage through storytelling and place naming.

These categories can overlap and are not mutually exclusive—and most of them are intrinsically linked to tangible or intangible heritage and apply in urban contexts as well.

So far a variety of different expressions has been presented and used to describe the bond that people develop with certain spaces, namely place attachment, sense of place, genius loci and place identity. Place attachment can be defined as:

> The symbolic relationship formed by people giving culturally shared emotional/affective meanings to a particular space or piece of land that provides the basis for the individual's or group understanding of and relation to the environment (Low 1992: 165).

The concept of place attachment refers to the bond that people develop with places (Low 1992; Lewicka 2008, 2010; Kyle et al. 2014). Historic sites have the potential to facilitate such attachment as there is a growing consensus that culture and cultural heritage can contribute to human well-being (Tweed and Sutherland 2007; Bandarin et al. 2011). Hidalgo and Hernandez (2001) cited the importance of a positive affective bond for maintaining a close association with place. For example Hernandez et al. (2007) studied residents' (university students and the general public residing in the Canary Islands off the north-western coast of Africa) attachment to three spatial contexts (i.e. island, city and neighbourhood) inferred that affective attachment develops prior to place identification. They also reported that emotional attachment can exist independent of place identification. Kyle et al. (2014) seek to conceptualize place attachment using identity theory, finding that there are slight differences in the definition and indicators used to measure place attachment.

'Sense of place' can be defined as a subjective emotional attachment that people feel towards a specific place (Yacobi 2004). It can be located at the intersection of three dimensions: physical space; conceived space, or the way in which planners and architects represent space; and the ideological space, which relates to the evolution of a place as influenced by its sociopolitical context (Yacobi 2004; Abu-Rabia 2010).

'Identity', when applied to a place, can either refer to the spatial aspects or—in psychology—can be understood as a feature of a person, not place. In this book, 'place identity' will be used in the sense of features that define a place's distinctiveness and unique character, in this sense very near to the concept of 'genius loci' (Norberg-Schulz 1982; Reuber 1993; Lewicka 2008; Wolfrum 2008a). According to Burke and Stets (1999), as an individual's identity is continually verified through the interaction with specific others, shared experiences, and settings, she or he begins to see the relationships, activity, and settings predictable and dependable (Kyle et al. 2014). According to Finnish geographer Paasi (2003: 479) identity is a social process, and

> 'Regional identity' is, in a way, an interpretation of the process through which a region becomes institutionalized, a process consisting of the production of territorial boundaries, symbolism and institutions. This process concomitantly gives rise to, and is conditioned by, the discourses/practices/rituals that draw on boundaries, symbols and institutional practices.

According to Lewicka (2008) there is no agreement in literature on how place attachment and place identity are related, sometimes they are used interchangeably, in other studies attachment is used as a part of identity. What both concepts have in common is a positive connotation (Lewicka 2005). In this research it was decided to use the phrases of 'place attachment' and 'identity' in the same sense without differentiating further.

People's attachment to place emerges from their desire to strive to preserve contexts for self-verification. The verification of place identities evolves through interaction with a certain space. Physical environment plays an important role in maintaining identity (Kyle et al. 2014). Australian New South Wales government (2004: 10) formulated that "*Heritage forms the backdrop of our identity*" and ascribes heritage an important role to play in understanding relationships, culture, and in fashioning future. Cultural heritage will be used in the broadest understanding of UNESCO (UNESCO 1972, 2011, 2013, 2014), as elaborated in Sect. 3.2. Cultural heritage combines space and endowed meanings or values (Wöhler 2008; Te Heuheu et al. 2012). In this sense, tangible cultural heritage ascribes a certain value to a building, ensemble, space or urban area, to a physical space, while an intangible cultural heritage includes traditions, living expressions, or social practices which are potentially but not necessarily attached to a certain physical space. Different authors have used the word 'heritage-scape', what would mean a heritage area which is endowed with a certain meaning (depending on the societal context and group), and thus in its very sense a 'construction' of space which contains certain heritage assets (Wöhler 2008; Garden 2009; Di Giovine 2011; Ronström 2014).

Talking about heritage—and how to protect or sustain it—reveals different definitions of related terms as there are four distinct preservation approaches

> 'Preservation' or 'conservation' refers to maintaining a site in its original condition to the greatest extent possible, taking only measures, such as repairing a leaking roof, that protect it from further damage. 'Restoration' refers to returning a structure to an earlier, often the original, state. In restoration work, keeping original elements in place, even when they are damaged, is preferable to replicating elements. 'Reconstruction' encompasses the building of a new structure based on historic designs (Alberts and Hazen 2010: 62).

All four concepts are applied in urban renewal projects. Actually, even for projects that are meant to upgrade certain urban areas (like urban centres) there is a variety of phrases: 'urban renewal', 'upgrading', 'regeneration', 'revitalization' or 'redevelopment' (cf. Chap. 4). While 'upgrading' has a clear economic connotation, other phrases are often used interchangeably, in particular 'regeneration' and 'renewal', which different authors use to describe current trends (Roberts and Sykes 2000; Chien-Yuan Lin 2007; Colantonio and Dixon 2011). Yeo and Han (2012) on the other hand uses 'regeneration' as an umbrella for all other terms. This research sticks to the words 'renewal' and 'regeneration', and uses them in the same sense. In the heritage context they are going beyond 'preservation' as heritage is included as a (social or economic) resource worth being protected but not necessarily the goal itself.

## 2.2 The Potential of Urban Heritage in Identity Formation

Aside of the discourses coming from the 'attachment' or 'identity' side, there is another one that emerged on the 'place' side, namely in planning and architectural criticism. As early as 1848 Ruskin formulated as one of his "Seven Lamps of Architecture" (1989) that buildings should respect the culture from which they have developed. In the late twentieth century the discourse was pursued to oppose uniform global planning and building activities. To Rossi (1984), the city is a collective memory, its forms are constant while its functions can change and vanish. Cities are the collective memory of people (Nerdinger 2008; Rautenberg 2011), what is not always considered in planning processes. While the understanding of public space throughout the 1960s and 1970s was more a functionalist one, actual discussions are increasingly considering historic and social values, public spaces are considered as important parts of identity (Knirsch 2004).

Today, cities are understood as a living entity, constantly under development and redefining itself and its identity over time (Constantinescu and Hărmănescu 2014; Stan 2014). Cultural heritage has become an integrative component for urban areas and sustainable cities (Turner 2012), as it showcases the 'contemporary of the uncontemporary', expressed in material forms as well as in immaterial realities, relations and linkages (Lefebvre 1991). Public urban space is shaped by various actors of different backgrounds, age, sex, ethnicity, etc. Its use differs over time or even in the course of the day. By using and shaping it, people are expressing their individual identity (Borden 2008), in return spaces can influence on the construction of identity of individuals."*Settings and context, spirit and feeling are all part of the elusive genus loci of the city*" (Turner 2013: 79).

Hassenpflug (2000) talks about spaces that are characterised by signs and symbols, thus ensuring their uniqueness. To him, spaces enable people to perceive the history of the past in the present, while at the same time contributing to shaping the future. For Baum (2010), 'urban' has four different dimensions: a built one; a functional one, comprising uses of public and private space; a social one in the sense of human interaction with its rules and regulations; and an atmospheric dimension, which is the character of a place, its history but also its contemporary development.

Berkes et al. (2009) stress the importance of maintaining, and at times restoring and cultivating, new cultural connections to space. Among the needs they identified to sustain the linkage between people and land—mainly for rural areas—are the maintenance of local and traditional knowledge, of cultural legacies, social institutions and networks. Scrutinizing their approach, it is worth asking why this should not apply to urban areas similarly, e.g. for maintenance of urban rivers, like in the Australian context where Aboriginal people attach cultural and religious values to the Darling river, linking well-being with cultural health and physical safety, as described by Gibson (2012).

'Place' is a social concept and the ways in which such a physical place or space is perceived, experienced, imagined and in the end maintained is tied to cultural values and beliefs (Gibson 2012). Jack (2012: 90) notes:

> Place can be said to come into existence when people give meaning to a part of the larger, undifferentiated space in which they live. While abstract knowledge about a place can be acquired relatively quickly, attachment to a place takes longer to develop.

The urban landscape is a spatial representation of social culture (Enache and Căplescu 2014)—of the past one(s) that constructed the space and the present one that is using and transforming it, and who are transmitting it then to future generations (Jaramillo Contreras 2012). Today, cultural change seems to be faster than at any time in history. As a consequence, social attitudes and values are altered globally; modernity led to a disengagement of direct links of local contexts

(Townroe 1996). Places are no longer clear supports of local identity (Giddens 1990; Morley and Robins 1995). Opinions on the value of places vary, so do emotional connections. Sense of place is a personal matter that might change and even get lost (Schofield and Szymanski 2011).

The urban functional segregation of the past decades has led to a decreased personal identification with the city and an increased identification with few clearly defined spaces (Grünberg 2004). Views on urban heritage, the attached values and local identities are manifold, and can even be conflicting (Lillehammer 2009). Cultural heritage is a construction, what is considered cultural heritage depends on social and political processes. The consideration as heritage of one societal group could imply its loss by another group that ascribed different meaning or use (Saretzki 2008). Aspects of social identity, knowledge, spirituality, recreation, and aesthetics are very likely to differ also between different cities (Grove 2009). Each city has a specific history and particular social, cultural, political and economic assets which change or overlap over time (Schmidt 2013); neither cultures nor their values are static (Paasi 2000; Taylor and Levine 2011). Many heritage values are set not by the market but by other types of social relations (Rojas 2002).

The ongoing and increasing global competition of cities (Amen et al. 2011) results—among others—in a growing recognition of urban features as unique selling propositions. Urbanity is undergoing a change from a normative concept towards an aesthetic perception of users and visitors. Urban density and complexity are no longer seen as a handicap but rather cultivated as urban self-promotion (Bittner 2010). Urban marketing and urban living quality are increasingly considered in planning and attached to heritage. Cultural heritage is an important part of the urban landscape; it forces the user to adapt to particular contexts but provides the surrounding for social interaction. It is why "maintaining cultural connections to the land and at times restoring and cultivating new connections" (Berkes et al. 2009: 129) was and is essential. Hatuka (2010) assigns collective memory the power to become a tool in modifying space. An intrinsic urban profile or character is composed of different layers that overlap over time, what turns historic does not vanish but becomes another layer (Hoffmann-Axthelm 1996; Vedru 2011; Turner 2013). Heritage is non-static and open for interpretation: "*It also becomes a piece of clay ready to be moulded into something we want it to be*" (Uzzell 2009: 326).

Taking a look at the linkage between heritage and identity from a social anthropology perspective, Filippucci (2009) finds that societies approach their past by means of heritage and then construct their identity based on it. Heritage is a 'social construct', linked to the social construct of identity. The meaning of heritage depends on the social group that defines it as part of its identity (Dormaels 2013). Baum (2010) explains the importance of identity with the intrinsic human need for orientation and security. The identity of a place permits us to identify with that particular area; places with an identity stand out from the crowd and attract people. Identity is therefore complex and multi-faceted, composed of built, social and societal layers. Dürr (2005) talks about a 'spatialization' of identity—certain actions are carried out in certain places. These spaces therefore become witnesses of distinct actions and are associated with them; as a consequence they become part of the urban or collective identity.

Identity is also a topic of growing importance in heritage studies. Albert (2013) witnesses this shift since heritage is no more seen as purely a tangible and static object, but rather as a cultural and social activity, constituting part for the shaping of identity:

> From the epistemic theoretical perspective, a change of paradigm has taken place from an identity that is immanent in an object and therefore static, to an identity that continuously develops and therefore also constructs heritage in a dynamic manner. In my opinion, this change of perspective in Heritage Studies should be followed (Albert 2013: 13).

Aspects of identity are important on different scales, for the values an individual attaches with a certain place or thing, and for strengthening the ties within groups, to develop a group identity (Weichhart 2004), e.g. in nation building or developing a distinct urban identity. Heritage is one specific interpretation of the past in the present. Only a small portion of past events ever makes it into the recorded or materially preserved past, and out of that another small part is used in the creation and reinforcement of group identities (Sommer 2009: 103). Remembrance is a subjective reconstruction of the past (Saretzki 2008), and one precondition for forming a cultural identity (West 2007). It is as well a precondition for developing a feeling of responsibility towards a city or urban area. To Norberg-Schulz (2013: 282), identification is the basis for man's sense of belonging

> It is therefore not only important that our environment has a spatial structure which facilitates orientation, but that it consists of concrete objects of identification. Human identity presupposes the identity of place. Identification and orientation are primary aspects of man's being-in-the-world.

The interest in cultural heritage that social and cultural sciences have developed over the past years can be regarded as a reaction to processes of globalization and modernization within the past around 20 years (Schmitt 2009) where cities and urban places are competing with each other on a global scale (Chien-Yuan Lin 2007). Amongst others it is expressed in a growing appreciation of values and potentials of traditional knowledge, culture and spirituality (Department of Economic and Social Affairs—Division for Social Policy and Development and Secretariat of the Permanent Forum on Indigenous Issues 2009).

Culture has never been static nor homogenous, although this is widely believed, resulting in attempts to freeze certain cultures or cultural expressions in time. Cultures are adapting, changing and without rigid borders, particularly in times of globalization and mobility (Department of Economic and Social Affairs—Division for Social Policy and Development and Secretariat of the Permanent Forum on Indigenous Issues 2009). Furthermore, Scazzosi (2011: 20) comments that:

> The process through which people give sense to places and elements and attribute values or disvalues to them, in order to make choices for the transformation pf places and implement and manage them, is extremely complex. It has not been much studied or experimented yet. It concerns the different kinds of relationships between the new and the old, in the light of the different cultural meanings and values that socio-cultural groups attribute to past

> (material and immaterial) heritage. But also in the light of conditions and requisites that contemporariness has put forward.

The interlinkages of identity, culture and heritage have found their way into international policies and declarations (cf. Charter for the Interpretation and Presentation of Cultural Heritage Sites and Nara Document, International Conference on Urban Culture 2007; ICOMOS 2008, 1994), Chap. 3 will elaborate further on this topic. Conservation in an international context is related to the debate on the 'authenticity' and 'integrity' of a site—debates which are power-laden as both guiding concepts can have different meanings in different cultural contexts. In addition, sites are multilayered so the point is for whom a site should be authentic and integer (Alberts and Hazen 2010). 'Sense of place', 'authenticity' and social values of space have become important research topics in heritage management in Australia, influenced by the Burra Charter. The charter also influenced on heritage practices in the UK that changed towards a stronger consideration of 'communal values' (Harrison 2011). The concept of identity also comprises recognition of different societal groups, on national and international scale happening, e.g. in the growing number of documents and policies on 'indigeneity' or the recognition of how closed natural assets and beauty are linked to cultural beliefs in different cultures (Te Heuheu et al. 2012; Thorsell 2012). In particular, urban centres and urban heritage are considered as manifestation of urban identity worldwide

> Twentyfirst-century heritage interpretation must be an informed and inclusive group activity, and expression of evolving community identity, facilitated by professionals and nonprofessionals alike. Moving from passive consumption of prepared presentations to enactment of identity and connection, this new form of heritage interpretation breaks through the confines of the tour and the site to become a form of discourse within the wider community (Silberman 2013: 30).

## 2.3 Previous Research on Urban Identities in Different Cultural Contexts

There are a number of studies from Euro-American or Western context, and a growing number of research and case studies on cities and areas in the Global South.

Even in—or because of—the globalizing world, place attachment is strong in Europe (Lewicka 2005). For instance European reconstruction after World War II can be considered as longing for lost urban identity (Tokya-Seid 2003a; Vinken 2010; Pellnitz 2013). This process seems to be taken up again within the last few years, e.g. in reconstruction processes like in the recent discourses on the historic centre of Frankfurt (Rodenstein 2010). From the 1960s on, middle European cities witnessed a transformation of urban structures towards a 'functional city', as a consequence the traditional form of social spaces was reformed in the post-industrial city (Bittner 2008).

Cultures of remembrance have been gaining importance in European cities since the 1980s, e.g. in the European Capital of Culture (Luger 2008; Roca et al. 2011). Subsequently, different programmes and policies have been developed and implemented, such as the 2008 programme Intercultural Cities Programme, seeking for diversity and a pluralistic city identity (Wood 2010). All over Europe, cities seek to define their specific image, referring to spatial and cultural particularities (Constantinescu and Hărmănescu 2014). In its conclusions, the "European Urban Charter II—Manifesto for a new urbanity" (2008: 15) states that:

> We know that our towns and cities have a long history and must be viewed from a long-term perspective of our cultures. We think that these roots in the past and in our collective memories are also an asset that helps us to project ourselves into the future on the basis of a strong identity. We are not proposing a single model of urban development. Our towns and cities have their own personalities. They are all different and their diversity is an opportunity for Europe.

In the UK for example, culture and heritage have been linked more closely since the 1990s. The complex of culture and heritage became of

> Particular importance to the successful growth and development of devolved areas for a myriad of reasons—not least the politics of identity. Devolution and articulation of cultural identity are intimately linked, evident by the immediate appointment and creation of ministerial posts with cultural responsibility, and major reviews of cultural policies, being instigated in the devolved countries. Scanning cultural policy statements, it can be clearly seen that the heritage's role within the politics of cultural identity and devolution was to provide a firm cultural context, community and physical manifestation of history and a sense of place for the emergent nations within the greater whole of the United Kingdom (Baxter 2009: 86).

Europe-wide rediscovery of historic centres can also be considered as a countermovement of an urban society that is losing its urban identity (Tokya-Seid 2003a). All over Europe, numerous historic cities are facing the challenge to protect built heritage without prohibiting change and becoming static. They are trapped between short-term economic interests that may alter the entire urban appearance and a musealization, as can be witnessed, e.g. in San Gimignano, Italy (Urban 2011), or Rothenburg ob der Tauber, Germany (Ashworth and Tunbridge 2000; Alberts and Hazen 2010). Traditional European cities undergoing revitalization are endangered to turn into open-air-museums presented as the embodiment of a collective memory (Eeckhout and Jacobs 2008; Gaines and Jäger 2009) instead of being a living entity.

This tendency is associated with urban marketing, which is increasingly based on the urban identity and the identification with a city, in particular with the city's urban heritage (Hilber 2004; Weichhart 2004; Ebert 2005; Luchsinger 2008; HerO 2011). The creation of an urban image is needed to foster urban marketing (Frank 2011) which itself often is based on renewal projects in run-down historic areas. After the major urban renewal actions in London, the iconic Docklands project, Thames River and its bank has become a major open space and heart of the town again (Farrell 2010), although the project was not free of conflicts, including some on the nature of its past heritage (Massey 1991). Like London, also other European cities define their identity based on water, even with brand names like 'sea-side town' or the 'pearl' of a certain river (Stiftung Lebendige Stadt 2005). In a study on German cities it was found out that particularly those cities undergoing an economic structural shift declared identity a political goal. Only the successful seem to have an identity (Helbrecht 2004).

A similar kind of 'branding' can be found on a global scale, in particular in tourist destinations, e.g. where Amsterdam is associated with red light district or Australia with Aborigines (Wöhler et al. 2010). On a global scale those cities that are associated with a certain image—including cultural and historic references—are comparably advantage, e.g. this is the case for cities like Venice or Rio de Janeiro. (Wolfrum 2008b).

Various studies have been carried out on different aspects of place attachment and identity all around Europe's cities. Hidalgo and Hernández (2001) interviewed people from different areas of Santa Cruz de Tenerife, Spain, and found a comparably larger grade of attachment to larger scales instead of neighbourhood and social attachment greater than physical one. In a later study on the same island, Hernández et al. (2007) found natives establish more intense links with their area compared to non-natives of the same nationality and immigrants, however, the later groups are attached to the place. Another study analysed the significance of urban open spaces for young people's social practises in the Canary Islands, Spain (Díaz-Rodríguez et al. 2015).

In the case of the North Pennines, UK, the local heritage was found to support the residents' sense of place by providing a source of pride and self-esteem (Hawke 2012). On the other hand, in studies on heritage reserves and villages in Estonia, done based on questionnaires handed to locals, Vedru (2011) found people to get used to their surroundings to the extent that they will no longer recognize values going beyond material ones. Social and cultural values sometimes seem to be more visible to outsiders than to locals.

In a study on the collective memory of the inhabitants of the cities of Lviv (Ukrainian with Polish history) and Wroclaw (Polish with German history), Lewicka (2008) found in place identity high in both cities, and place memory loaded with national contents, with the place origin and most recent times and events recalled best. However, they differ in terms of scale as place attachment in Lviv was dominated by national identity and in Wroclaw by local (district, city) one. Therefore she hypothesizes that place may be construed either in a top–down manner, as a national, ethnic, or religious symbol, or in a bottom–up way, as an autonomous unique entity.

The importance of preserving historic cities is widely recognized in Europe, including the protection of its aesthetic values. However, urban sprawl is identified as a growing threat, resulting in a loss of the urban or regional identity, as de Noronho Vaz et al. (2012) point out in the case of the Algarve region in Portugal. Simultaneously, people's acting radius has enlarged over time (Weichhart 2009), resulting in a changing perception. Today, Europe is witnessing new trends in 'placemaking' and post-consumerism structures, e.g. by transition town movements or urban gardening, aiming at the creation of turning spaces into places with a specific—and mostly community-based use (Andrews and Urbanska 2010). Sociocultural activities and events are increasingly reintegrated into daily life and become part of urban lifestyle and identity of certain groups, e.g. urban gardening does not only serve supply functions but fulfils a certain 'back to the roots' feeling of gardeners—in particular in major cities like New York (Eizenberg 2010).

Most studies found differences in objects and kinds of identification of different user groups with a city, e.g. inhabitants with Hispanic background perceive their hometowns in the US in a different way than groups with a US background (Dürr 2005); or international migrants in Cologne, Germany, that have a comparably different spatial reference system which is on a more or local quarter scale and related to partly different objects (Reuber 1993; Gebhardt et al. 1995; Espahangizi 2011). Other studies deal with specific meanings endowed, e.g. oral memory of Roma in Spain (West 2007) or battlefields, a research that the authors explicitly understand as not being about the past but about attitudes and understandings of the past in the present (Carman and Carman 2009).

Not surprisingly, sense of place and identity topics are of particular interest in areas which are not free of conflicts over land and over values ascribed to it, e.g. there is a comparably large number of case studies on the topic from Israel and Palestine (Yacobi 2004, 2010; Abu-Rabia 2010; Fenster 2010). Abu-Rabia (2010) investigated on construction of territorial belonging and memory by Bedouin-Arabs in the Negev desert. In his findings he points out how the 'sense of place' differs between competing groups, and how it is constructed through spatial practices of memory and belonging. Fenster (2010) analysed different memories and symbolism attached to certain places by Jews and Palestinians in Israel, and witnesses a growing interest in the links between memory, belonging and commemoration in the recent years. Handal emphasized on the linkages between tourism and identity and potential conflicts among the various actor groups, using the case study of Bethlehem (2006).

In an essay on multicultural Britain and its heritage, sociologist and principal figure in cultural studies Hall (1999) raised several questions which can be summarized as follows:

- Whose heritage are we actually talking about?
- Who is it for?
- And who is concerned by it?

These questions can be transferred to postcolonial contexts in many countries of the Global South, where aspects of heritage and attached values are intertwined with different or even conflicting views on distinct spaces by different actor groups (Lagae 2010), raising questions of cultural heritage as a manifestation of colonial times. For instance, in the case study of Lubumbashi, DR Congo, Lagae (2010) refers to the question 'whose heritage' by analysing and comparing sites of former 'colonizers' and 'colonized' that act as *lieux de mémoire* for different communities that coexist but not necessarily interact. Hewitt (2012) finds culture and cultural identities increasingly politicized, and still popular Western or global stereotypes find cultures and societies outside their range as underdeveloped, traditional, or 'backward'.

With the arrival of the Europeans, Latin America became a 'laboratory' for European (urban) planning and theories. It is what Adams calls a 'constructed identity' (2002: 19), however, an expression of the local and urban history. To her, the construction of an identity is more than freezing some area in time, but rather a socially compatible reuse. That matches the growing concern to preserve historic urban centres that can be witnessed in many Latin American cities, intertwined with certain nostalgia (Hiernaux 2013). In Latin America, built heritage is increasingly becoming a resource for the reconfiguration of urban spaces (Lacarrieu 2013). In their article on the upgrading history of the historic centre of Recife, Brazil, the authors claim "*nowadays, collective identity and memory are essential values that must be present in any urban planning task*" (de Albuquerque Lapa and Almeida de Melo 2007: 37).

In the relatively young states of southeast-Asia, historic cities have a more negative connotation as they are associated with colonialism or poverty (Tokya-Seid 2003b). Nevertheless, this colonial heritage is used by different countries to create a certain image or national pride, e.g. in South Korea where government tries to define the country's national identity through Seoul's urban postcolonial landscape (Podoler 2010). In the case of Southeast Asia and in particular Indonesia, Evers (2007) talks about an 'archaeology of meaning'; built artefacts from previous eras which survived the ongoing urbanisation process and which do not necessarily still have a meaning for the current population. Ahmad (2006) calls cultural resources of Southeast Asian countries 'items of national pride', and finds them still rooted in vibrant and largely traditional communities while Yap (2012) notices a modernization trend in Southeast Asian cities, replacing traditional quarters with international-style malls and towers. In the end, cities regret the losses of fabric and subsequently identity too late.

However, other researchers found different kinds and levels of attachment to places in different Asian countries, e.g. functional and emotional attachment to main traditional shopping streets in the city centre of Kuala Lumpur, Malaysia (Ujang 2012), while China has already erased most of its history and built whole new cities and quarters with new artificial or 'instant history', representing the identity of the class of young and well-travelled Chinese citizens (Mars 2008). Michel (2010) found discourses on urban renewal all over the metropolises in Southeast Asia, from small-scale projects that consider heritage and identity aspects up to large-scale ones focussing comparably more on economic aspects. Referring to a global scale, Di Giovine (2011) elaborates on UNESCO's role in 'valorising' and 'creating' the heritage of Angkor Wat in Cambodia, while Nyaupane et al. (2015) researched on the linkages between tourism, religion and heritage by analysing the Buddhist World Heritage site of Lumbini, Nepal

Globalization is incorporated into the fabric of cities also to satisfy the expectation of international tourists, resulting in changes in the perception of spaces, uses, etc., e.g. as Ellingsen (2010) has noticed in Kathmandu. Ronström (2014) links heritage and tourism when stating that festival and heritage provide the destinations while tourism provides the visitors. Tourism has the potential to support cultural heritage and culture conservation in developing countries (Luger 2008), but the complex relationship between cultural heritage and global tourism may end up in a 'tourismification' (Salazar 2010) of heritage, where the cultural assets are rather seen as unique selling propositions that are used to attract visitors than as living—and changeable—heritage for the local communities:

> Ironically, pioneering projects of originality and uniqueness have been successfully replicated to the point where they no longer express the sense of a locally distinctive identity that was the intention of their creators and promoters (Salazar 2010: 133).

In the worst case, interpretation of heritage in global tourism—in particular in developing countries—can have the effect of disembedding local sense of identity, e.g. in the case of 'glocalized' heritage of Yogyakarta and central Java in Indonesia (Salazar 2008, 2010, 2012). Zahnd (2005) worked on the potential of revitalisation and innovative uses in historic quarters of Yogyakarta and Semarang, Indonesia, illustrating the discrepancy between modernity and tradition within the Indonesian urban planning and urban development.

Contemporary understandings of heritage are using elements of the past to represent shared values as a basis for a future vision of a nation, city or community. Ireland (2012) finds this concept particularly powerful in Twenty-first century post-colonial nations with their culturally heterogeneous populations. In other cases, cities are trying to develop an identity, e.g. Dubai, which seeks to brand itself as the global destination for the wealthy of the world, trying to counteract the partly bad image of Arab countries in the Western world. The strategy is therefore to create a particular modality of Muslim modernity without completely denying the past and culture, but by adapting it to become more globally 'acceptable' (Haines 2011). Culture is not static, and is increasingly influenced by global trends, for emigrants as well as immigrants, e.g. in a case study on Chinese dwellers in Canada that found an 'Asianiziation' of Vancouver and subsequent 'Vancouverization' of Asia (Lowry and McCann 2011).

In a review of 81 recent studies on sustainable urban renewal, Zheng et al. (2014) found a growing number of publications over the past years indicating the growing interest in the topic. However, they identified a lack in terms of mechanism to achieve sustainable urban renewal and claim a need for more comprehensive approaches that analyse more than one or two aspects (cf. Chap. 4 on regeneration). In an interview published in the Journal of International Affairs (2012), architect Rem Koolhaas described that:

> If you look back a 100 years, you find that there was still such a thing as Indian architecture, Thai architecture, Chinese architecture, African architecture, Dutch architecture, and Russian architecture. But now, almost all of these languages have disappeared, and are subsumed in a larger and seemingly universal style. The process has been like the disappearance of a spoken language. Remnants of these differences still exist. For example, a high-rise in Singapore is inhabited in a very different way from a high-rise in the suburbs of Paris or a high-rise in China. Each of these cultures, which once had its own form of speaking, is not trying to resurrect its old language, but is interested in defining and asserting its uniqueness again.

To sum it up, there is a variety of past studies on different cities and areas that dealt with different aspects of place attachment and the linkages between heritage and identity. Studies stem from countries all around the globe with a growing number from Asia. However, studies on place attachment mostly do not consider heritage as one aspect that is of importance for feeling attached to a city. As a consequence, these studies mostly dealt with attachment on different scales, ranging from neighbourhood to city, with none of them asking about the historic city centre explicitly. In return, studies on heritage only recently shifted from researching on 'scientifically' ascribed values to those ascribed by local actor groups and dwellers.

In their book on aspects of urban living quality Baur et al. (2010) define a set of indicators to assess what makes a city 'liveable'. In their opinion standard rankings focus too much on economic aspects, while they explicitly include assets and indicators like 'proud of the city', or 'feeling at home', combining aspects of the built environment with intangible assets while explicitly mentioning cultural heritage as one asset—all aspects of place attachment or identity.

## 2.4 Conceptual and Theoretical Embedding of the Research Topic

As described above, there is multiple research and literature on how people attach to places, using different nomenclature, namely on place attachment, urban or place identity, different aspects of collective memory, tangible and intangible values. Their use and definitions change over time and depend on the scientific discipline. However, they coincide in the very core, as they all deal with certain meanings, values or feelings attached to distinct places or occurrences. Such interplay between time and space is what makes historic cities and city cores so special. Such historic centres are changing or disappearing continuously—although being 'of value' as argued by a multitude of researchers, organisations, policymakers and others.

Perception and the perceived reality depend on the perceiver and is thus unique to each individual. Places are part of such realities and are a topic analysed by philosophers, psychologists, sociologists, human geographers, and urban planners (Dürr 2005; Barthel-Bouchier 2012; Casakin et al. 2015). Like in any other discipline, (human) geography has witnessed certain research topics and related theories emerging and declining. As mentioned previously, this research cannot be ascribed to a single geographical research area only, but is going beyond. The approaches and paradigms in German and international geography are manifold and have changed over time, e.g. as Ehlers (2007) has listed. In recent years the research in urban geography was much more actor oriented, with space treated as a container where certain actions are located but without regarding it as determining factor.

This research work is based at the interface of urban and (new) cultural geography, and as case studies from cities in the Global South are tackled, also aspects of development geography are part of it. Subsequently, the different concepts and paradigms will be highlighted before deducing a methodology.

With the cultural turn that occurred in the last decades of the twentieth century in (German-speaking) humanities and social sciences, the symbolic dimension of space gained recognition (Ehlers 2007; Freytag 2014; Lossau 2014). Dörfler (2013) describes the disregard of space in (German-speaking) geography until it came back on the agenda via the debate on globalization and related topics emerging in the 1990s. Rothfuß and Dörfler (2013) claim a yet untapped potential of (re-)introducing a spatial reference in geography research and theory, by linking the discourse on space emerging in the 'spatial turn' debate (cf. Schlögel 2003; Klaus and Drüeke 2010) to qualitative research methodology.

The Construction of space has been a research topic within German-speaking human geography since around 2000 (Weichhart 2010). It derives from actual cultural and social geography dealing with links between societal and spatial aspects, with human production of spaces by their present and past uses (Freytag 2014), marking spatial boundaries based of societal differentiation (Reuber 2014). Nature and culture are increasingly seen as linked to each other, thus moving away from traditional Western scientific paradigms (Lippuner 2014). Cultural geography also comprises questions of cultural governance (cf. Mattissek and Prossek 2014, on the concept of governance) and global culture governance (Schmitt 2009, 2011) which is concerned with social actors, mechanisms and conditions of 'culture' production, including cultural expression or intangible values, symbols and also sense of place.

Geography has also been concerned with research on regional identities and how they are constructed for decades. The topic has gained a new momentum in the course of 'new regionalism', a kind of countermovement to globalization (cf. Paasi 2009, 2013, who researches on regional identity and regionalisms in the context of Europe). The spatiality of culture is a geographical concern (Pratt 2012). The importance of communication in space-making and constructing identities was also stressed by system theory, which was developed by Luhmann (1997, 2000). According to system theory, society is based on communication which is then creating reality as well as identity. It therefore can also serve to understand 'space' which is constituted in a societal process (Pott 2007b; Rampley 2009; Jönhill 2011; Freytag 2014), as also dealt within new cultural geography which is also concerned with the linkages of space and identities (Gebhardt et al. 2007). In Luhmann's approach, one main hypothesis is that society is an all-encompassing communication system solely consisting of communication, including actions, but excluding human beings and static objects like buildings (Jönhill 2011). To Luhmann communications and not actions define social systems. Luhmann (2000) also described art as a social system. Pott (2007b) claims that Luhmann's system theory is predestined for the social-geographical analysis of spaces that are constituted in a linguistic approach, as the theory is grounded in the concept of communication. Different authors have used this approach to analyse the construction of heritage cities as tourism destinations (cf. studies of Vanderstraeten 2005; Pott 2007a, b; Wöhler 2010; Wöhler et al. 2010).

Furthermore, aspects of tourism geography are included in the methodology as it deals with the social construction and production of tourist space (cf. Wöhler et al. 2010). Concepts related to the research topic including global tourism, location branding and competition amongst destinations are strongly linked to the question of identity formation (cf. Häußermann and Siebel 1993, on instrumentalisation of large-scale cultural and sports events to shape urban images in Europe; Poytner 2009, on the Olympics in London; Steinbrink et al. 2011, on the context of urban renewal and festivals in the Global South by the example of 2010 World Cup in South Africa; or Scharr and Steinicke 2012, on impacts of Olympic Games on Sochi) and 'destination branding' (cf. Glatter and Weber 2010, on branding urban quarters as 'in' in tourist guide books, thus influencing on shaping urban identities).

The institutionalization of cultural protection and urban cultural heritage has been tackled both in heritage studies and tourism geography (Wöhler 2008; Butina 2011). In international cultural preservation, Wöhler defined (2008) processes of 'heritageification' (cf. Chap. 3), liberating locally rooted cultural heritage from its sociohistorical context by assigning it a new meaning, which in turn is integrated in a global system of meaning in a process of 'achronisation' (Butina 2011).

Development research is concerned insofar as the case study cities are all located on the Global South. For a long time "*hegemonial theories of globalisation and postcoloniality*" (Ong 2011: 8) have shaped the debate on cities in Asia and other developing countries (Roy 2009). Euro-American cities have been found a suitable 'urban role model'. However, today many cities in the Global South have become centres of enormous changes, including economic growth and cultural vitality (Ong 2011). Much research on cities of the Global South deals with questions of social disparities and poverty, such as marginal settlements or slum upgrading—publications like Mike Davis' 'Planet of Slums' (2006) or Abdoumaliq Simone's 'City Life' (2010) called for international attention and importance. One of the milestone publications on urbanisation-related issues in developing countries was David Drakakis-Smith's book on "Third World Cities" (2000). Related topics, also a focal area within German-speaking urban geography, are megacities and mega-urbanisation, particularly in Asia (Kraas and Mertins 2008; Kraas and Nitschke 2008; Kraas 2006, 2010) but also in Latin America (Borsdorf and Coy 2009). Increasingly, aspects of urban renewal in developing and emerging countries have become research topics (Coy 2007; Kraas 2010). 'Culture' has become a very important topic in urban context, in particular in renewal or redevelopment schemes (Montgomery 2003, 2004). Increasingly, studies are dealing with heritage sites as spaces where social networks are created and maintained (Murzyn-Kupisz and Dzialek 2013), acknowledging that material and immaterial are inseparably combined (Weichhart 2009).

Currently, the understanding of urban assets and heritage is moving away from the Western knowledge-centred societies with its more analytical perspective, particularly in cultural heritage whose values are defined differently in traditional societies (Barth 2002; Koch 2013). Insofar post-colonialism is one important political trend allowing for increased appreciation of different views of the past, e.g. about importance of certain heritage, also the birth of 'heritage studies' as new field (Carman and Stig Sørensen 2009). Attitude of people towards the past and the question of how such attitudes are formed is a major area of heritage research. The construction of cultural heritage is a process involving actors from local to world society (Wöhler 2008). Nevertheless, so far research has focussed on particular aspects instead of reflecting on the interdependences and interlinkages between them (Stig Sørensen 2009).

Different authors have stated research demands concerning various aspects of heritage, conservation as well as place identity.

Heritage in the understanding of this research is composed of tangible and intangible ones; both of them together are part of urban identity construction. To Roodhouse (2010), there is a need for research on effects of culture on local development and human capital, in particular on linkages to intangible assets. There are comparably more efforts to protect tangible objects considered as cultural heritage, than intangible ones. Collective memory contributing to urban identity and represented in the symbolic aspects of tangible heritage is comparably less researched while too often the experts' view on its values is considered as more important than the local communities' ones (Jaramillo Contreras 2012). In this context, the analysis of meanings attached to spatial constructions and material aspects is of particular interest (Weichhart 2010), especially the local actor groups' ones.

Alberts and Hazen (2010) found a lack of research on heritage in terms of defining the concepts of authenticity and integrity while appreciating the uniqueness of individual sites for different reasons: different cultural contexts, the multiple layers of sites in terms of time and groups that have shaped them and the different expectations various actor groups may have. This research gap should be addressed and particularly geographers are found to be well placed in developing a greater understanding on the complex and multi-level processes of global heritage conservation (Alberts and Hazen 2010). However, different authors claim a need for interdisciplinary approaches to address cultural impacts, despite potential difficulties (Satterfield et al. 2014), e.g. environmental psychologist Uzzell (2009) who calls heritage studies per se interdisciplinary. Development research is also closely related to interdisciplinarity as it addresses complex phenomena that require multi-perspective approaches (Novy and Howorka 2014)

This research also has a potential practical aspect as research on processes of identification can contribute to more efficient and sustainable conservation practices (Vinken 2011). Shortcomings of existing approaches on urban regeneration might be overcome—among others—through a greater understanding of how people interact with the urban heritage (Tweed and Sutherland 2007). There is a lack of understanding about the complex and multi-level interactions between people and the built environment. A more integrated view would also support a stronger consideration of social dimensions in planning and serve practical action (Tweed and Sutherland 2007). Beyond academia, the relevance of a holistic view is also proven by numerous national and global policies and frameworks that have emerged in recent years especially. They have in common a more and more holistic approach, tackling aspects of place attachment, urban identity or cultural heritage, e.g. global recommendations on 'Traditional Culture and Folklore', 'Historic Urban Landscapes', intangible cultural heritage or cultural participation (UNESCO 1989, 2011, 2014; UNESCO Institute for Statistics 2012).

A number of studies on the perception of different areas and sites has been carried out, e.g. a case study of recognition of Jongmyo Shrine in Seoul, done by using interviews and desk study (Yeo and Han 2012) to analyse if cultural heritage is a tool for globalization of a city or a means of achieving sustainable urban regeneration. As a result, the authors claim the need to rethink the importance of social values of and for local communities. The value that communities assign to heritage varies over time as well as among actor groups (Rojas 2002; Brown et al. 2013). So far research has focused mainly on aspects of place attachment and social capital. However, cultural capital is not a less important resource in people's life, and deserves more attention in research (Bourdieu 1989; Lewicka 2005).

Casakin et al. (2015) criticize that place attachment and place identity are only dealt with on a neighbourhood scale in most urban and environmental studies. Among others, Hidalgo and Hernandez (2001) as well as Lewicka (2010), have researched on relationships between scale of place (apartment, neighbourhood, city) and strength of attachment to the place. However, their studies left aside both the city centre as physical place, and cultural heritage as object of place attachment.

Keitumetse (2009, in the case of Botswana) identified a scarcity of methods for investigating the changing attitudes of communities towards cultural heritage, as this becomes a tourism product. Among others, she used qualitative interviews with people living close to heritage sites or who interact with to include local perception in heritage tourism research. Scazzosi (2011) defines the need to do further research on the extremely complex processes through which people give sense to places and attribute values. In her opinion it has not been studied much yet, including the different cultural meanings and values that sociocultural groups attribute to past (material and immaterial) heritage.

Finally, Filippucci (2009) sees the need for comparative heritage studies, comparing the Euro-American with other ways "in which societies imagine, materialize and make the past known and visible to themselves and claim it in processes of identity formation (p. 20)"—in societies that are now dealing with the international idiom of 'heritage' as an imported cultural influence. Garden (2009) claims that up to now there is no clear or widely used methodology that is typically applied to heritage sites. Previous research mostly focused on either intangible aspects, on materials or tangible aspects. She finds both of these two approaches insufficient to account for the multiple functions of heritage sites as they are not able to depict the sites' complexities and ongoing changes.

Comparative studies (Rojas and Lanzafame 2011) in urbanism are increasingly popular in recent years. Its advantage is that it allows for researching how variables work differently in a range of settings, and allows the encounter of cross-cutting issues (Gough 2012). As a conclusion, to assess the complex interaction of global heritage discourses with local constructions of place attachment based on the tangible and intangible assets of historic centres, a mix of different methods seems adequate

> It is the interplay between people and things (including texts) that makes the field of heritage so unique and theoretically exciting. In order to fully do justice to this quality of heritage, a sustained 'comparison' or dialogue between, qualitative and non-qualitative methods is in order, helping to develop methodologies for connecting qualitative with quantitative data (Filippucci 2009: 324).

## 2.5 Own Analytical Model

The analytical model applied is composed of different steps referring to a variety of authors and concepts which are rooted in both geography and heritage studies. The research is done on different scales, ranging from the global reference frame of 'heritage-making' and the ascription of values, to the urban and urban centre level of the three case study cities, down to personal perception of tangible and intangible values, and finally again to a broader scale when concluding with a comparison of the case studies against the background of the global frame.

Overall the research is divided in four major steps: heritagefication, heritage-scape, identification and achronisation. Figure 2.1 shows an overview of the particular scales, goals, and applied methods of each research step. This division permits the analysis of different scales with different methods. All four steps refer to different theoretical backgrounds that will be described in more detail below.

The starting point of the research and first perspective is a supranational one, looking at the global reference frame which is determining global down to national discourses, policies, paradigms and best practices, here called heritagefication. Heritagefication as defined by Wöhler (2008) is the process of making (cultural) heritage. It is strongly linked to Pott's and Wöhler's approaches to assess the construction of (urban) tourism spaces (Pott 2007a, b), which itself is rooted in system theory. Research on urban tourism and the identity of heritage tourism destinations is logically based on the perspectives of actors in tourism, and their perception of and communication about the city. In this sense, following the system theory approach, any inhabitant of a city, whether native or not, is nothing but someone who is part of the communication about the city itself. As the focal point is the place attachment and construction of identity of locals and not tourists, the perspectives were modified, adding other approaches of doing research on urban heritage. The content of the communication might differ; however, to assess this, a similar methodology can still be applied. This allows for adapting such a model to research on the place perception and construction of identity in historic city centres of a certain local actor group.

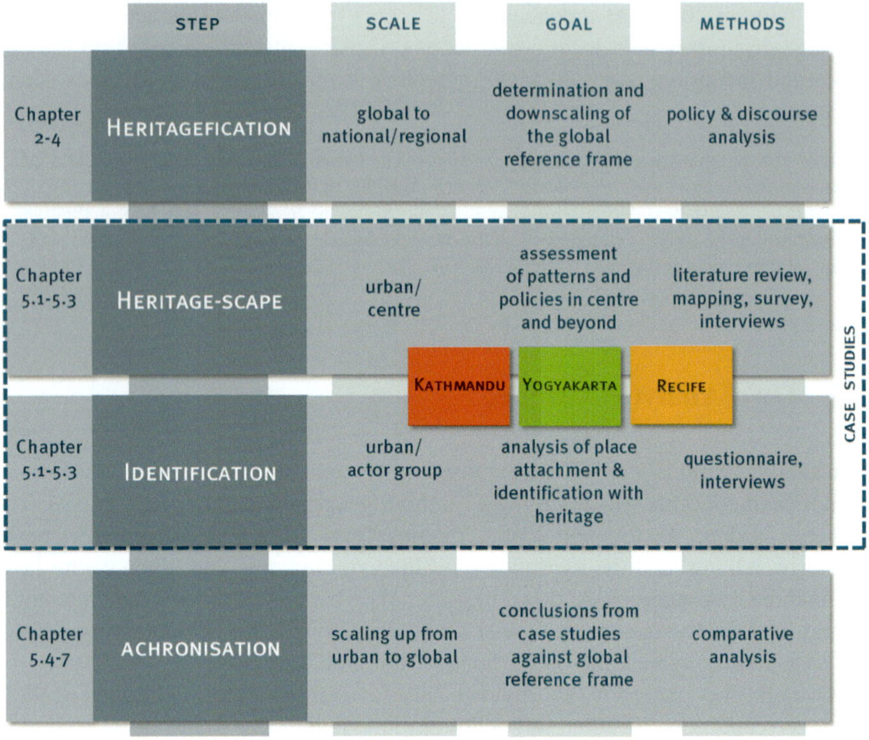


**Fig. 2.1** Research framework, scales, goals and related methods

In a second step, the heritage-scape of the case study cities is analysed. Garden (2006, 2009) has been among the first authors using this word as both a descriptor and a methodology for analysing the living processes that circulate at heritage sites (Di Giovine 2011). Additionally, the word—inspired by 'landscape'—indicates that the multilayered and multi-temporal diversity of assets a certain site is composed of. Garden (2006, 2009) developed the three-step 'heritage-scape' methodology: composed of boundaries (defining a heritage site's 'boundaries', in contrast to the outside areas); cohesion (sense of place, connecting visible and invisible components of site); and visibility (perception of tangible and intangible features that refer to the past, attached to physical elements) as well as their links and interplay. Such stepwise approach corresponds to the mosaic character of cities, with a whole variety of contrasting subspaces. Different assets, such as functional, social or structural ones allow subdividing the urban landscape. The same is done by the breakdown into heritage—non heritage. Therefore, this step (somewhat comparable to 'heterogenization' in the analysis framework Pott (2007a) is using in his research on urban tourism in historic cities) analyses the past and present urban layout as well as urban policies, and formal as well as informal development before focussing again on the heritage (in particular the urban centre) as the potential key issue for the construction of an urban identity.

One main assumption is that the historic centre serves as a means of identification. The same geographical space (Ronneberger 2000; 'containerraum' c.f. Dörfler 2013; Freytag 2014) can cause very different feelings about it. Its perception as well as the attached values may be totally different among different individuals or groups. Declaring some (assets of a) historic centre valuable must not necessarily imply that locals perceive it the same way. Therefore in this step the perception of the historic centres of the three case study cities is analysed, based on the assumption that the centre is not just as any other part of the town but that there is a distinct place attachment which may even be different to the values endowed by formal conservation authorities on national or global scale. In this step methodology stems from studies on place attachment and urban identity, carried out by Lewicka (2005, 2008), Hidalgo and Hernández (2001) and Hernández et al. (2007). However, the questionnaires were adapted, as most of these studies stem from European context and were not focussed on the historic centre.

Finally, a comparison of the values endowed to the historic centres by interviewees and focal group and the actual processes as well as policies of urban planning and development will reveal how far the urban reality is considering the values found in the previous step. This last step is called achronisation, referring to Butina (2011). She is using the word to describe a process of instilling a space with symbolic charge and meaning, going hand in hand with the global 'meaning-making system' of cultural heritage (heritagefication). Therefore, first the three case studies will be compared to draw conclusions on their similarities and differences in terms of processes and values ascribed to the historic centres. Finally, the case studies outcomes will be evaluated against the background of the global perspectives of heritagefication, allowing for conclusions on the impacts and appropriateness of such global concepts for the individual case of historic city centres in the Global South.

## 2.6 Methodology

There is a variety of methods that have been used in previous investigations on urban, heritage and/or identity topics. Different methods have been used to survey people's perception and feelings of places, using comparative studies, quantitative questionnaires (c.f. studies on European cities by Lewicka 2008, 2010) or visual tools like images of a city (c.f. Salesses et al. 2013, using geo-tagged images to measure the perception of uniqueness, safety and class in US and Austrian cities).

In her thesis on urban places that are undergoing a conversion, Baum (2008) distinguishes between spatial aspects (e.g. location, building patterns or quality of open spaces), functional aspects (e.g. accessibility, uses, private and public spaces), social and atmospheric aspects (e.g. identification, history, atmosphere), using a mixed method approach. While the spatial analysis is executed based on methods from planning and geography, empirical analysis uses qualitative and quantitative methods.

The construction of urban identities and place attachment in different urban contexts has been studied before, e.g. by Lewicka (2005, 2008) who researched on attachment in cities with Polish roots or inhabitants. Lewicka (2005, 2008, 2010) as well as Kyle et al. (2014) used standardized interviews to assess place attachment and neighbourhood ties in different case studies—to do so they developed different evaluation scales.

Other investigations making use of questionnaires on place attachment (Hidalgo and Hernández 2001; Hernández et al. 2007; Lewicka 2008, 2010), proclaim interviews as means of investigation on attitudes to heritage and identity (Stig Sørensen 2009) or do analyses of past and present policy instruments in combination with key actor interviews (Manzi and Jacobs 2009; Yeo and Han 2012). Waterton et al. (2006) promote discourse analysis for the heritage topic as they found a common sense and distinctly Western understanding of what heritage entails on a global scale, reflected in legislation, charters and the value system of experts.

Ujang (2012) uses a mix of qualitative and quantitative data (field surveys and interviews, in this case with users of the shopping street) to assess attachment and place identity, while Ellingsen (2010) focusses on qualitative methods, different kinds of interview types, in his study on territoriality of different ethnic groups in Kathmandu. Field surveys were also one method used by Schmitt (2011), who analysed and compared different World Heritage sites for his work on global cultural governance. His methodology comprises qualitative interviews, participatory observation, own surveys and document analysis. In his study on discourses of regional identity in Finland, Paasi (2013) relies on the analysis of strategic regional plans and expert interviews.

In this research, a mix of quantitative and qualitative methods is applied, following a number of previous studies using such a mix or papers that recommend such an approach (Reuber 1993; Townend and Whittaker 2011; Dannecker and Vossemer 2014; Englert and Dannecker 2014; Slezak 2014; Ujang 2012). The methods used in each step of investigation are as well described in the right column of Fig. 2.1.

The global reference scale is assessed through an analysis of global discourses, in policies, charters and legislations and decreed by various national and international organisations (c.f. studies by Schmitt 2011; Gfeller 2013; Veldpaus et al. 2013). A particular emphasis is placed on the period after 1972, when the UNESCO Convention Concerning the Protection of the World Cultural and Natural Heritage was adopted.

One of the hypotheses is that global policies are originally rooted in Western concepts, and that there is a global production of heritage sites sense and meaning which is slowly changing over time (c.f. Schmitt 2009, 2011, on global cultural governance). Based on this assumption, qualitative methods including a discourse analysis of heritage charters and documents as well as an analysis of previous studies dealing with cultural heritage were chosen for the first part of the investigation.

To gain a deeper insight on processes, goals, policies and governance of urban planning and heritage conservation in general in the case study cities, semi-structured expert interviews were carried out following an interview guideline. After having developed the expert interview guideline it was revised and pretested to avoid culturally non-sensitive questions. Revision was done by researchers familiar with at least one of the three research environments, including such based in the countries and such based in Europe (c.f. Dannecker and Vossemer 2014, on qualitative methods in development research; Müller-Mahn and Verne 2014, on development research). Overall ten interviews were done in Yogyakarta, twelve in Kathmandu and fifteen in Recife, with the interview duration differed between 20 min and one and a half hours. All of the selected interview partners had a certain relation to the case study areas; they came from regional or urban planning authorities, monument preservation authorities, private foundations dealing with the urban tangible or intangible heritage, entrepreneurs being based in the area, the urban history museum, local university researchers dealing with the urban area in various ways, or an NGO conducting projects on urban issues.

The expert interviews were recorded; additionally notes were taken, after the interviews a transcription was done for further analysis (Bohnsack 2008). For all interviews the same transcription system was used. The analysis categories were set based on the expert questionnaire itself. During the analysis of the interviews categories and coding were revised and adapted, based on the studies of Hopf (Hopf and Weingarten 1993; Kuckartz 2007; Hopf 2008). The analysis and interpretation of interviews are done based on coding (Englert and Dannecker 2014) and with MAXQDA software (cf. Annex IV with the coding system). However, interpretation of results in terms of how the respondents expressed their opinions was primarily based on a comparison with other interviewees from the same city. It has been decided purposefully not to do a comparative study of how strong interviewees expressed their satisfaction or rejection of certain policies or processes, and how they acted non-verbally, as diction and straightforwardness of language depends is culturally specific. It therefore differs between the three cities and prohibits a direct comparison of expressions used, as, e.g. in Javanese context a direct 'no' is regarded as rude. Therefore, disagreement is demonstrated in a very polite and indirect way. Extensive prior experiences in all three countries and cities, however, permitted an appropriate evaluation and comparison of the interviews.

The interviews were supported by a survey of corresponding legislations, local governance system and previous scientific publications with related topics, including theses from universities in the three case study cities accessed in the university libraries. This collection also supported a comparison of global policies discourse analysis (c.f. Schicho 2014) to local discourses.

To assess the place attachment and identification with the centre, a questionnaire was prepared, reviewed, tested and then handed out (the questionnaire is provided in Annex III). Table 2.1 is providing an overview of interview and questionnaire numbers per city. The questionnaire itself was designed in such a way, that it allowed being used in the different cities while only changing the city name itself. Overall the questionnaire comprised of twenty questions, starting with general ones on the respondent's background, followed by a set of questions on the urban history and place memory, e.g. asking places considered as important for the respondent or

**Table 2.1** No. of interviews, questionnaires and photos for photo documentation obtained in case study cities

| Data | case studies | | |
|---|---|---|---|
| | Kathmandu | Yogyakarta | Recife |
| Expert interviews | 10 | 12 | 15 |
| Questionnaires | 120 | 80 | 81 |
| Photo documentation | 1,478 | 1,426 | 1,475 |

the urban history. The next block of questions then investigated the personal ties to the city and the historic centre, leading over to the perception of the centre, and then processes and projects. Finally, respondents were asked about their vision for the future urban development in their city, and in the last section about their attachment to different places on various scales. Only one question was a purely open one, while the others were mainly giving different answer options, e.g. from “absolutely agree” to “absolutely disagree”, in few cases with the voluntary option to given an additional comment.

As this research deals with different case studies, it was decided not to select the inhabitants of the city centre as peer groups, as the sociocultural population and use structure of the centre areas is very different, and also not all areas are equally inhabited. University students were selected as a sample for this study, a method applied, e.g. in recent research on place attachment and place identity in Israeli cities (Casakin et al. 2015). All three cities comprise of universities of supraregional importance that run graduate, postgraduate and Ph.D. courses on various topics. To narrow down the peer group, only students of at least postgraduate level were chosen, studying various subjects with a close link to sustainability and/or planning. That way it could be assumed that the concept of sustainable development was known, also allowing for the investigating on the respondents’ vision of a sustainable development for his/her city. In addition, it was very likely that the respondents, as having access to higher education, will be future decision-makers of the three cities and regions, thus their opinions also permits assumptions on guiding planning principles the city authorities might follow in future.

Participants were approached before or after attending courses. They were informed about the research objectives and asked to fill the questionnaire voluntarily and anonymously, in the presence of the researcher and only for the researcher. It took between 25 and 40 min to complete the questionnaires. The analysis of the questionnaire results was done with SPSS. Overall, between 80 and 120 filled questionnaires were obtained. This quite different number of responses results from different factors: the number of overall students in the respective courses, and the number of students available (due to unscheduled vacation time in Brazil and a large number of Nepalese students doing field work during the field research stay, in addition to a complicated inner-university situation with the faculty head office being closed for months).

The interviews and questionnaires were accompanied by site inspections and mapping surveys of the research areas and its vicinities. Based on available maps, the different uses within the area, the condition of buildings and places, and user groups were mapped and documented. Street names were taken from maps bought in the respective cities or obtained online from the municipalities. In Yogyakarta and Kathmandu sometimes street names differed between different maps or between local names and formal names. In these cases it was tried to use the street and location names found on road signage or in documents from urban authorities. During the survey overall 1,475 photos of streetscapes, buildings and activities, during different daytimes and days of week were taken in Recife; another 1,478 in Kathmandu and 1,426 in Yogyakarta (cf. Table 2.1).

After analysing the case study findings, the three case studies are compared with each other, to allow the drawing of conclusions on a more abstract scale. Such an approach was followed in different studies on urban regeneration (cf. Delmelle 2015, who compared regeneration in four US cities), as well as place attachment (cf. Hidalgo and Hernández 2001; Lewicka 2008), and also permits to conclude on particularities of the different cities.

## 2.7 Selected Case Study Areas

To carry out a research on the construction of urban identities in historic city centres in the Global South, it was decided to have different case studies instead of only one to allow for a comparison. For this book three cities were selected as case study areas: Yogyakarta, Indonesia; Kathmandu, Nepal; and Recife, Brazil. The three selected cities obviously differ in many aspects, but in terms of urban heritage and identity they have much in common: all three cities still comprise of a historic centre of supraregional publicity which is under pressure due to ongoing urban change.

In Latin America, Recife, the city with the highest number of urban development plans in the whole of Brazil will be studied. The historic centre is under pressure of tremendous changes due to economic development and population increase. At the same time, the debate on the urban cultural heritage and ascribed values is a very vivid one, especially the discourse on intangible heritage like Afro-descendant culture.

Kathmandu in Nepal is currently facing rapid pressures and tremendous changes due to population increase, a lack of economic resources combined with weak planning and governance systems. Although the unique historic centre around the Durbar Square is protected by law, it suffers from these processes. In the central area of Yogyakarta, Indonesia, bigger renewal projects have already been carried out, and the first outcomes can be perceived. However, economic development and increasing importance of tourism sector is putting pressure on the area.

Both Asian cities are based on Buddhism and Hinduism planning paradigms and thus partly comparable in terms of urban patterns. Recife and Yogyakarta both were influenced by Dutch colonial rule that can still be perceived in the urban outline.

An overview on similarities and differences between the three cities is provided in Table 2.2. There, a large number of similarities are revealed. All three cities played and still play a role for the regional or even national history and culture; they are administrative hubs and seats of regional or national governments. More importantly, they are facing quite comparable urban development processes, in particular growing urban pressure due to increasing population and (comparably high) economic growth rates, resulting in densification and verticalisation of the urban areas.

**Table 2.2** Comparison of case study cities

| | Kathmandu | Yogyakarta | Recife |
|---|---|---|---|
| **Administration and ongoing processes** | | | |
| City of (supra) regional economic and administrative importance | ✔ | ✔ | ✔ |
| Urban agglomeration consisting of different municipalities | ✔ | ✔ | ✔ |
| Urban pressure due to population and/or economic growth | ✔ | ✔ | ✔ |
| Ongoing urban densification and verticalization processes | ✔ | ✔ | ✔ |
| **Urban history and culture** | | | |
| Urban colonial past which is still visible in the urban layout | o | ✔ | ✔ |
| City depicts important part/ phase of national history | ✔ | ✔ | ✔ |
| City considered as national or regional cultural hub | ✔ | ✔ | ✔ |
| Strong focus on preservation of intangible values | o | ✔ | ✔ |
| **Urban heritage** | | | |
| Historic urban historic center | ✔ | ✔ | ✔ |
| Urban cultural heritage of national importance | ✔ | ✔ | ✔ |
| Urban world cultural heritage site | ✔ | o | – |
| Intangible world heritage site | – | o | ✔ |

**Legend:** ✔ correct; o partly correct; – not correct

All three historic city centres are still preserved, being considered at least partly as heritage on regional or national scale. In the case of Kathmandu, overall seven sites of the valley are inscribed as Cultural World Heritage "Kathmandu Valley". Yogyakarta Palace Area had been listed on the Indonesian tentative list of world Heritage, while the *Frevo* of Recife is listed on the Representative List of the Intangible Cultural Heritage of Humanity (cf. Sect. 3.5). A more detailed description of the case study cities, their urban history, built environment, and governance system is given in Chap. 5.

## 2.8 Research Constraints

Doing a research in different cities located not only in different countries, but even continents, obviously bears a number of risks when designing a methodology to be applied in all places.

As explained before, the number of interviews and questionnaires varied among the cities. The number of interviews and the background of interviewees differed due to unequal accessibility of authorities and due to differences in administrative systems, e.g. kind of authorities and number of employees. Furthermore, not all authorities were in favour of answering questions, in other cases—in particular in Indonesia—different formal permits were requested that were not accessible in all cases. Overall, the study tried to access those authorities and institutions that were dealing with urban planning and urban tangible and intangible heritage. Logically, the authority names, size and levels were not the same due to different government and governance structures as well as different natural environments.

Also, the number of questionnaire respondents varied as well as their scientific background, due to the different educational systems in the three universities. They were students of different M.Sc. or Ph.D. programmes dealing with the sustainability concept, but the names of the MSc programmes were not the same, nor was the distribution among level of education. One major reason for this is that programmes in Brazil and Nepal only allowed a limited number of students per year, while this was not the case in Indonesia, resulting in a higher number of filled questionnaires from there. Other unexpected difficulties—as explained above—resulted from availability of students in university itself which was not always given.

Only comparably low shares of the questionnaire respondents are inhabitants of the historic centres: 6.7 % in Yogyakarta, and only 2.5 % in Kathmandu and Recife. Instead of asking for the absolute distance between their places of residence and the centre areas the questionnaire asked for time needed to go there and means of transportation, as the abstract distance is not a good comparative indicator for accessibility. The majority of Kathmandu's and Yogyakarta's respondents need between 16 and 30 min to get to the historic centre, while the peer group in Recife needs up to 15 min more on average (in this case either using bus or car). In Kathmandu walking or motorbikes are the main means of transportation, in

Yogyakarta motorbikes are preferred. Overall, the centre areas are not too far away of the respondents' residences, not preventing visits to the area.

All three research areas have different native languages. As the researcher was able to read and speak Portuguese, interviews in Recife were done in Portuguese, with also the questionnaire translated to Portuguese and handed in a bilingual version. Many students had hesitations concerning English, as the proficiency is comparably lower than in the other countries. In Nepal and Indonesia, questionnaires were distributed in English as university education from postgraduate level on is taking place at least to a large extent, if not completely in English. In Nepal, it was possible to carry out all interviews in English, also most legal documents and reports are available in English. In Yogyakarta, some interviews were done in English, others in Bahasa Indonesia with the help of a translator. It was also possible to get some legal documents translated, as often they are not translated. Thus language barriers did not prohibit conducting research.

As the field work took place in three locations time schedule was tight and did not permit too extensive stays in each location. Probably another researcher focussing on only one of these cities still may be far more expert in this location. Nevertheless the comparative concept of this research justifies the scientific approach chosen, which in no case took place in an unknown location. The researcher was familiar with all three cities before doing the field research itself, between two and seven previous stays served to become acquainted with the area and to build up strong networks of resource persons. These preparatory visits were followed by stays of up to two months dedicated to field research only—interviews, questionnaires, mappings and surveys. Beforehand, visits were used to collect literature and to gain overviews of the cities. Impressions from previous stays were also used to back up the field surveys which logically only took place within a certain period, not being representative for uses throughout the year, due to climate, religious calendars or other festivals. In addition, the definition of key persons and some of the interview partners as well as the delimitation of research areas were done during the preparatory visits, mainly in 2011/2012, while field research was carried out in 2013. Having described the constraints, it can, however, be concluded that it was possible to overcome them and to legitimately do a comparison of the three cities.

The questionnaire itself was designed for a comparative study, with only the questions on distinct places of remembrance and intangible values being tailored to the different cities (see Annex III with the questionnaire). The other questions were generalized to allow a full comparative analysis. Obviously the peer group chosen for the questionnaires is not representative for the whole urban population. Surely they have a better education and probably a historical awareness at least as high as the urban average. Therefore, the results obtained in the interviews cannot be generalized in terms of absolute numbers and valuation. It, however, permits generalizing overall positive or negative valuation and allows for a comparison of the three case study cities, what has been the intention of the study.

In the field the questionnaire itself proved to be quite long and exhaustive, taking more time than initially expected and resulting in the fact, that the last question was answered considerably less often than the others. Therefore, this question was omitted in the evaluation.

## References

Abu-Rabia S (2010) Memory, belonging and resistance: the struggle over place among the Bedouin-Arabs of the Naqab/Negev. In: Fenster T, Yacobi H (eds) Remembering, forgetting and city builders re-materialising cultural geography. Ashgate, Farnham, UK, pp 65–83

Adams B (2002) Preservação Urbana: Gestão e Resgate de uma História. O patrimônio de Florianópolis, Editora da UFSC, Florianópolis

Ahmad AG (2006) Cultural heritage of Southeast Asia: preservation for world recognition. Malaysian Townplan Journal 4(1):52–62

Albert M-T (2013) Heritage studies—paradigmatic reflections. In: Albert M-T, Bernecker R, Rudolff B (eds) Understanding heritage: perspectives in heritage studies. de Gruyter, Berlin/Boston, pp 9–17

Alberts HC, Hazen HD (2010) Maintaining authenticity and integrity at cultural world heritage sites. Geogr Rev 100(1):56–73

Amen M, Toly N, McCarney P, Segbers K (2011) Sighting or slighting cities in international relations. In: Amen M, Toly NJ, McCarney PL, Segbers K (eds) Cities and global governance: new sites for international relations. Ashgate, Farnham, pp 13–32

Andrews C, Urbanska W (2010) Inspiring people to see that less is more. In: Starke L, Mastny L (eds) State of the world 2010: transforming cultures: from consumerism to sustainability. A worldwatch institute report on progress toward a sustainable society. Norton, Washington DC, pp 178–184

Ashworth GJ, Tunbridge JE (2000) The tourist-historic city. Retrospect and prospect of managing the heritage city. Adv Tour Res Ser. Elsevier, Oxford

Bandarin F, Hosagrahar J, Sailer Albernaz F (2011) Why development needs culture. J Cult Heritage Manage Sustainable Dev 1(1):15–25. doi:10.1108/20441261111129906

Barth F (2002) Anthropology of knowledge. Curr Anthropol 43(1):1–11

Barthel-Bouchier D (2012) Heritage and social change: anticipating future trends. In: Negussie E (ed) Changing world, changing views of heritage: heritage and social change. Proceedings of the ICOMOS Scientific Symposium 2010. ICOMOS, Dublin, Ireland, pp 82–86

Baum M (2008) Urbane Orte. Ein Urbanitätskonzept und seine Anwendung zur Untersuchung transformierter Industrieareale. Universitätsverlag Karlsruhe, Karlsruhe

Baum M (2010) Urbane Orte. In: Baur R, Feuz M, Gasser Derungs C, Gmünder A (eds) meine, deine schönste stadt der welt. Merkmale urbaner Lebensqualität. Lars Müller Publishers, Baden, pp 152–156

Baur R, Feuz M, Derungs CG, Gmünder A, Hausheer T, Jann M, Krass P, Mv Lupin, Maag T, Tgetgel U, Zwissler M (2010) meine, deine schönste stadt der welt. Lars Müller Publishers, Baden, Merkmale urbaner Lebensqualität

Baxter I (2009) Means maketh the end. The context for the development of methodologies to assess the state of the historic environment in the UK. In: Stig Sørensen ML, Carman J (eds) Heritage studies. Methods and approaches. Routledge, London, New York, pp 85–102

Berkes F, Kofinas GP, Chapin III SF (2009) Conservation, community, and livelihoods: sustaining, renewing, and adapting cultural connections to the land. In: Chapin III SF, Kofinas GP, Folke C (eds) Principles of ecosystem stewardship: resilience-based natural resource management in a changing world. Springer, New York, pp 129–148

Bittner R (2008) Cities of Perfection: Stadtfluchten oder Stadtversprechen. In: Wolfrum S, Nerdinger W (eds) Multiple City. Stadtkonzepte 1908|2008. Jovis, Berlin, pp 234–238

Bittner R (2010) Der Reiz der Stadt. In: Baur R, Feuz M, Gasser Derungs C, Gmünder A (eds) meine, deine schönste stadt der welt. Merkmale urbaner Lebensqualität. Lars Müller Publishers, Baden, pp 132–134

Bohnsack R (2008) Rekonstruktive Sozialforschung. Einführung in qualitative Methoden. 7 edn. Barbara Budrich, Opladen

Borden I (2008) Performance, risk and the public realm. In: Wolfrum S, Nerdinger W (eds) Multiple City. Stadtkonzepte 1908|2008. Jovis, Berlin, pp 154–158

Borsdorf A, Coy M (2009) Megacities and global change: case studies from Latin America. Erde 140:1–20

Bourdieu P (1989) Social space and symbolic power. Sociol Theory 7(1):14–25

Brakman A (2011) Maastricht-Lanakerveld: the place to be? In: Schofield J, Szymanski R (eds) Local heritage, global context: cultural perspectives on sense of place. Heritage, culture and identity. Ashgate, Farnham, pp 120–128

Brown K, O'Neill S, Fabricius C (2013) Social science understandings of transformation. In: SC and UNESCO (ed) Changing global environments, world social science report 2013. OECD Publishing and UNESCO Publishing, Paris, pp 100–106

Burke PJ, Stets JE (1999) Trust and commitment through self-verification. Soc Psychol Quart 62:347–366

Butina B (2011) From the seven wonders of the ancient world to the UNESCO world heritage: political and economic aspects of institutionalised cultural preservation. Hrvatski Geografski Glasnik 73(2):141–154

Carman J, Carman P (2009) The intangible presence. Investigating battlefields. In: Stig Sørensen ML, Carman J (eds) Heritage studies. Methods and approaches. Routledge, London, New York, pp 202–315

Carman J, Stig Sørensen ML (2009) Heritage studies. An outline. In: Stig Sørensen ML, Carman J (eds) Heritage studies. Methods and approaches. Routledge, London, New York, pp 11–28

Casakin H, Hernández B, Ruiz C (2015) Place attachment and place identity in Israeli cities: the influence of city size. Cities 42(B):224–230. doi:10.1016/j.cities.2014.07.007

Castello L (2010) Rethinking the meaning of place. Ashgate, Farnham

Chien-Yuan Lin J (2007) Urban regeneration and sustainable development. Nature-Conservation-led Approach for Green Island in Taiwan. Lincoln Institute of Land Policy, Cambridge

Colantonio A, Dixon T (2011) Urban regeneration & social sustainability. Best practice from European cities. Real estate issues. Wiley-Blackwell, Chichester, West Sussex

Constantinescu C, Hărmănescu M (2014) Recondition the urban texture through public interventions. In: Crăciun C, Bostenaru Dan M (eds) Planning and designing sustainable and resilient landscapes. Springer, Dordrecht, Heidelberg, New York, London, pp 243–257

Council of Europe, The congress of local and regional authorities (2008) European urban charter II—Manifesto for a new urbanity. Council of Europe. https://wcd.coe.int/ViewDoc.jsp?id=1302971&Site=Congress&BackColorInternet=e0cee1&BackColorIntranet=e0cee1&BackColorLogged=FFC679#RelatedDocuments. Accessed 11 Aug 2009

Coy M (2007) Innenstadtentwicklung und Innenstadterneuerung in São Paulo—Akteure, Wahrnehmungen, Interessenkonflikte. In: Rothfuß E, Gamerith W (eds) Stadtwelten in den Americas., vol Passauer Schriften zur Geographie 23. Passavia Universitätsverlag, Passau, pp 57–69

Dannecker P, Vossemer C (2014) Qualitative Methoden in der Entwicklungsforschung. In: Dannecker P, Englert B (eds) Qualitative Methoden in der Entwicklungsforschung. Mandelbaum, Wien, pp 153–175

Davis M (2006) Planet of slums. Verso, London

de Albuquerque Lapa T, Almeida de Melo RF (2007) Interventions in historic areas, mobility and urban conservation: the case study of Bairro do Recife District. City Time 2(3–4):37–46 de Noronha Vaz E, Cabral P, Caetano M, Nijkamp P, Painho M (2012) Urban heritage endangerment at the interface of future cities and past heritage: a spatial vulnerability assessment. Habitat International 36:287–294

Delmelle EC (2015) Five decades of neighborhood classifications and their transitions: A comparison of four US cities, 1970–2010. Appl Geogr 57:1–11

Department of Economic and Social Affairs—Division for Social Policy and Development, Secretariat of the Permanent Forum on Indigenous Issues (2009) State of the world's indigenous peoples. United Nations Publication, New York

Di Giovine M (2011) UNESCO's heritage-scape: a global endeavour to produce 'peace in the minds of men' through tourism and preservation. In: Singh RPB (ed) Heritagescapes and cultural landscapes. Shubhi Publications, Delhi, pp 57–86

Díaz-Rodríguez MC, Armas-Díaz A, García-Herrera LM, Garía-Hernández JS (2015) Spaces of transition: young people's social practises in Santa Cruz de tenerife (Canary Islands, Spain). Erde 146(1):79–89

Donnelly B (2014) New ideas need old buildings. http://sustainablecitiescollective.com/donnellyb/249976/new-ideas-need-old-buildings. Accessed 10 Oct 2014

Dörfler T (2013) Raumbezogene Qualitative Sozialforschung. Konzeptionelle Überlegungen zwischen Geographie und Soziologie. In: Rothfuß E, Dörfler T (eds) Raumbezogene Qualitative Sozialforschung. Springer, Wiesbaden, pp 33–60

Dormaels M (2013) the concept behind the word: translation issues in definitions of heritage. In: Albert M-T, Bernecker R, Rudolff B (eds) Understanding heritage: perspectives in heritage studies. de Gruyter, Berlin/Boston, pp 107–115

Drakakis-Smith D (2000) Third world cities. Routledge introductions to development series, 2nd edn. Routledge, New York

Dürr E (2005) Identitäten und Sinnbezüge in der Stadt: Hispanics im Südwesten der USA. Lit, Münster

Ebert C (2005) Identitätsorientiertes Stadtmarketing. In: Meffert H, Burmann C, Koers M (eds) Markenmanagement. Identitätsorientierte Markenführung und praktische Umsetzung. Gabler, Wiesbaden, pp 565–587

Eeckhout B, Jacobs S (2008) (Dys)Functionalism in a Post(sub)urban landscape. In: Wolfrum S, Nerdinger W (eds) Multiple city. Stadtkonzepte 1908|2008. Jovis, Berlin, pp 34–37

Ehlers E (2007) Deutsche Kulturgeographie im 20. Jahrhundert. Geographische Rundschau 59(7–8):4–11

Eizenberg E (2010) Remembering forgotten landscapes: community gardens in New York and the reconstruction of cultural diversity. In: Fenster T, Yacobi H (eds) Remembering, forgetting and city builders Re-materialising cultural geography. Ashgate, Farnham, UK, pp 7–26

Ellingsen W (2010) Ethnic appropriation of the city. The territoriality of culture in Kathmandu. LAP Lambert, Saarbrücken

Enache C, Căplescu OA (2014) Media landscape. In: Crăciun C, Bostenaru Dan M (eds) Planning and designing sustainable and resilient landscapes. Springer, Dordrecht, Heidelberg, New York, London, pp 259–273

Englert B, Dannecker P (2014) Praktische und ethische Aspekte der Feldforschung. In: Dannecker P, Englert B (eds) Qualitative Methoden in der Entwicklungsforschung. Mandelbaum, Wien, pp 233–265

Espahangizi R (2011) Migrantische Unternehmen in Köln. In: Löw M, Terizakis G (eds) Städte und ihre Eigenlogik. Ein Handbuch für Stadtplanung und Stadtentwicklung, vol Interdisziplinäre Stadtforschung, Band 11. Interdisziplinäre Stadtforschung. Campus Verlag, Frankfurt am Main, pp 57–66

Evers H-D (2007) The end of urban involution and the cultural construction of urbanism in Indonesia. Internationales Asienforum 38(1–2):51–65

Farrell T (2010) Shaping London. The patterns and forms that make the metropolis, Wiley, Chichester, West Sussex

Fenster T (2010) One Place—different memories: the case of Yaad and Miaar. In: Fenster T, Yacobi H (eds) Remembering, forgetting and city builders Re-materialising cultural geography. Ashgate, Farnham, UK, pp 85–105
Filippucci P (2009) Heritage and methodology. A view from social anthropology. In: Stig Sørensen ML, Carman J (eds) Heritage studies. Methods and approaches. Routledge, London, New York, pp 319–325
François E, Schulze H (2005) Deutsche Erinnerungsorte. Beck, München
Frank S (2011) Stadtmarketing. In: Löw M, Terizakis G (eds) Städte und ihre Eigenlogik. Ein Handbuch für Stadtplanung und Stadtentwicklung, vol Interdisziplinäre Stadtforschung, Band 11. Interdisziplinäre Stadtforschung. Campus Verlag, Frankfurt am Main, pp 37–46
Frehse F (2013) Zeiten im Körper: das Potenzial der Lefebvre'schen Methode für die (lateinamerikanische) Stadtforschung. In: Huffschmid A, Wildner K (eds) Stadtforschung aus Lateinamerika: Neue urbane Szenarien: Öffentlichkeit—Territorialität—Imaginarios. Transcript, Bielefeld, pp 145–169
Freytag T (2014) Raum und Gesellschaft. In: Lossau J, Freytag T, Lippuner R (eds) Schlüsselbegriffe der Kultur- und Sozialgeographie. Eugen Ulmer, Stuttgart, pp 12–24
Gaines J, Jäger S (2009) Albert speer & partner. Ein Manifest für nachhaltige Stadtplanung: think local, act global. Prestel, München
Garden M-CE (2006) The heritagescape: looking at landscapes of the past. Int J Heritage Stud 12 (5):394–411. doi:10.1080/13527250600821621
Garden M-CE (2009) The heritagescape. Looking at heritage sites. In: Stig Sørensen ML, Carman J (eds) Heritage studies. Methods and approaches. Routledge, London, New York, pp 270–291
Gebhardt H, Schweizer G, Reuber P (eds) (1995) Zuhause in der Großstadt: Ortsbindung und räumliche Identifikation im Verdichtungsraum. Geographisches Institut der Universität zu Köln, Köln
Gebhardt H, Mattissek A, Reuber P, Wolkersdorfer G (2007) Neue Kulturgeographie? Geographische Rundschau 59(7–8):12–19
Gfeller AE (2013) Negotiating the meaning of global heritage: 'cultural landscapes' in the UNESCO World Heritage convention, 1972–92. J Global Hist 8 (03):483–503. doi:10.1017/S1740022813000387
Gibson L (2012) 'We are the river': place, wellbeing and aboriginal identity. In: Atkinson S, Fuller S, Painter J (eds) Wellbeing and place, vol Farnham. Ashgate, pp 201–215
Giddens A (1990) The consequenses of modernity. Polity Press, Cambridge University Press
Glatter J, Weber D (2010) Die mediale Konstruktion des Stereotyps Szeneviertel in Reiseführern. In: Wöhler K, Pott A, Denzer V (eds) Tourismusräume. Zur soziokulturellen Konstruktion eines globalen Phänomens. transcript, Bielefeld, pp 43–65
Gough KV (2012) Reflections on conducting urban comparison. Urban Geogr 33(6):866–878
Greenwald R (2013) Why is 'authenticity' so central to urban culture? The Atlantic Media Company. http://www.theatlanticcities.com/neighborhoods/2013/01/why-authenticity-so-central-urban-culture/4543/
Greverus I-M (2008) Lebendige Orte oder: Die Spuren der Berührung, Gedanken zu einer Ästhetik urbaner Vielfalt. In: Wolfrum S, Nerdinger W (eds) Multiple city. Stadtkonzepte 1908|2008. Jovis, Berlin, pp 139–141
Grove JM (2009) Cities: managing densely settled social-ecological systems. In: Chapin SF III, Kofinas GP, Folke C (eds) Principles of ecosystem stewardship: resilience-based natural resource management in a changing world. Springer, New York, pp 281–294
Grünberg W (2004) Orientierungssuche. Stadtidentität zwischen Markt, Event und Orientierungssuche. In: Hilber MLE, Ayda (ed) Stadtidentität—Der richtige Weg zum Stadtmarketing. Orell Füssli Verlag AG, Zürich, pp 51–64
Haines C (2011) Cracks in the Façade: landscapes of hope and desire in Dubai. In: Roy A, Ong A (eds) Worlding cities: Asian experiments and the art of being global. Studies in urban and social change. Wiley-Blackwell, Chichester, pp 160–181
Halbwachs M (1980 [1950]) The collective memory. Harper & Row, New York

Hall S (1999) Whose heritage? Un-settling 'the heritage', re-imagining the post-nation. Third Text 13(49):3–13. doi:10.1080/09528829908576818

Handal J (2006) Rebuilding city identity through history: the case of Bethlehem-Palestine. In: Zetter R, Butina Watson G (eds) designing sustainable cities in the developing world. Ashgate, Aldershot, pp 51–68

Harrison R (2011) 'Counter-mapping' heritage, communities and places in Australia and the UK. In: Schofield J, Szymanski R (eds) Local heritage, global context: cultural perspectives on sense of place. Heritage, Culture and Identity. Ashgate, Farnham, pp 79–98

Hassenpflug D (2000) Citytainment oder die Zukunft des öffentlichen Raums. In: Matejovski D (ed) Metropolen: Laboratorien der Moderne, vol Schriftenreihe., des Wissenschaftszentrums Nordrhein-Westfalen, Band 5Campus Verlag, Frankfurt/New York, pp 308–323

Hatuka T (2010) The fragility of memory and its remedy through spatial practices. In: Fenster T, Yacobi H (eds) Remembering, forgetting and city builders Re-materialising cultural geography. Ashgate, Farnham, UK, pp 194–199

Häußermann H, Siebel W (1993) Die Politik der Festivalisierung und die Festivalisierung der Politik. In: Siebel W (ed) Festivalisierung der Stadtpolitik. Stadtentwicklung durch große Projekte. Springer, Wiesbaden, pp 7–31

Hawke SK (2012) Sense of place in changing communities: the plurality of heritage values. In: Negussie E (ed) Changing world, Changing views of heritage: heritage and social change. Proceedings of the ICOMOS scientific symposium 2010. ICOMOS, Paris, pp 37–44

Helbrecht I (2004) Stadtmarketing. Vom Orakel zum Consulting. Identitätspolitiken in der Stadt. In: Hilber MLE, Ayda (ed) Stadtidentität—Der richtige Weg zum Stadtmarketing. Orell Füssli Verlag AG, Zürich, pp 165–174

Hernández B, Hidalgo MC, Salazar-Laplace ME, Hess S (2007) Place attachment and place identity in natives and non-natives. J Environ Psychol 27:310–317

HerO (Heritage as Opportunity) (2011) The road to success—A new approach to the management of historic towns. Policy recommendations. City of Regensburg, planning and building division. World Heritage Management, Regensburg

Hewitt K (2012) Culture, hazard and disaster. In: Wisner B, Gaillard JC, Kelman I (eds) The Routledge handbook of hazards and disaster risk reduction. Routledge, London and New York, pp 85–96

Hidalgo MC, Hernández B (2001) Place attachment: conceptual and empirical questions. J Environ Psychol 21:273–281

Hiernaux D (2013) Die historischen Stadtzentren Lateinamerikas: Auf dem Weg zu einer kreolischen Gentrifizierung? In: Huffschmid A, Wildner K (eds) Stadtforschung aus Lateinamerika: Neue urbane Szenarien: Öffentlichkeit—Territorialität—Imaginarios. Transcript, Bielefeld, pp 377–395

Hilber ML (2004) Der richtige Weg zum Stadtmarketing. In: Hilber MLE, Ayda (ed) Stadtidentität —Der richtige Weg zum Stadtmarketing. Orell Füssli Verlag AG, Zürich, pp 17–26

Hillier B, Hanson J (1984) The social logic of space. Cambridge University Press, Cambridge

Hoffmann-Axthelm D (1996) Anleitung zum Stadtumbau. Campus, Frankfurt, New York

Hopf C (2008) Qualitative interviews—ein Überblick. In: Flick U, Kardorff Ev, Steinke I (eds) Qualitative sozialforschung. Handbuch. Rowohlt, Reinbek bei Hamburg, pp 349–360

Hopf C, Weingarten E (1993) Qualitative sozialforschung. Klett-Cotta, Stuttgart

ICOMOS (1994) The nara document on authenticity. Nara, Japan

ICOMOS (2008) Charter for the interpretation and presentation of cultural heritage sites ICOMOS. http://www.international.icomos.org/charters/interpretation_e.pdf. Accessed 21.08.2014

International Conference on Urban Culture Beijing Declaration concerning Urban Culture (2007). In: International conference on urban culture, 2nd international forum on city planning Beijing 2007

Ireland T (2012) Excavating globalisation from the ruins of colonialism: archaeological heritage management responses to cultural change. In: Negussie E (ed) Changing world, changing views of heritage: heritage and social change. Proceedings of the ICOMOS scientific symposium 2010. ICOMOS, Paris, pp 18–26

Jack G (2012) The role of place attachments in Wellbeing. In: Atkinson S, Fuller S, Painter J (eds) Wellbeing and place, vol Farnham. Ashgate, pp 89–103

Jacobs J (1961) The death and life of great American cities. Random House, New York

Jaramillo Contreras M (2012) Beyond the protection of material cultural heritage in times of conflict. Protecting Cultural Heritage in Times of Conflict. ICCROM, Rome

Jive'n G, Larkham PJ (2003) Sense of place, authenticity and character: a commentary. J Urban Des 8(1):67–81. doi:10.1080/1357480032000064773

Jönhill JI (2011) Inclusion and exclusion—a guiding distinction to the understanding of issues of cultural background. Systems Research and Behavioral Science 29(4):387–401

Jung MM (2010) RAUMimage—ImageRÄUME. Marketing von (urbanen) Räumen als Instrument von Gemeinwesenentwicklung. Schriftenreihe des europäischen Masterstudiengangs "Gemeinwesenentwicklung, Quartiersmanagement und Lokale Ökonomie" an der Hochschule München. AG SPAK, Neu-Ulm

Keitumetse S (2009) Methods for investigating locals' perceptions of a cultural heritage product for tourism. Lessons from Botswana. In: Stig Sørensen ML, Carman J (eds) Heritage studies. Methods and approaches. Routledge, London, New York, pp 201–216

Klaus E, Drüeke R (2010) Inklusion und Exklusion in medialen Identitätsräumen. In: Sedmak C, Drüeke R, Schweiger G (eds) Klaus E. Identität und Inklusion im europäischen Sozialraum VS Research, Wiesbaden, pp 113–132

Knirsch J (2004) Stadtplätze. Architektur und Freiraumplanung, Alexander Koch, Leinfelden-Echterdingen

Koch G (2013) Studying heritage in the digital era. In: Albert M-T, Bernecker R, Rudolff B (eds) Understanding heritage: perspectives in heritage studies. de Gruyter, Berlin/Boston, pp 169–181

Koolhaas R (2012) The invention and reinvention of the city. J Int Aff 65(2):113

Kraas F (2006) Megacities: dimension and consequences of global change—and the need for shifted priorities in perception, research and planning. In: Kraas F, H. Gaese, Mi Mi Kyi (eds.) (ed) Megacity Yangon: transformation processes and modern developments. Second German-Myanmar workshop in Yangon/Myanmar 2005 vol 7. Southeast Asian Modernities, Berlin, pp 21–29

Kraas F (2010) Urbanisierung als weltweite Herausforderung. In: Debiel T, Messner D, Nuscheler F, Roth M, Ulbert C (eds) Globale Trends 2010. Frieden, Entwicklung, Umwelt. Stiftung Entwicklung und Frieden, Frankfurt, pp 181–199

Kraas F, Mertins G (2008) Megastädte in Entwicklungsländern: Vulnerabilität. Informalität, Regier- und Steuerbarkeit Geographische Rundschau 60(11):4–10

Kraas F, Nitschke U (2008) Megaurbanisierung in Asien: Entwicklungsprozesse und Konsequenzen stadträumlicher Reorganisation. Informationen zur Raumentwicklung Themenheft „Raum- und Stadtentwicklung in Asien" (8–2008):447–456

Kuckartz U (2007) Einführung in die computergestützte Analyse qualitativer Daten. VS Verlag für Sozialwissenschaften, Wiesbaden. doi:10.1007/978-3-531-90664-5_4

Kyle GT, Jun J, Absher JD (2014) Repositioning identity in conceptualizations of human-place bonding. Environ Behav 46(8):1018–1043

Lacarrieu M (2013) Kulturerbe, Materialität und die Frage nach dem Städtischen. In: Huffschmid A, Wildner K (eds) Stadtforschung aus Lateinamerika: Neue urbane Szenarien: Öffentlichkeit—Territorialität—Imaginarios. Transcript, Bielefeld, pp 419–439

Lagae J (2010) From "patrimoine partagé" to "whose heritage"? critical reflections on colonial built heritage in the City of Lubumbashi, democratic republic of the Congo. In: Fenster T, Yacobi H (eds) Remembering, forgetting and city builders Re-materialising cultural geography. Ashgate, Farnham, UK, pp 175–191

Lefebvre H (1991) The production of space. Blackwell, Oxford

Lewicka M (2005) Ways to make people active: the role of place attachment, cultural capital, and neighborhood ties. J Environ Psychol 25:381–395. doi:10.1016/j.jenvp.2005.10.004

Lewicka M (2008) Place attachment, place identity, and place memory: restoring the forgotten city past. J Environ Psychol 28:209–231. doi:10.1016/j.jenvp.2008.02.001

Lewicka M (2010) What makes neighborhood different from home and city? Effects of place scale on place attachment. J Environ Psychol 30(1):35–51. doi:10.1016/j.jenvp.2009.05.004
Lillehammer G (2009) Making them draw. The use of drawings when researching public attitudes towards the past. In: Stig Sørensen ML, Carman J (eds) Heritage studies. Methods and approaches. Routledge, London, New York, pp 253–269
Lippuner R (2014) Natur und Landschaft. In: Lossau J, Freytag T, Lippuner R (eds) Schlüsselbegriffe der Kultur- und Sozialgeographie. Eugen Ulmer, Stuttgart, pp 38–51
Lossau J (2014) Kultur und Identität. In: Lossau J, Freytag T, Lippuner R (eds) Schlüsselbegriffe der Kultur- und Sozialgeographie. Eugen Ulmer, Stuttgart, pp 25–37
Löw M (2008) Soziologie der Städte. Suhrkamp, Frankfurt am Main
Löw M (2011) Lokale Ökonomie—Lebensqualität als Standortfaktor. In: Löw M, Terizakis G (eds) Städte und ihre Eigenlogik. Ein Handbuch für Stadtplanung und Stadtentwicklung, vol Interdisziplinäre Stadtforschung, Band 11. Interdisziplinäre Stadtforschung. Campus Verlag, Frankfurt am Main, pp 29–35
Low SM (1992) Symbolic ties that bind. In: Altman I, Low SM (eds) Place Attachment, vol 12. Human behavior and environment. Springer, New York and London, pp 165–185. doi:10.1007/978-1-4684-8753-4_8
Lowry G, McCann E (2011) Asia in the mix: urban form and global mobilities—Hong Kong, Vancouver, Dubai. In: Roy A, Ong A (eds) Worlding cities: Asian experiments and the art of being global. Studies in urban and social change. Wiley-Blackwell, Chichester, pp 182–204
Luchsinger C (2008) Zweierlei Städte. In: Wolfrum S, Nerdinger W (eds) Multiple city. Stadtkonzepte 1908|2008. Jovis, Berlin, pp 279–282
Luger K (2008) Welterbe-Tourismus. Ökonomie, Ökologie und Kultur in weltgesellschaftlicher Verantwortung. In: Luger K, Wöhler K (eds) Welterbe und Tourismus: Schützen und Nützen aus einer Perspektive der Nachhaltigkeit. StudienVerlag, Innsbruck, pp 17–42
Luhmann N (1997) Die Kunst der Gesellschaft. Suhrkamp, Frankfurt a.M
Luhmann N (2000) Art as a social system, vol (originally published in Germany 1995 under the title "Die Kunst der Gesellschaft"). Stanford University Press, Stanford
Lynch K (1960) The image of the city. MIT Press, Cambridge
Manzi T, Jacobs K (2009) From a 'society of fear' to a 'society of respect': the transformation of Hackney's Holly Street Estate. In: Imrie R, Lees L, Raco M (eds) Regenerating London. Governance, sustainability and community in a global city, vol Reprint 2010. Routledge, Milton Park, pp 273–288
Mars N (2008) Cities without history. In: Wolfrum S, Nerdinger W (eds) Multiple City. Stadtkonzepte 1908|2008. Jovis, Berlin, pp 274–278
Massey D (1991) A Global Sense of Place. Marxism Today June 1991:24–29
Mattissek A, Prossek A (2014) Regieren und Planen. In: Lossau J, Freytag T, Lippuner R (eds) Schlüsselbegriffe der Kultur- und Sozialgeographie. Eugen Ulmer, Stuttgart, pp 198–211
Michel B (2010) Global city als Projekt: Neoliberale Urbanisierung und Politiken der Exklusion in Metro Manila. Transcript, Bielefeld
Montgomery J (2003) Cultural quarters as mechanisms for urban regeneration. Part 1: Conceptualising cultural quarters. Planning, practice & research Vol. 18, No. 4:293–306. doi:10.1080/1561426042000215614
Montgomery J (2004) Cultural quarters as mechanisms for urban regeneration. Part 2: A review of four cultural quarters in the UK, Ireland and Australia. Plan Pract Res 19(1):3–31. doi:10.1080/0269745042000246559
Morlcy D, Robins K (1995) Spaces of identity: global media. Electronic landscapes and cultural boundaries. Routledge, New York
Mueller-Haagen I, Simonsen J, Többen L (2014) Die DNA der Stadt.: Ein Atlas urbaner Strukturen in Deutschland. Schmidt, Mainz
Müller-Mahn D, Verne J (2014) Entwicklung. In: Lossau J, Freytag T, Lippuner R (eds) Schlüsselbegriffe der Kultur- und Sozialgeographie. Eugen Ulmer, Stuttgart, pp 94–107
Murzyn-Kupisz M, Dzialek J (2013) Cultural heritage in building and enhancing social capital. J Cult Herit Manage Sustain Dev 3(1):35–54. doi:10.1108/20441261311317392

Nerdinger W (2008) Stadt als gebautes und kollektives Gedächtnis. In: Wolfrum S, Nerdinger W (eds) Multiple city. Stadtkonzepte 1908|2008. Jovis, Berlin, pp 74–76
Nora P (1989) Between memory and history: les lieux de mémoire. Representations 26 (Special issue: memory and counter-memory):7–24
Norberg-Schulz C (1982) Genius Loci. Landschaft, lebensraum, baukunst. Klett, Stuttgart
Norberg-Schulz C (2013) The phenomenon of place. In: Larice M, Macdonald E (eds) The urban design reader (Routledge urban reader series), 2nd edn. Routledge, Milton Park, pp 272–284
Novy A, Howorka S (2014) Transdisziplinarität und Wissensallianzen. In: Dannecker P, Englert B (eds) Qualitative Methoden in der Entwicklungsforschung. Mandelbaum, Wien, pp 20–37
NSW Heritage Office (2004) Heritage & sustainability. A discussion paper.23
Nyaupane GP, Timothy DJ, Poudel S (2015) Understanding tourists in religious destinations: A social distance perspective. Tour Manag 48:343–353. doi:10.1016/j.tourman.2014.12.009
Ong A (2011) Worlding cities, or the art of being global. In: Roy A, Ong A (eds) Worlding cities: Asian experiments and the art of being global. Studies in urban and social change. Wiley-Blackwell, Chichester, pp 1–26
Paasi A (2000) Territorial identities as social constructs, Hagar. Int Soc Sci Rev 1(2):91–113
Paasi A (2003) Region and place: regional identity in question. Prog Hum Geogr 27(4):475–485. doi:10.1191/0309132503ph439pr
Paasi A (2004) Place and region: looking through the prism of scale. Prog Hum Geogr 28(4):536–546. doi:10.1191/0309132504ph502pr
Paasi A (2009) The resurgence of the 'Region' and 'Regional Identity': theoretical perspectives and empirical observations on regional dynamics in Europe. Review of International Studies 35 (Supplement S1):121–146. doi:10.1017/S0260210509008456
Paasi A (2013) Regional planning and the mobilization of 'regional identity': from bounded spaces to relational complexity. Reg Stud 47(8):1206–1219. doi:10.1080/00343404.2012.661410
Pellnitz A (2013) Sehnsuchtsstädte und Stadtsehnsüchte: Urbanität, Natur und Schönheit. In: Bosshard MT, Döhling J-D, Janisch R et al. (eds) Sehnsuchtsstädte: Auf der Suche nach lebenswerten urbanen Räumen. transcript, Bielefeld, pp 29–43
Petermann S (2014) Erinnern und Gedenken. In: Lossau J, Freytag T, Lippuner R (eds) Schlüsselbegriffe der Kultur- und Sozialgeographie. Eugen Ulmer, Stuttgart, pp 212–226
Podoler G (2010) Seoul: City, Identity and the Construction of the Past. In: Fenster T, Yacobi H (eds) Remembering, forgetting and city builders Re-materialising cultural geography. Ashgate, Farnham, UK, pp 121–140
Pott A (2007a) Orte des Tourismus: eine raum- und gesellschaftstheoretische Untersuchung. Transcript, Bielefeld
Pott A (2007b) Sprachliche Kommunikation durch Raum. Das Angebot der Systemtheorie. Geographische Zeitschrift 95 (1/2—Sprachorientierte Forschungsansätze in der Humangeographie nach dem linguistic turn):56–71
Poytner G (2009) The 2012 Olympic Games and the reshaping of East London. In: Imrie R, Lees L, Raco M (eds) Regenerating London. Governance, sustainability and community in a global city, vol Reprint 2010. Routledge, Milton Park, pp 132–148
Pratt AC (2012) The cultural and creative industries: Organisational and spatial challenges to their governance. Erde 143(4):317–334
Rampley M (2009) Art as a social system: the sociological aesthetics of Niklas Luhmann. Telos 2009(148):111–140. doi:10.3817/0909148111
Rautenberg M (1998) L'émergence patrimoniale de l'ethnologie: entre mémoire et politiques publiques. In: Poulot D (ed) Patrimoine et modernité. L'Harmatan, Paris, pp 279–291
Rautenberg M (2010) Stereotypes and emblems in the construction of social imagination. Outlines Crit Pract Stud 2:126–137
Rautenberg M (2011) Industrial heritage, regeneration of cities and public policies in the 1990s: elements of a French/British comparison. Int J Heritage Stud 18(5):513–525. doi:10.1080/13527258.2011.637945

Reuber P (1993) Heimat in der Großstadt. Eine sozialgeographische Studie zu Raumbezug und Entstehung von Ortsbindung am Beispiel Kölns und seiner Stadtviertel. Kölner Geographische Arbeiten, vol 58. Geographisches Institut der Universität zu Köln, Köln

Reuber P (2014) Territorien und Grenzen. In: Lossau J, Freytag T, Lippuner R (eds) Schlüsselbegriffe der Kultur- und Sozialgeographie. Eugen Ulmer, Stuttgart, pp 182–195

Roberts P, Sykes H (eds) (2000) Urban Regeneration. A Handbook. Sage, London

Roca Z, Oliveira JA, Roca MdN (2011) Claiming territorial identity and local development: from wishes to deeds. In: Roca Z, Claval P, Agnew J (eds) Landscapes, identities and development. Ashgate, Surrey, pp 319–334

Rodenstein M (2010) Forgetting and remembering: Frankfurt's *Altstadt* after the second world war. In: Fenster T, Yacobi H (eds) Remembering, forgetting and city builders Re-materialising cultural geography. Ashgate, Farnham, UK, pp 159–174

Rojas E (2002) Urban heritage conservation in Latin America and the Carribean. A task for all social actors. Sustainable development department technical papers series:25

Rojas E, Lanzafame F (2011) Tracking preservation work in ten historic centers. In: Rojas E, Lanzafame F (eds) City development: experiences in the preservation of ten world heritage sites. Inter-American Development Bank, Washington, D.C., pp 1–10

Ronneberger K (2000) Container des 21. Jahrhunderts. In: Matejovski D (ed) Metropolen: Laboratorien der Moderne, vol Schriftenreihe des Wissenschaftszentrums Nordrhein-Westfalen, vol 5. Campus Verlag, Frankfurt/New York, pp 324–342

Ronström O (2014) Consequences of world heritage production: the heritage town. In: Hølleland H, Solheim S (eds) Between dream and reality: debating the impact of world heritage Listing. Primitive tider special edition. Reprosentralen, Oslo, pp 7–17

Roodhouse S (2010) Cultural quarters: principles and practice, 2nd edn. Intellect Ltd., Bristol

Rossi A (1984) The architecture of the city. MIT Press, Cambridge

Roy A (2009) The 21st-century metropolis: new geographies of theory. Reg Stud 43(6):819–830. doi:10.1080/00343400701809665

Ruskin J (1989) The seven lamps of architecture. Dover Publications, New York

Salazar NB (2008) "Enough stories!" Asian tourism redefining the roles of Asian tour guides. In: Doquet A, Evrard O (eds) Tourisme, Mobilités et Alterités Contemporaines, vol Civilisations, vol.LVII, n°1–2. Institut de Sociologie, Université libre de Bruxelles, Bruxelles pp 207–222

Salazar NB (2010) The glocalisation of heritage through tourism: balancing standardisation and differentiation. In: Labadi S, Long C (eds) Heritage and globalisation. Routledge, New York, pp 130–147

Salazar NB (2012) Shifting values and meanings of heritage from cultural appropriation to tourism interpretation and back. In: Lyon S, Wells EC (eds) Global tourism. Cultural heritage and economic encounters, vol 30, Society for economic anthropology monographs. Altamira, Lanham, pp 21–41

Salesses P, Schechtner K, Hidalgo CA (2013) The collaborative image of the city: mapping the inequality of urban perception. PLoS ONE 8(7). doi:10.1371/journal.pone.0068400

Saretzki A (2008) Postmoderne Mutationen—Aktualisierung des Kulturerbes als Entortung. In: Luger K, Wöhler K (eds) Welterbe und Tourismus: Schützen und Nützen aus einer Perspektive der Nachhaltigkeit. StudienVerlag, Innsbruck, pp 59–69

Satterfield T, Gregory R, Klain S, Roberts M, Chan KM (2014) Culture, intangibles and metrics in environmental management. J Environ Manage 117:103–114

Scazzosi L (2011) Limits to transformation in places' identity: theoretical and methodological questions. In: Roca Z, Claval P, Agnew J (eds) Landscapes, identities and development. Ashgate, Surrey, pp 9–24

Scharr K, Steinicke E (eds) (2012) Sotschi 2014 troubles in paradise? Olympische Winterspiele zwischen Alpen und Kaukasus. Innsbruck University Press, Innsbruck

Schicho W (2014) Diskursanalyse. In: Dannecker P, Englert B (eds) Qualitative Methoden in der Entwicklungsforschung. Mandelbaum, Wien, pp 127–152

Schlögel K (2003) Im Raume lesen wir die Zeit. Über Zivilisationsgeschichte und Geopolitik. 2. Auflage 2007 edn. Fischer Taschenbuch, Frankfurt am Main

Schmidt JA (2013) Urbane Sehnsüchte entwickeln und Entscheidungsprozesse beflügeln. In: Bosshard MT, Döhling J-D, Janisch R et al. (eds) Sehnsuchtsstädte: Auf der Suche nach lebenswerten urbanen Räumen. Transcript, Bielefeld, pp 209–229

Schmitt TM (2009) Global cultural governance. Decision-making concerning world heritage between politics and science. Erdkunde 63(2):103–121

Schmitt TM (2011) Cultural governance. Zur Kulturgeographie des UNESCO-Welterberegimes. Erdkundliches Wissen., vol 152. Franz Steiner, Stuttgart

Schofield J, Szymanski R (2011) Sense of place in a changing world. In: Schofield J, Szymanski R (eds) Local heritage, global context: cultural perspectives on sense of place. Heritage, culture and identity. Ashgate, Farnham, pp 1–11

Schubert D (2014) Jane Jacobs und die Zukunft der Stadt. Diskurse—Perspektiven—Paradigmenwechsel. Beiträge zur Stadtgeschichte und Urbanisierungsforschung—17. Franz Steiner, Stuttgart

Seifert J (2011) Stadtbild, Wahrnehmung, Design. Kevin Lynch revisited. Birkhäuser, Gütersloh, Berlin

Sen A (2007) Die Identitätsfalle. Warum es keinen Krieg der Kulturen gibt, Beck, München

Senil N (2011) Reorganising space and time. Revue de Géographie Alpine. J Alp Res 99(2):2–11. doi:10.4000/rga.1442

Sentosa LS (2001) Genius loci within Balinese dwellings environments: the unlikely scenarios of urban development in Bali. Habitat Int 25(2):255–272. doi:10.1016/S0197-3975(00)00020-5

Sharifi A, Murayama A (2013) Changes in the traditional urban form and the social sustainability of contemporary cities: a case study of Iranian cities. Habitat Int 38:126–134

Silberman NA (2013) Heritage interpretation as public discourse: towards a new paradigm. In: Albert M-T, Bernecker R, Rudolff B (eds) Understanding heritage: perspectives in heritage studies. de Gruyter, Berlin/Boston, pp 21–33

Simone A (2010) City life from Jakarta to Dakar. Movements at the Crossroads. Routledge, New York

Slezak G (2014) Von Wahrnehmung und Erfahrung. Überlegungen zu explorativen Beobachtungsverfahren. In: Dannecker P, Englert B (eds) Qualitative Methoden in der Entwicklungsforschung. Mandelbaum, Wien, pp 176–196

Sommer U (2009) Methods used to investigate the use of the past in the formation of regional identities. In: Carman J, Stig Sørensen ML (eds) Heritage studies. Methods and approaches. Routledge, London, New York, pp 103–120

Sörensson E (2008) Making a living in the world of tourism: livelihoods in backpacker tourism in urban Indonesia. Umeå University, Umeå, Sweden

Stan A (2014) Urban culture, urban cultural landscape. In: Crăciun C, Bostenaru Dan M (eds) Planning and designing sustainable and resilient landscapes. Springer, Dordrecht, Heidelberg, New York, London, pp 107–115

Steinbrink M, Haferburg C, Ley A (2011) Festivalisation and urban renewal in the Global South: socio-spatial consequences of the 2010 FIFA World Cup. S Afr Geogr J 93(1):15–28. doi:10.1080/03736245.2011.567827

Stadt Stiftung Lebendige (ed) (2005) Stadt am Wasser: Neue Chancen für Kommunen und Tourismus, Schriftenreihe Lebendige Stadt, vol 4. Frankfurter Societäts-Druckerei GmbH, Frankfurt/Main

Stig Sørensen ML (2009) Between the lines and in the margins. Interviewing people about attitudes to heritage and identity. In: Stig Sørensen ML, Carman J (eds) Heritage studies. Methods and approaches. Routledge, London, New York, pp 164–177

Taylor WM, Levine MP (2011) Prospects for an ethics of architecture. Routledge, London and New York

Te Heuheu T, Kawharu M, Tuheiava RA (2012) World heritage and indigeneity. World heritage, vol 62. UNESCO, Paris

Thorsell J (2012) Scenic natural beauty. How can it be judged? World Heritage, vol 63. UNESCO, Paris

Tokya-Seid M (2003a) Bristol-Köln. In: Böhme H, Körte A, Tokya-Seid M (eds) WOHNEN—BAUEN—PLANEN. Erneuerung historischer Kernstädte in Südostasien und Europa im historisch-architektonischen Vergleich. Wissenschaftliche Buchgesellschaft, Darmstadt, pp 19–74
Tokya-Seid M (2003b) Einleitung. In: Böhme H, Körte A, Tokya-Seid M (eds) WOHNEN—BAUEN—PLANEN. Erneuerung historischer Kernstädte in Südostasien und Europa im historisch-architektonischen Vergleich. Wissenschaftliche Buchgesellschaft, Darmstadt, pp 9–15
Townend S, Whittaker K (2011) Being accounted for: qualitative data analysis in assessing 'place' and 'value'. In: Schofield J, Szymanski R (eds) Local heritage, global context: cultural perspectives on sense of place. Heritage, culture and identity. Ashgate, Farnham, pp 65–77
Townroe PM (1996) Urban sustainability and social cohesion. In: Pugh C (ed) Sustainability, the environment and urbanisation, vol Neudruck 2002. Earthscan, London, pp 179–195
Turner M (2012) World heritage and sustainable development. World Heritage Rev Sustain Dev 65:8–15
Turner M (2013) UNESCO recommendation on the historic urban landscape. In: Albert M-T, Bernecker R, Rudolff B (eds) Understanding heritage: perspectives in heritage studies. de Gruyter, Berlin/Boston, pp 77–87
Tweed C, Sutherland M (2007) Built cultural heritage and sustainable urban development. Landscape Urban Plan 83:62–69
Ujang N (2012) Place attachment and continuity of urban place identity. Procedia—Social Behav Sci 49:156–167
UNESCO (1972) Convention concerning the protection of the world cultural and natural heritage. UNESCO. http://whc.unesco.org/en/conventiontext/. Accessed 25 July 2014
UNESCO (1989) Recommendation on the safeguarding of traditional culture and Folklore. UNESCO, Paris
UNESCO (2011) Recommendation on the historic urban landscape. UNESCO. http://portal.unesco.org/en/ev.php-URL_ID=48857&URL_DO=DO_TOPIC&URL_SECTION=201.html. Accessed 16 Jun 2012
UNESCO (2013) Operational guidelines for the implementation of the world heritage convention. UNESCO World Heritage Centre. http://whc.unesco.org/archive/opguide13-en.pdf. Accessed 25.07.2014
UNESCO (2014) Lists of intangible cultural heritage and Register of best safeguarding practices. UNESCO. http://www.unesco.org/culture/ich/index.php?lg=en&%20pg=00011#tabs. Accessed 20.07.2014
UNESCO Institute for Statistics (2012) Measuring cultural participation. 2009 Framework for Cultural Statistics Handbook No. 2. UNESCO-UIS, Montreal
Urban F (2011) Denkmalschutz und Stadtentwicklung—Einführung. Informationen zur Raumentwicklung 3/4 (Denkmalschutz und Stadtentwicklung):I–III
Uzzell D (2009) Where is the discipline in heritage studies? A view from environmental psychology. In: Carman J, Stig Sørensen ML (eds) Heritage Studies. Methods and Approaches. Routledge, London, New York, pp 326–333
Vanderstraeten R (2005) System and environment: notes on the autopoiesis of modern society. Syst Res Behav Sci 22(6):471–481
Vedru G (2011) Memory and the value of place in Estonia. In: Schofield J, Szymanski R (eds) Local heritage, global context: cultural perspectives on sense of place. Heritage, culture and identity. Ashgate, Farnham, pp 53–64
Veldpaus L, Pereira Roders AR, Colenbrander BJF (2013) Urban heritage: putting the past into the future. Hist Environ 4(1):3–18
Vinken G (2010) Zone Heimat. Altstadt im modernen Städtebau. Deutscher Kunstverlag, Berlin, München
Vinken G (2011) Lokale Sinnstiftung—Die Bedeutung der Denkmale. In: Löw M, Terizakis G (eds) Städte und ihre Eigenlogik. Ein Handbuch für Stadtplanung und Stadtentwicklung, vol Interdisziplinäre Stadtforschung, Band 11. Interdisziplinäre Stadtforschung. Campus Verlag, Frankfurt am Main, pp 73–82

Wagner G (2008) Kulturelles Gedächtnis versus Verhandlung. Das Welterbe und die Konstruktion glokaler Identität. In: Luger K, Wöhler K (eds) Welterbe und Tourismus: Schützen und Nützen aus einer Perspektive der Nachhaltigkeit. StudienVerlag, Innsbruck, pp 87–102

Waterton E, Smith L, Campbell G (2006) The utility of discourse analysis to heritage studies: the burra charter and social inclusion. International Journal of Heritage Studies 12(4):339–355

Weichhart P (2004) Regionalentwicklung. Identitätsmanagement für Orte. In: Hilber MLE, Ayda (ed) Stadtidentität—Der richtige Weg zum Stadtmarketing. Orell Füssli Verlag AG, Zürich, pp 129–138

Weichhart P (2009) Humangeographie—quo vadis? In: Musil R, Staudacher C (eds) Mensch. Raum. Umwelt. Entwicklungen und Perspektiven der Geographie in Österreich. Österreichische Geographische Gesellschaft, Wien, pp 63–77

Weichhart P (2010) Raumkonstruktionen, "Turns" und Paradigmen. In: Wöhler K, Pott A, Denzer V (eds) Tourismusräume. Zur soziokulturellen Konstruktion eines globalen Phänomens. Transcript, Bielefeld, pp 21–39

Werlen B (1997) Gesellschaft, Handlung und Raum: Grundlagen handlungstheoretischer Sozialgeographie. Steiner, Stuttgart

Werlen B (2014) Globalisierung. In: Lossau J, Freytag T, Lippuner R (eds) Schlüsselbegriffe der Kultur- und Sozialgeographie. Eugen Ulmer, Stuttgart, pp 54–66

Werlen B, Lippuner R (2007) Regionale Kulturen und globalisierte Lebensstile. Geographische Rundschau 59(7–8):22–27

West C (2007) Gedächtnis—Kultur—Raum. Zur Identitätskonstruktion der Roma ('Gitanos') in Spanien. Geographische Rundschau 59(7–8):44–52

Wöhler K (2008) Heritagefication: Zur Vergegenwärtigung des Kulturerbes. In: Luger K, Wöhler K (eds) Welterbe und Tourismus: Schützen und Nützen aus einer Perspektive der Nachhaltigkeit. StudienVerlag, Innsbruck, pp 43–58

Wöhler K (2010) "Wir wissen nicht, was wir tun sollen"—Expertisierte Raumaneignung und -konstruktion. In: Wöhler K, Pott A, Denzer V (eds) Tourismusräume. Zur soziokulturellen Konstruktion eines globalen Phänomens. transcript, Bielefeld, pp 157–188

Wöhler K, Pott A, Denzer V (2010) Formen und Konstruktionsweisen von Tourismusräumen. In: Wöhler K, Pott A, Denzer V (eds) Tourismusräume. Zur soziokulturellen Konstruktion eines globalen Phänomens. Transcript, Bielefeld, pp 11–19

Wolfrum S (2008a) Genius Loci|Ort und Atmosphären. In: Wolfrum S, Nerdinger W (eds) Multiple city. Stadtkonzepte 1908|2008. Jovis, Berlin, p 132

Wolfrum S (2008b) Stadt als Bild|Urban Icon. In: Wolfrum S, Nerdinger W (eds) Multiple city. Stadtkonzepte 1908|2008. Jovis, Berlin, p 252

Wood P (ed) (2010) Intercultural cities. Towards a model for intercultural integration. Council of Europe, Strasbourg

Yacobi H (ed) (2004) Constructing a sense of place: architecture and the zionist discourse. Ashgate, Aldershot, UK and Burlington, VT

Yacobi H (2010) Memory, recognition and the architecture of a diasporic place: the case of Netivot, Israel. In: Fenster T, Yacobi H (eds) Remembering, forgetting and city builders Re-materialising cultural geography. Ashgate, Farnham, UK, pp 27–45

Yap KS (2012) The challenges of promoting productive, inclusive and sustainable urbanisation. In: Yap KS, Thuzzar M (eds) Urbanisation in Southeast Asia: issues and impacts. Institute of Southeast Asian Studies, Singapore, pp 10–77

Yeo HK, Han SM (2012) Heritage conservation as urban regeneration policy in globalizing cities: social exclusion and gentrification at the vicinity of the Jongmyo World Heritage site in Seoul, Korea. In: Negussie E (ed) Changing world, changing views of heritage: heritage and social change. Proceedings of the ICOMOS scientific symposium 2010. ICOMOS, Paris, pp 56—62

Zahnd M (2005) Traditionelle Stadtquartiere in Semarang und Yogyakarta, Indonesien. Möglichkeiten der Revitalisierung und innovativen Nutzung des historischen Städtebaus. Universität Stuttgart, Stuttgart

Zheng HW, Shen GQ, Wang H (2014) A review of recent studies on sustainable urban renewal. Habitat Int 41:272–279. doi:10.1016/j.habitatint.2013.08.006

# Chapter 3
# Heritage and Conservation in Changing Environments

**Abstract** Global cultural heritage debates and definitions from international organizations such as UNESCO are rooted in a predominantly Euro-American understanding. Material, original state and authenticity are the major justifications for being considered as cultural heritage, not suiting different non-material focused understandings of heritage that allow for continuous change as intrinsic part of the heritage object itself. Since the 1990s historic cities and urban heritage have become one of the focal points of global heritage policies, headed by international organizations like UNESCO. A shift from single buildings towards an (urban) landscape approach has occurred, in parallel with the growing consideration of intangible heritage and approaches linking tangible and intangible heritage, such as the Historic Urban Landscape Approach. In both cases the process was triggered outside the Euro-American sphere, with the aim to adapt the global heritage understanding to other cultural environments and to introduce heritage concepts which suit the cultural understandings and actual urban challenges of the non-Western world. Key concerns of the international heritage debates are the understanding of urban areas as a layering of cultural and natural values and attributes, historic and actual ones and acknowledging the potential of urban heritage on urban development. However, in reality rapid urban change, economy-driven new urban developments, and non-suitable urban policies can impact negatively on historic cities, particularly in the Global South, where cities in addition have to cope with rapid change, high levels of informality and administrative shortcomings.

**Keywords** World heritage · UNESCO · Conservation policies · Intangible values · Urban heritage

This chapter will highlight global discourses on cultural heritage that are setting the frame for national and regional heritage policies and also sometimes conflicting understanding(s) of heritage itself. Starting with an overview of the roots of cultural heritage conservation, the concept of 'UNESCO World Heritage' will be illustrated, as it is impacting strongly on the global reference frame of the heritage debate. This



S. Sandholz, *Urban Centres in Asia and Latin America*,
The Urban Book Series, DOI 10.1007/978-3-319-43735-4_3

section includes a review of the criteria for defining heritage and is followed by a revision of the debate on representativeness and euro-centricity of the World Heritage List.

After dealing with the tangible heritage, the comparably recent concept to protect intangible heritage will be highlighted. This is followed by an illustration of tangible and intangible World Heritage sites in the three case study countries of Brazil, Indonesia and Nepal. Subsequently, the chapter focusses on urban heritage in particular, on its roots and current paradigms like the Historic Urban Landscape approach.

## 3.1 Roots of Cultural Heritage Conservation

The point of origin of modern heritage understanding can be seen in Renaissance Europe, an epoch highly interested in the antiques of the ancient Greek and Roman world. Since that period, European heritage conservation practices began to spread around: "*The Papal government created the post of 'Overseer over Protection of all Antiquities' which marked the entry into formal configurations of urban conservation endeavors*" (Birabi 2007: 40).

Throughout the following centuries, more laws and policies were decreed with the aim to protect heritage, from single monuments to urban fabric, e.g. the launch of the Swedish Antiquities Ordinance in 1666, being responsible for urban historic ensembles, the establishment of Prussian 'Ober-Bau-Departement' to protect historic urban areas, or the 1789 pro-heritage protection Decree of Revolutionary France (Birabi 2007). Since the late eighteenth century, heritage then became a public and societal concern: the heritage 'valorization' and wish to preserve historic remains emerged, going hand in hand with the development of national states and subsequently nationalism movements. This is true for different parts of the world, leading to the foundation of various legal institutions and measures (Carman and Stig Sørensen 2009).

Victorian England witnessed the foundation of the Society for the Protection of Ancient Buildings in 1877 to oppose ongoing tendencies of restoring buildings by adding elements instead of just doing repair works. In the same year, the co-founder William Morris published a manifesto emphasizing on the importance of authenticity of the historic fabric, on the look and feel of the original materials:

> Thus, the emphasis on the authenticity of materials must be seen in context and its origins in the late nineteenth century English Art and Crafts movement, a romantic vision of decaying ruins that led to reverence for authentic historic object (Kwanda 2010: 6f).

The topic of authenticity has continued being a focal point of the heritage debate since these days, remaining "*the back-bone for urban heritage conservation into the 20th Century*" (Birabi 2007: 41).

In the late nineteenth century, the first conventions related to cultural heritage protection, mainly focusing on times of conflicts, were developed (Alberts and

Hazen 2010). The transition from the nineteenth to the twentieth century also meant a shift from consumption of urban heritage resources and ad hoc maintenance of the urban heritage to more long-term conservation and management processes, entering the agendas of formal planning and resulting in guidelines and charters (Birabi 2007). As early as in 1903, the Austrian theoretician on the preservation of monuments Riegl (1903) dealt with the socially constructed character of cultural heritage and found its selection depending on specific present-day needs (Schmitt 2009). His book "*Der moderne Denkmalkultus, sein Wesen und seine Entstehung*" (The modern cult of monuments: its character and origin) is still regarded as one of the major sources of heritage debate, impacting beyond language borders. Riegl and Dehio (1905) are considered to be two of the founding fathers of modern conservation, due to their proposal of restoration measures that respect the historic fabric being a duty of modern heritage conservation.

International conservation conferences were held with the aim to come up with more comprehensive charters and recommendations, beginning with the 1904 Sixth International Congress of Architects in Madrid. The conference proceedings not only proposed unified principles for conserving urban heritage but also gave recommendations for how to conserve monuments and their beauty (Locke 1904). During the Hague Conference 1907, principles of heritage preservation in case of an armed conflict were defined (Birabi 2007), which unfortunately could not avoid massive destructions of historic fabric all over Europe during World War I.

After having faced many losses during the war years, the most affected nations co-operated in reconstructing historic ensembles. It was during these years when the idea of an international agency responsible for heritage conservation emerged. In 1926, the International Museums Office was opened in Paris, and in 1930, the first International Conference on the Study of Scientific Methods for the Examination and Preservation of Works of Art took place in Rome, while 1-year later the First International Congress of Architects and Technicians of Historic Monuments took place in Athens. The Athens Charter for the Restoration of Historic Monuments (First International Congress of Architects and Technicians of Historic Monuments 1931) for the first time ever suggested binding standards for conservation. Among others, the charter proposed the establishment of "International organizations for Restoration on operational and advisory levels" and of suitable national legislations. It already mentioned urban heritage, the adequate uses of heritage buildings to ensure the continuity of their life, and that areas surrounding historic sites should respect the heritage (Birabi 2007; Vinken 2010).

Throughout the 1930s, Europe witnessed the first comprehensive conservation programmes, rooted in different ideological backgrounds but always going along with a formal standardization based on the canon of '*Heimatschutz*', e.g. proposed by Rudorff (1897), one of the founding fathers of natural and local history conservation. Members of the German-based '*Heimatschutz*' movement opposed against the visible changes of the new industrial era, in architecture and urban planning, and promoted traditions and their superiority. This movement had strong influence on the modern definition of heritage, including the value of the object itself, its impression, and its embedding in the scenery. Ensembles and the protection not only of landmarks like churches or castles but also farm and middle-class buildings aroused public interest. Beauty and scenery were of utmost importance and even overruled artistic value or preservation of historic fabric of single buildings (Vinken 2010).

World War II then again led to massive destructions, in Europe and beyond. As a response, then, in 1945, the United Nations Educational, Scientific and Cultural Organization (UNESCO) was created, with the mission to establish peace "on the basis of humanity's moral and intellectual solidarity" (UNESCO 2014c). As the name already indicates, the UNESCO initial mission was the formation of networks among nations on the topics of education as prerequisite of development, scientific and trans-boundary cooperation especially in the field of natural resources and hazards, freedom of expression, and most important in this context, fostering intercultural understanding by protecting heritage and supporting cultural diversity. Cultural heritage of a diverse world was defined as a human right, with its protection as prerequisite for a lasting peace.

This statement can be regarded as the basis of the World Heritage idea realized then in the 1970s, at the same time it fostered the establishment of conservation as an organizational field from mid-twentieth century onwards. Heritage became a global concept of growing importance. Barthel-Bouchier (2012) summed this up by stating that basically the two central concepts of science and human rights have served to legitimate heritage conservation in the second half of the twentieth century. From being focused on isolated monuments, the definition of heritage later expanded to whole landscapes, and also intangible cultural patterns and practices.

Beginning with the 1954 Hague Convention (The Convention for the Protection of Cultural Property in the Event of Armed Conflict), an impressive list of conventions, charters, recommendations and other heritage policy documents was published (Barthel-Bouchier 2012), passed by various mainly international or supra-national organizations. They show that the scope of heritage has changed a lot over time. Following up on the 1931 congress, the Second Congress of Architects and Specialists of Historic Buildings that took place in Venice 1964 formulated the International Charter for the Conservation and Restoration of Monuments and Sites, known as the Venice Charter. Ahmad (2006) has stated that development of conservation principles in the second half of the twentieth century is often regarded as the most significant international achievement of conservation activities, with the Venice Charter from 1964 as the most important document:

> Since the Venice Charter 1964, the scope of heritage has broadened from a concern for physical heritage such as historic monuments and buildings to groups of buildings, historic urban and rural centres, historic gardens and to non-physical heritage including environments, social factors and, lately, intangible values (Ahmad 2006: 293).

For the first time, the term 'authenticity' was introduced in an international context in the charter's preface, but without theoretical explanation in any of the following 16 articles (Falser 2010).

The congress also agreed on the need to establish the International Council on Monuments and Sites (ICOMOS), founded in 1965. Together with the International Union for Conservation of Nature (IUCN, established 1948, responsible for natural sites) and the International Centre for the Study of the Preservation and Restoration of Cultural Property (ICCROM, established 1956), ICOMOS is the advisor to UNESCO for World Heritage, being responsible for the field of cultural and mixed heritage. The scope of heritage defined by the Venice Charter as historic monuments was interpreted slightly differently by UNESCO and ICOMOS, but agreeing in principle that heritage was no longer confined to historic monuments and buildings and should be extended to include groups of buildings and historic quarters. UNESCO defined heritage as 'cultural property', while ICOMOS called heritage as 'monuments and sites' (Ahmad 2006).

During the late 1970s and the 1980s, the focus of international charters, recommendations and resolutions was more towards the refinement of principles at national and regional levels (Ahmad 2006). The 1972 UNESCO Convention concerning the Protection of the World Cultural and Natural Heritage merged two distinct areas: the preservation of cultural sites, and the conservation of nature (Peña Moreno 2013). Since then a wide range of cultural, environmental and sustainable development issues have been addressed at the global level—the subsequent chapter will follow up on this.

Within Europe the 'European Architectural Heritage Year' in 1975 can be considered a milestone for the consideration of heritage. Subsequently, the preservation of cultural heritage returned on the agenda of national governments, after having been of less importance in the post-war decades. In Germany, this is especially true, while already having lost a large amount of its tangible heritage during World War II; even more buildings were destroyed in the years after 1945, often with the idea of a more modern city in mind where the old fabric would only be obstructive. Over the years, these approaches have changed, rooted in the 1975 movement. The European initiative of formulating charters like the Declaration of Amsterdam 1975 on the European Architectural Heritage was going beyond former nation-wide approaches and extended the heritage scope to groups of buildings and their surroundings: old quarters, areas of towns and villages of historic or cultural interest, and even historic parks and gardens. This initiative to broaden the horizon of both heritage definition and spatial borders, was soon followed by other developed countries around the world (Ahmad 2006), while it took longer for most developing countries (cf. Sects. 3.3, 3.6, and 3.9).

Figure 3.1 gives an overview on the main documents, indicating the main changes and focal points of the heritage debate (in addition see Annex I with a short overview on the documents and their content). The Athens Charter of 1931 and the Venice Charter of 1964 are especially regarded as milestones for international cultural heritage preservation, path-breaking for the elaboration and ratification of the World Heritage Convention in 1972, which itself was enabled by the creation of UNESCO in 1945. Other conventions followed, notably the 1994 Nara Document on Authenticity and the 2003 Convention for the Safeguarding of Intangible Heritage (see Sects. 3.3 and 3.4). Basic concepts however stem from even earlier

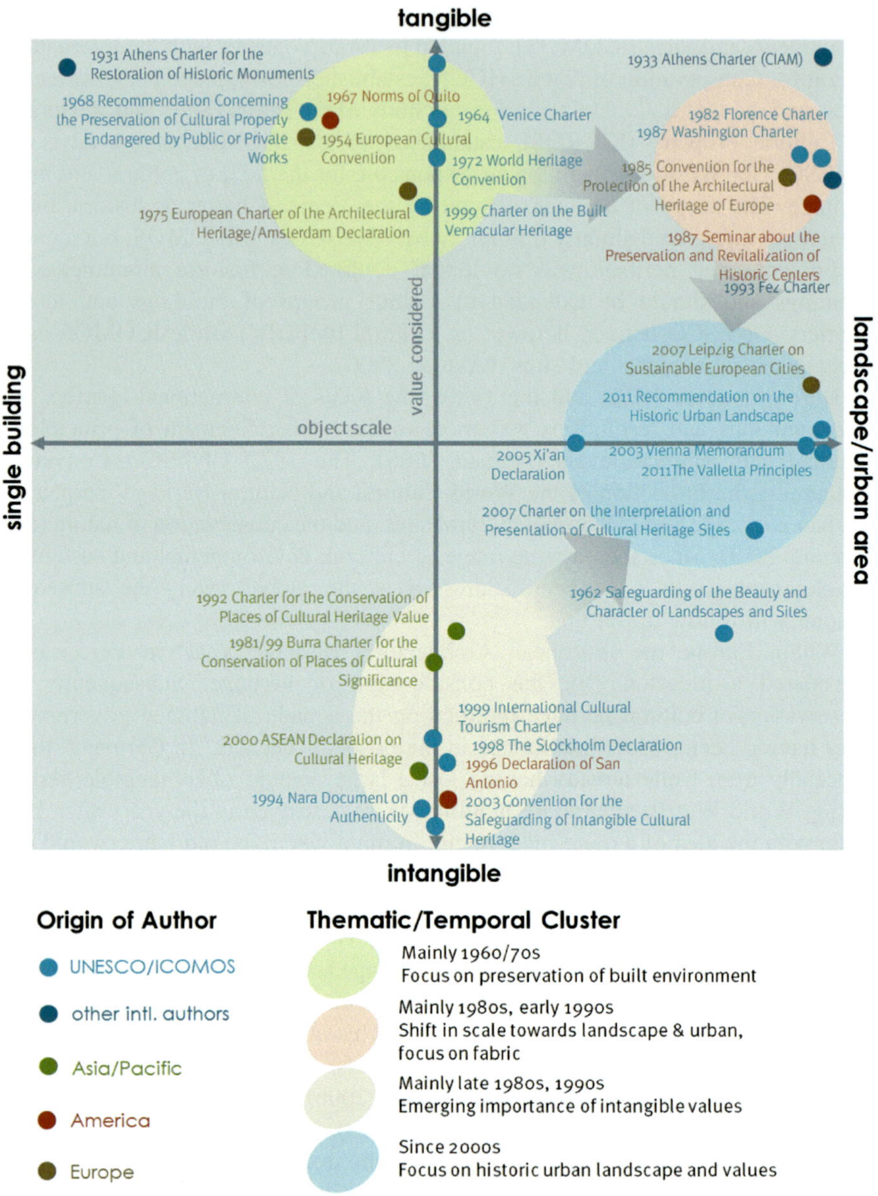


**Fig. 3.1** Overview of documents, charters and recommendations of supranational importance on cultural and intangible heritage (not including documents focusing on museums, archaeological sites, moveable properties, illicit trade, underwater heritage, artefacts, tourism, cultural routes). *Sources* Ahmad (2006), Birabi (2007), Falser (2010), Bandarin (2011), Ruland (2011), Veldpaus et al. (2013), The Getty Conservation Institute (2015)

times, especially the debate on the authenticity of heritage and the discussion on adequate conservation methods least harmful for historic fabric, and are still impacting on today's discourses.

In sum, two parallel movements towards each other can be ascertained: one concerning tangible cultural heritage, moving from a focus on single buildings or smaller entities like ensembles towards an urban and/or landscape approach. On a global scale it emerged after the 1931 charter. The second movement is a more recent one which has emerged in the 1990s only, focussing on the conservation of intangible heritage. Since the end of the 1990s and the early 2000s both movements are pooled, resulting in a more ample approach covering urban areas with their tangible and intangible heritage altogether.

## 3.2 The Emergence of UNESCO and International Cultural Heritage Policies

The idea to create a global movement for heritage protection emerged after World War I. It was later put into practice by UNESCO, founded after the even more destructive World War II. Since then, UNESCO and its partnering organizations are shaping the global heritage debate. The key event leading to the establishment of the World Heritage concept was the construction of the Aswan High Dam in Egypt. The initial plan to flood the precious temples of Abu Simbel and the Upper Nile Valley called international attention, and resulted in the 1959 international safeguarding campaign led by UNESCO (Prott 1993; Francioni 2012). International fundraising and technical as well as scientific support helped to dismantle and reassemble the temples on dry ground. This momentum of international attention and support kicked of a debate on the importance of heritage sites going beyond the mere responsibility of individual state parties. It was accelerated when being linked to the US-led movement on nature conservation. 'Mankind' became a perceptible scale, influenced also by the first space missions, despite cold war and political blocks (Wagner 2008).

In 1965, White House Conference in Washington suggested the foundation of an international 'World Heritage Trust' to safeguard natural and historic sites for the global community, followed by a similar appeal from IUCN in 1968. In 1972 the General Conference of UNESCO adopted the 'Convention concerning the Protection of the World Cultural and Natural Heritage'. Initially, ratified by 20 state parties, the number of states signing up rose rapidly to more than 60 in the early 1980s and 191 today (state in July 2014), with this number, its degree of execution is surpassed by no other convention in the world (Butina 2011; von Droste 2012). It is this postulation of a common cultural and natural heritage that Schmitt (2009: 119) calls "*an idealistic moment in international politics in the mankind*" while Alberts and Hazen (2010: 57) describe it as "*internationalization of the heritage movement*".

Today, 1007 sites belonging to 161 state parties are inscribed on the World Heritage List, with 780 of them as cultural heritage (state in July 2014). The most unique feature of the World Heritage approach is that it links natural and cultural heritage and their preservation (UNESCO 1972; Meskell 2013). Since then the concept has strengthened, matching the emerging discourses on the need to recognize the interaction of man and nature. Initially, a site was inscribed either as a cultural or a natural property. Cultural sites are in definition either monuments, groups of buildings or sites as stated in article 1 of the convention. Later, modifications meanwhile permit the inscription of mixed cultural and natural heritage if a site fulfils criteria of both and the inscription of cultural landscapes (formally inscribed as cultural sites). In definition cultural landscapes represent "*combined works of nature and man*", as stated in article 47 of the Operational Guidelines (UNESCO 2013b). The Operational Guidelines for the Implementation of the World Heritage Convention are neighbouring the Convention and have been slightly changed and adapted a few times since their initial formulation in 1977, and last time in 2013. The guidelines contain criteria for the inscription of properties on the World Heritage List and for the provision of international assistance under the World Heritage Fund. Such funding usually is limited to only a few sites in developing countries as the budget is limited. Being inscribed on the list is thus more a marketing and networking benefit for a site than a direct economic one.

The one requirement all sites on the list have to fulfil is the so-called outstanding universal value (OUV). Article 49 of the Operational Guidelines defines:

> Outstanding Universal Value means cultural and/or natural significance which is so exceptional as to transcend national boundaries and to be of common importance for present and future generations of all humanity. As such, the permanent protection of this heritage is of the highest importance to the international community as a whole (UNESCO 2013b).

UNESCO defines the OUV according to ten overall criteria, among them six for the cultural and four for the natural heritage sites. Table 3.1 lists the six criteria for cultural properties. Each site that is proposed to be inscribed on the list has to fulfil at least one of them. The majority is inscribed under two or more criteria.

To be considered as property on the World Heritage List, sites must meet the conditions of 'integrity' and/or 'authenticity', as both of them are key concepts and critical points in discussions of threats to any World Heritage site. "*The notion of integrity refers to the goal of maintaining all the critical elements of a site intact*" (Alberts and Hazen 2010: 60).

It is mentioned in the Operational Guidelines, as a kind of indicator for the wholeness and intactness of the heritage site, but the concept of authenticity is not easy to define and leaves room for interpretation:

> In particular, authenticity is socially constructed, so it has different meanings in different cultural contexts. This brings certain problems associated with enforcing common standards, but it also offers flexibility in taking the characteristics of each individual site into consideration when making preservation decisions (Alberts and Hazen 2010: 62).

**Table 3.1** Criteria for outstanding universal value of cultural world heritage sites (*Source* operational guidelines, UNESCO 2013b)

| Nominated cultural properties shall | |
|---|---|
| (I) | represent a **masterpiece of human creative genius** |
| (II) | exhibit an important **interchange of human values**, over a span of time or within a cultural area of the world, on developments in architecture or technology, monumental arts, town-planning or landscape design |
| (III) | bear a unique or at least exceptional **testimony to a cultural tradition or to a civilization** which is living or which has disappeared |
| (IV) | be an outstanding **example of a type of building, architectural or technological ensemble or landscape** which illustrates (a) significant stage(s) in human history |
| (V) | be an **outstanding example of a traditional human settlement, land-use, or sea-use** which is representative of a culture (or cultures), or human interaction with the environment especially when it has become vulnerable under the impact of irreversible change |
| (VI) | be **directly or tangibly associated with events or living traditions, with ideas, or with beliefs, with artistic and literary works** of outstanding universal significance. (The Committee considers that this criterion should preferably be used in conjunction with other criteria) |

It was particularly this authenticity concept that was leading to later discussions on the euro-centricity of the World Heritage and its policies, demanding for judgments on authenticity in the belief system of the local culture:

> The World Heritage Convention is an international legal instrument which calls on States Parties and the international community as a whole to do their utmost to protect sites of outstanding universal value for future generations (Ringbeck and Rössler 2011: 211).

Any development of World Heritage sites has to be in line with the conservation of those aspects or values the site was inscribed for. The World Heritage Committee's mission is more a monitoring and support of the state parties where the site is located. In the process of its implementation, the Convention has succeeded in bringing together different conservation concepts and approaches to create globally acceptable international standards in heritage conservation practice. Subsequently, also capacity building and building curricula in conservation of both, natural and cultural heritage, became part of UNESCO's mission (Cameron and Rössler 2013), mainly carried out by ICCROM, IUCN and ICOMOS:

> The six scientific criteria for cultural heritage were not, however, sufficient to the task of selection. As the list grew through the 1980s and early 1990s, it became increasingly evident that it was heavily weighted toward specific types of sites, namely sites of art and architectural merit such as castles, churches and cathedrals, and toward a specific region, namely Europe (Barthel-Bouchier 2012: 83).

Against this background, UNESCO tried and is still trying to balance the list. Subsequently, the OUV concept has changed a lot since it was implemented (Schmitt 2009). Sites inscribed have changed from the artefacts of 'classical' high cultures like the Roman Empire or ancient Egypt, towards an inclusion of sites linked to 'indigenous' cultures, to modern architecture of the twentieth century, disasters of civilization, and sites representing the industrial or global history like African fortifications commemorating slavery. Schmitt (2009: 119) calls this "a result of the need to respect the List's claim of serving the world society".

It has impacted on the understanding of OUV, changing its sense and meaning, and subsequently also influencing global debates on heritage.

The UNESCO World Heritage programme was, and still is, a reference for what is worth preserving for future generations—a long-term approach with a sustainable development perspective (Schmitt 2009; Ringbeck and Rössler 2011). With the World Heritage, UNESCO also established a new standard of site management: Any natural or cultural site has to establish a proper monitoring and management plan as a basic requirement for the site protection (Cameron and Rössler 2013), including also the establishment of surrounding buffer zones.

During the last quarter of the twentieth century, a shift towards landscape scale occurred, from individual buildings or sites to complex landscape values. In 1984, the category of cultural landscapes was put on the agenda by authorizing IUCN, ICOMOS and IFLA with developing guidelines for mixed and natural/cultural properties; 9-years later the first one was inscribed (von Droste 2012), rising to the number of 85 by now (state July 2014). Today, all landscapes are considered to have associative and intangible values that are requisite for their understanding (Smith 2013).

An integration of cultural and natural conservation efforts started in the 1990s when the Operational Guidelines were revised, adopting a single set of evaluation criteria, now focussing on a nature-culture continuum rather than seeing natural and cultural items separated (von Droste 2012). Initially, the World Heritage Convention recognized only archaeological sites, monuments and historic town centres as cultural heritage reflecting the dominant influence of the European traditions of archaeology, art history, and architectural conservation. By the turn of the millennium, the range of World Heritage cultural categories had included cultural landscapes, canals, and routes, as well as modern, rural and industrial architecture (Gfeller 2013).

To support the convention, in 2002 the World Heritage Committee developed a set of 'Strategic Objectives': 'Credibility', 'Conservation', 'Capacity-Building' and 'Communication'. In 2007, the objective of 'Community' was added—one reason for adding the fifth 'C' was to emphasize the importance of local, community-driven values. According to Te Heuheu and Kawharu et al. (2012: 10), it is crucial for a community to "*be given an opportunity to share the way they traditionally see, feel and listen to the Universe, and to do so through secured and sustainable processes*".

Throughout the past 10 years, aspects of community and indigeneity became prominent on local, national and international levels, promoting for instance the importance of intangible and associative values, represented in (oral) traditions and landscape interpretation. This way, UNESCO and its policies give advice to state parties and site managers on preservation approaches and appropriate measures, being aware of the limited influence an inscription on the World Heritage has for the long-term protection of a site (Alberts and Hazen 2010; Te Heuheu et al. 2012). Being relevant for natural sites exclusively in the beginning, the aspect of integrity of a site was applied to cultural heritage in 2005, referring to the intactness of a site, how resistant it is to threats, and how well the features and processes express its OUV (Perry and Falzon 2014).

Among the most important recent documents on heritage, there are specifically two that prove the shift towards cultural diversity and intangible heritage: the 2003 Convention for the Safeguarding of Intangible Heritage, and the 2005 Convention on the Protection and the Promotion of the Diversity of Cultural Expressions. Another important document published almost simultaneously, the Vienna Memorandum (2005), which focuses on protecting cultural sites in the face of modern development, states that even in living historic cities, authenticity and integrity should not be compromised (Alberts and Hazen 2010; Barthel-Bouchier 2012).

The history of World Heritage is also the history of shifting concepts, first developed and implemented on a global scale, then trickling-down to regional and local scales (Gfeller 2013). UNESCO and its partners have been the referent power for others, with the World Heritage List as the core of a 'metacultural' production of symbols for global society (Schmitt 2009). The World Heritage is being increasingly promoted as the focal point of international development, with the consequence that other UN agencies and international organizations like the World Bank are promoting heritage as a cultural resource for achieving socioeconomic development. Culture is becoming a tool for development, on a national and international level, and for many donor agencies (Negussie 2012). World Heritage governance systems have undergone remarkable changes as well since the 1970s, "*which can be aptly described as increasing professionalization, scientificization, bureaucratization and also NGOization*" (Schmitt 2009: 109).

Kwanda (2010: 1) states that at present, there are two paradigms on the notion of conservation:

- one is the classical conservation theory rooted in nineteenth century Europe, considering that the object has a value existing independently of people that should not be threatened to any change;
- the second one is following a more people-centred approach, developed since the 1980s and assuming that "*heritage is inevitably rooted more about people as creators of heritage, who attributes meaning, and selects what is to become heritage from the past*".

Alberts and Hazen (2010: 71) conclude:

> One of the major strengths of the World Heritage program is its ability to operate in a huge diversity of individual contexts. Despite all of UNESCO's efforts to generate and implement guidelines for preserving sites of outstanding value to all people, the protection of the sites, and therefore decisions taken about appropriate preservation measures, remain the responsibility of state parties. Developing effective guidelines for state parties is critical. Although such notions as authenticity and integrity have considerable shortcomings in terms of their vagueness and differences of opinion in how to define them, these factors provide sufficient flexibility for the framework to be useful across a wide variety of sites that are part of the program. Effective dialogue to ensure that such concepts remain meaningful in diverse contexts, and for all stakeholders, remains essential to maintaining and improving their efficacy.

## 3.3 Discourses on the Representativeness and Euro-Centricity of the World Heritage List

The previous chapters have already elaborated on the mostly Western origins of today's heritage movements. The visible result was the double-unbalanced list of World Heritage with a disproportionately high number of European and North American sites and a majority of cultural sites, such as castles, churches and cathedrals. This can also be understood as a sign of policies favouring Western conservation measures, as the basic concepts of integrity and authenticity are rooted there.

Von Droste (2012) calls the geographically balanced composition of the list one out of five major challenges the World Heritage Committee had to face during the initial years, beside the affirmation of the Committee's prerogatives vis-á-vis state parties, the interaction between nature and culture, a proper site management and public information and involvement. To counteract the euro-centric and imbalanced list, he asks for a new understanding of authenticity. Authenticity is deeply rooted in European understanding of heritage as described in Sect. 3.1; on a global scale its Western notion was constituted with the Venice Charter in 1964 and executed after 1977 when the World Heritage Convention became operational for cultural heritage (Kwanda 2010; von Droste 2012). Albert (2013) identified a lack of cause analyses or action strategies dealing with the unbalanced developments, although euro-centrism was so evident.

The biased definition of authenticity then caused difficulties for non-European cultural contexts. For example, Southeast Asian experts claimed that the Venice Charter was too close to European cultural values, and thus hard to apply to societies outside of Europe and European-based cultures. However, the new international organizations somehow acted like "*privileged interpreters of the past*" (Kwanda 2010: 9), producing a "*global canonization of the heritage of mankind*" (Schmitt 2009: 112). Their interpretation was then adopted and executed by conservation authorities.

European models of heritage practices were spread during colonial times in the latter part of the nineteenth century and the first half of the twentieth century. This is especially true for Africa and Asia, while Latin America had been following European planning and administration even earlier:

> Moreover, even areas not under colonial administration began to import the Euro-American heritage model and having heritage and a national museum became as much a part of a country's national paraphernalia as the flag and the anthem (Carman and Stig Sørensen 2009: 16).

In many former colonies, the administrative system was generated by the colonial masters, after their national models or reconstructed after the pre-independency system during the global conservation movement after the 1970s (Kwanda 2010). As a consequence, conservation had an ideological background, defined and executed in Western way, and therefore "*it is striking in this respect that a large number of sites outside Europe are concerned with European colonial history*" (Schmitt 2009: 112).

Post-colonialism became one important political trend allowing for increased appreciation of different views of the past, broadening the view of heritage and related values (Carman and Stig Sørensen 2009). Fundamental transformations of practice occurred then in many of these countries throughout the 1980s, when indigenous rights discourses emerged, raising awareness on ownership and control of their heritage in many post-colonial nations (Ireland 2012).

Twenty years after launching the World Heritage, the vast majority of sites were cultural ones in Europe, indicating the list's early focus on ancient monuments. The six criteria for cultural heritage were proven not sufficient for the task of selection (Barthel-Bouchier 2012; Albert 2013; Meskell 2013). In 1994, two breakthrough documents were published then by UNESCO and its partner organizations: The Nara Document on Authenticity (1994) which will be explained more in depth in the following chapter, and the Global Strategy for a Representative, Balanced and Credible World Heritage List adopted by the World Heritage Committee. The strategy aimed at a more holistic and representative list, broadening heritage definitions and opening it especially to living and traditional cultures—broadening the material understanding of heritage:

> Based on the scientific principles of classification and categorisation, the Global Strategy is essentially a quota sampling mechanism meant to guarantee that effort is made to identify under-represented examples of the full range of possible sites. Thus, preference should be given to under-represented regions such as Africa and Asia, and to under-represented forms such as industrial, scientific, or underwater heritage (Barthel-Bouchier 2012: 83)

Beside this, UNESCO decided to establish capacity building for nations without sites or tentative lists to empower them.

In 2000, Cairns Declaration decided to limit the annual inscriptions to 30, allowing only one per country except for natural sites, tackling both weak points. Subsequently, the ICOMOS "Filling the Gaps" report (2004) ought to identify under-represented categories contributing to the further development of the Global Strategy.

Another paradigm shift was the introduction of the concept of 'Community' and subsequently community values as the fifth strategic objective by WHC in 2007 (cf. Sect. 3.1). This was, and is not, only facilitating the inclusion of local community values, but also the concept of indigeneity, acknowledging the importance of intangibility and associative values (Te Heuheu et al. 2012), and thus facilitating the recognition of indigenous cultures and cultural landscape assets.

The dominance of Western understanding of heritage, as expressed and executed in the World Heritage Convention, has been both booth and bane for non-European sites and cultures. On the one hand sites and their inherent values were neglected or seen as inferior; on the other hand at least the growing awareness of heritage values from the 1970s onwards attracted global attention. From the 1990s on, many efforts were made to adapt and broaden the global understanding of heritage and conservation to match the underlying varieties of cultural concepts (Schmitt 2009; Te Heuheu et al. 2012).

Today, European cultural sites still continue to dominate in terms of nominations and inscriptions as can be seen in Fig. 3.2. The new 'upstreaming process', suggested in 2009 to assist underrepresented countries in preparing these dossiers and identifying the OUV criteria, is supposed to counteract this (Meskell 2013). Nevertheless, in 2012, still 38 % of the newly inscribed sites were located in Europe and North America indicating the strategy's limitations and the comparable advantages of these countries in preparing the very complex application documents. To many countries a good deal of the international recommendations are very challenging or even irrelevant, taking into consideration their cultural and socioeconomic realities (Kulikauskas 2007). In June 2014, the Okavango Delta in Botswana was inscribed on the World Heritage List as the 1,000th site. It is not inappropriate assuming that exactly this spot, representing a natural heritage site—currently accounting for only 197 out of 1007 properties—coming from the continent with the lowest number of inscriptions on the list was chosen to stand for this symbolic number.

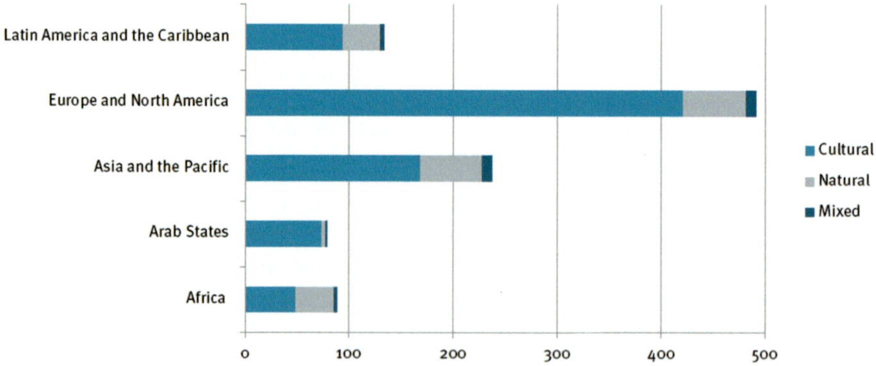


**Fig. 3.2** Number of world heritage properties by region in 2015 (*Data source* UNESCO 2015d)

The first site ever deleted from the World Heritage List was the Dresden Elbe Valley in Germany; due to the construction of a four-lane bridge the World Heritage Committee suspected the loss of the site's OUV. UNESCO set a precedent showing that the inscription on the World Heritage List can be withdrawn. This was it done with a European site—while the vast majority of sites inscribed on the List of World Heritage in Danger is from emerging or developing countries—so again there might be some purpose behind, on a more global political level than the local arguments for or against the cancellation of World Heritage status.

Asia in comparison to Europe and Northern America has different traditions affecting conservation principles (Kwanda 2010). Non-Western, oral cultural traditions, such as the Hindu cosmology and the Western knowledge-centred views of the world are competing fundamental principles, with one of them imposed within the conservation movement until only very recently (Kwanda 2010; Koch 2013). Kwanda (2010: 9) notes:

> Buildings are built, repaired, restored, rebuilt, and extended continuously by the people with their skills, rituals, and knowledge for fulfilling their needs in everyday life. This tradition of conservation to prolong the life of buildings in Asia leads to the common practices of continuous renewal as opposed to the Eurocentric notion of material authenticity that insists on the notion of minimum intervention and reversibility.

Kwanda (2010) describes the conflict between Western notion and the Asian understanding of heritage, with its tradition of continuous renewal of its perishable historic structures and closer link to naturalistic and spiritual values (cf. Rigg 1997). The same is true for many African countries, while in Latin America one has to basically distinguish between the colonial sites, that followed a mostly Spanish or Portuguese model, and indigenous communities and their values and traditions. Traditional architecture is means of communicating intangible values; its renewal is part of the spirituality and worshipping of the people. He therefore claims a need for redefining authenticity, broadening the limiting Western understanding and including the so-called living authenticity found in local leaders or craftsmen and rituals to be preserved. Jaramillo Contreras (2012) argues that intangible heritage deserves the same protection as built heritage, as it is of highest importance for local identity especially in poor (developing) countries.

The recent years since the end of the twentieth century have witnessed a shift from the object-centric to the subject-centric conservation, facilitating the recognition of a plurality of tangible and intangible values and functions, and trying to bridge the differences between a European monumental tradition and non-Western concepts of culture (Kwanda 2010; Irr 2011; Weise 2012). Meskell (2013: 492) goes on to comment:

> Despite the valid critiques of the World Heritage List and its Eurocentrism, the recognition and value that inscription bestows is remarkably still desired by almost all the nations of the world, regardless of political or religious affiliations, economic status, or historical trajectory. That fact, in itself, offers a powerful lens onto the potentials of something called heritage in political cultural, economic, and spiritual terms.

## 3.4 Consideration of Intangible and Immaterial Heritage

In the past, cultural heritage was often identified by taking into account the aesthetic and technical values perceived by experts exclusively, but not the symbolic aspects that it may represent. Few efforts were made to protect the collective memory of communities, which apparently contributes to an object's value (Jaramillo Contreras 2012).

The difficulties and shortcomings of the very much fabric-focused conservation approaches became evident when policy documents on national or regional level began to tackle the issue. In this context, the Australian Burra charter in its 1999 revision (the first draft stems from 1979, elaborated by Australia ICOMOS, the 1999 revision was then adopted by Australian Heritage Council and regional authorities) has to be mentioned as a path-breaking document for the global conservation discourses, although being drafted for the national context. The charter was the first one outlining standards for using cultural significance for managing and conserving sites, adapting international principles to national values and needs, paying particular attention to aboriginal culture.

The first international document giving consideration to these difficulties was the 1994 Nara Document on Authenticity (introduced in the last chapter), proposing that assessments of authenticity should encompass matters relating to "*form and design, materials and substance, use and functions, traditions and techniques, location and setting, and spirit and feeling, and other internal and external factors*" (article 13).

Furthermore, it stated that authenticity had to be understood as a relative concept including intangible attributes, such as usage, tradition and spirit. The key message was that cultural heritage needed to be judged within its respective cultural context (von Droste 2012). It was drafted at the Nara Conference on Authenticity in Relation to the World Heritage Convention, organized by the government of Japan, in cooperation with UNESCO, ICCROM and ICOMOS. To Kwanda (2010), it represents a change from euro-centric heritage definitions focusing on the material originality, towards a culturally centred approach inheriting intangible values. Authenticity and values do not reside in fabric and its proper maintenance only but incorporate local traditions and values:

> Temporal change is not the only influence on authenticity. The World Heritage program also accepts that different cultures interpret the concept in different ways, and it has convened several conferences to discuss such ambiguities. The first and most significant of these, the Nara Conference on Authenticity, gave rise to The Nara Document on Authenticity, which outlines how the term 'authenticity' should be interpreted (Alberts and Hazen 2010: 61).

The Nara document was then mainstreamed into national and regional policies, e.g. in the UNESCO Asia-Pacific Awards programme awarding outstanding conservation projects since 2000, including explicitly such with a focus on intangible heritage (Kwanda 2010).

In 2003, the Convention for the Safeguarding of the Intangible Cultural Heritage was adopted, entering into force 3-years later. With this document, it became possible to safeguard and acknowledge oral traditions, such as dance and song, literature, theatre and rituals on a global scale (Butina 2011). In article 2/1 it defines intangible cultural heritage as:

> The practices, representations, expressions, knowledge, skills – as well as the instruments, objects, artefacts and cultural spaces associated therewith – that communities, groups and, in some cases, individuals recognize as part of their cultural heritage (UNESCO 2003).

It thus enabled the formal protection of performing arts, social practices, rituals, festive events and traditional craftsmanship.

After a long time considering only the tangible values of a site, this fundamental shift formed part of a growing recognition of non-Western views of heritage. Namely, Asian countries faced difficulties with the very strict concepts of authenticity and the utmost importance of preserving historic fabric as they were competing with the traditional understanding of conservation measures and the close link between place, fabric, maintenance activities and attached values. Japanese Shinto shrines or Nepalese Hindu temples for instance were maintained by local communities for many generations; periodic maintenance was understood as an aspect of worshipping. Maintenance, however permitted changes and repair works, harming the original fabric and thus affecting the heritage value in European understanding but not in the local one, as the spiritual values were not affected but maintained. The 2003 convention at last allowed for protection of such spiritual assets. In the course of the policy document, UNESCO created the Lists of intangible cultural heritage and the Register of best safeguarding practices. Today, 336 elements (UNESCO 2016) are listed as intangible heritage, with the register of safeguarding practices comprising of eleven projects, with six of them from non-Western countries. Figure 3.3 visualizes the distribution of intangible heritage elements among the world regions, indicating a comparably lower regional bias compared to the World's Cultural Heritage—except for the complete absence of Northern American inscriptions.

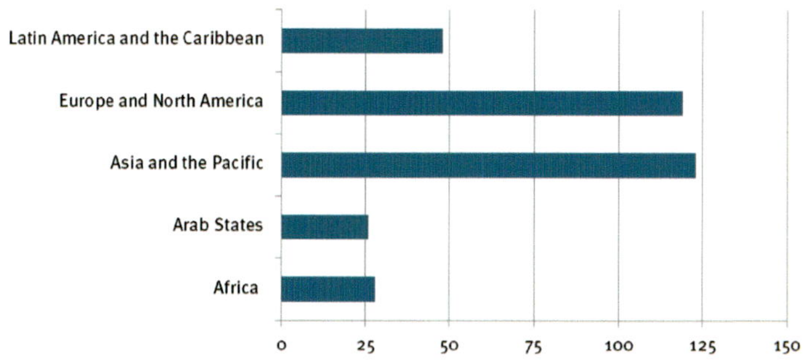


**Fig. 3.3** Number of elements on representative list of the intangible cultural heritage of humanity per region (*Source* UNESCO 2016)

The 2008 ratification of the ICOMOS Charter for the Interpretation and Presentation of Cultural Heritage Sites formed the most recent step towards acknowledging immaterial values and interpretive practices. It emphasizes on community-based activities as well as on participation, counteracting the dominance of expert interpretation over the values communities assign to their heritage. One of the seven objectives therefore is to "safeguard the tangible and intangible values of cultural heritage sites in their natural and cultural settings and social contexts" (ICOMOS 2008).

Participation in heritage conservation and interpretation is increasingly seen as an aspect of civic engagement (Silberman 2013). In terms of authenticity, the charter refers to the spirit of the Nara document, when stating that traditional social functions of the site, and the cultural practices and dignity of local residents and associated communities should be respected, e.g. when planning any cultural activities. It also claims for continuing research, consultation as well as knowledge exchange, on local as well as international level.

Over time, the initially quite narrow scope of the 1972 World Heritage Convention on clearly defined sites within national territories (as adopted from earlier national policies and interpretations) has been widened. Firstly, it was extended to larger sites, like whole city centres or settlements and their vicinities, slowly anticipating the shift towards cultural landscapes. They were added to the World Heritage Convention in 1992, drawing attention to the interaction between humans and their natural environment, offering a means by which to (re)negotiate the meaning of heritage on a global scale. Not only did it bridge the traditional nature–culture divide, but it also introduced the notion of intangible cultural heritage a decade before the adoption of the 2003 Intangible Cultural Heritage Convention (Gfeller 2013). Since then, the amalgamation of tangible and intangible values got acknowledged on a global scale and by global actors, such as UNESCO or ICOMOS (Smith 2013; Turner 2013; Alberts and Hazen 2010). Twenty-first century heritage interpretation is becoming more inclusive, fostered by global policies, trickling-down to national ones, at the same time also initiated and fostered by national efforts—somehow a bottom–up and a top–down approach at the same time.

## 3.5 Tangible and Intangible World Heritage Sites in Brazil, Indonesia and Nepal

**Brazil** ratified the World Heritage convention in 1977—the first site inscribed was the Historic Town of Ouro Preto in 1980. Actually, the country comprises of nineteen World Heritage sites, and eighteen on the tentative list (UNESCO 2014a). Despite the large number of different landscapes and their scenic beauty, only seven natural properties are inscribed. Since 2012, the cultural landscape of Rio de Janeiro is inscribed, the other eleven sites are cultural heritage properties, mainly urban centres or historic squares. They reflect different phases of Brazil's cultural history, from colonial times to independency and modern architectural movements manifested, e.g. in the urban planning of Brasilia. Overall, six historic centres or towns were inscribed, among them three colonial cities in the Northeast of the country: Sao Luis, Olinda and Salvador.

Indonesia joined the World Heritage initiative years later, with the acceptance of the convention in 1989. Today four cultural and four natural sites are inscribed, and another 26 sites form part of the tentative list (UNESCO 2014b). The first four sites were inscribed in 1991: two temple compounds in central Java in the vicinity of Yogyakarta (Prambanan and Borobodur), plus two national parks. Besides that, other sites showing human evolution and a cultural landscape were inscribed as cultural sites; so far no urban heritage found its way into the list. Since 1995, the Yogyakarta Palace Complex is listed on the tentative list; however, no further efforts to inscribe it on the list have been made. Among the 26 sites on the tentative list, natural sites like national parks are predominating.

The World Heritage Convention had been ratified by Nepal already in 1978, only 1 year later the first two sites were inscribed on the list: the Sagarmatha National Park with Mount Everest and the Kathmandu Valley. Since then, two more sites have been inscribed—one cultural and one natural site. Currently, 15 sites are listed on the tentative list, with all of them being cultural sites. For some years the Kathmandu Valley was put on the List of World Heritage in Danger, but it was removed in 2006 while making some buffer zones and boundary modifications (UNESCO 2014e).

The history of intangible heritage is a much more recent one. The Convention for the Safeguarding of the Intangible Cultural Heritage (2003) entered into force in April 2006. In 2008, the first 90 'elements' were inscribed into The Representative List of the Intangible Cultural Heritage of Humanity—intangible elements that help demonstrate the diversity of this heritage and raise awareness about its importance. Since then (2009–2013) another 192 elements were inscribed.

Out of these, four of these elements come from Brazil, in 2012 the 'Frevo', performing arts of the Carnival of Recife', got listed as an "artistic expression comprising music and dance, performed mainly during the Carnival of Recife" (UNESCO 2014d).

Four intangible heritage elements from Indonesia are listed, among them the Indonesian Batik (2009) and the traditional Wayang puppet theatre (2008). Despite not mentioning a specific location in the justification for inscription, Yogyakarta is well known for both of these, having flourished at the Royal Court.

There are no Nepalese sites yet inscribed as Intangible Cultural Heritage.

## 3.6 Current and Future Challenges of Cultural Heritage Conservation

As described in the previous chapters, the understanding of cultural heritage has changed over the past decades, becoming more ample and diverse (Carman and Stig Sørensen 2009). Conservation approaches had to adapt to the changing circumstances, ranging from the shift in heritage understanding to global urban change in general. Globalization has impacted on urban planning paradigms worldwide, resulting in a singular contrast between the aim to follow a global urban look and the wish to preserve the locally specific characteristics. "*There is a growing realization that heritage is not simply a top–down conservation effort, or a potentially lucrative resource to be exploited for short-term economic gain*". (Silberman 2013: 30).

Apart of the aspects described in the previous chapters, also discourses on urban governance, global climate change, environmental aspects, the sustainability debate and globalization have penetrated the heritage debate(Amen et al. 2011; Albert 2013; Turner 2012). It is no longer a stand-alone topic, like an isolated monument not rooted in its surrounding, but rather one asset of a city or landscape, linked to its natural and built environment. New topics emerged, such as heritage as *a "means of governing living populations*" (Carman and Stig Sørensen 2009: 20). The procedural character of heritage interpretation is gaining importance, taking into consideration the present uses and attached values of a site.

The sustainability debate that arose after the Brundtland Report (1987) and its famous definition of sustainable development as a "*development that meets the needs of the present without compromising the ability of future generations to meet their own needs*" has also impacted on the heritage debate. Subsequently, heritage was considered a non-renewable resource in various policies, emphasizing at the need to preserve it. As a consequence, sustainability requires an integrative approach including culture and development (Turner 2012). The social pillar of the sustainability understanding is closely linked to cultural heritage or cultural landscapes and the preservation of community values, while the preservation of natural heritage is connected to goals of environmental sustainability (Conradin et al. 2015).

Environmental aspects have been key aspects of the heritage debate for more than a century now, as protection of natural environments was one of the roots of natural heritage conservation (Conradin et al. 2015). In 1980s, the aspects of ecological damage found an entry into the natural and cultural heritage debate. It contributed to the broadening of categories (Irr 2011), e.g. to the inclusion of cultural landscapes and their fragile components, like the irrigated rice terraces of Ifugao, Philippines, which are threatened by deforestation and climate change impacts:

> We live in a time when a younger generation is asking tough questions about the systems put in place during the modernist period, particularly during the second half of the twentieth century. In terms of the built environment, this generation is questioning the legal

framework for design and development, the definition of professionalism the privileging of abstract learning. The role of experts in the operation of social, cultural and economic systems is being re-examined. Within the cultural heritage field, the interest in cultural landscape theory and practice can be seen as part of this shift towards a more ecological worldview. There is a change in the motives and interest of those who choose to become involved with historic places (Smith 2013: 62).

Climate change impacts are likely to exacerbate environmental and related risks for any given area, e.g. for fragile cultural landscapes or coastal cities, thus also affecting heritage sites. UNESCO itself has come up with working groups emphasizing on Climate Change and World Heritage in (2006). On the other hand, aspects of historic urban landscape are now increasingly seen from a sustainability point of view. Especially, European historic urban cores are regarded as role models, due to their variety of usage within the core area, allowing for shopping and leisure without dependency on individual transport but promoting public transportation and pedestrian areas.

The climate change topic is closely linked to aspects of risk and risk reduction, also within the heritage context. Initiatives of risk preparedness for cultural heritage emerged throughout the 1990s, on local, regional and international level. Stovel (1998: 13f) claims that the loss of cultural heritage in a disaster is an issue as post-disaster recovery and rebuilding livelihoods "*depend very much on efforts to retrieve and strengthen those heritage elements and symbols that have traditionally given meaning, order and continuity to life*".

According to Stovel (1998), so far there are only a few countries where the conservation and risk-preparedness fields routinely collaborate, e.g. the Netherlands, Switzerland, but also non-Western countries like Sri Lanka. Any heritage site is prone to natural and man-made risks and disasters. On the other hand, tangible and intangible heritage can play a significant role in reducing disaster impacts on livelihoods, related technologies, practices, skills and knowledge systems and at the same time can support reducing risks from disasters at all phases of the process (readiness, response and recovery), and hence in contributing to sustainable development in general as the Strategy for Risk Reduction at World Heritage Properties (UNESCO 2007) states.

Built cultural heritage as well as the intangible values and rituals attached to it are deeply political and an instrument of power. We are living in times where the targeted destruction of heritage has become a weapon of war again—60 years after the Hague Convention. Leftovers of earlier civilizations are destroyed to erase selected part of the people's past. In this context heritage is far more than fabric. The systematic destruction of cultural heritage, e.g. in Timbuktu, in Afghanistan (Taylor and Levine 2011) or now in Syria, Iraq and surrounding countries have triggered responses on a global scale (UN Security Council 2015). Irina Bokalova, Direcor General of UNESCO, talked about "*cultural cleansing*" taking place in Iraq after the destruction of Mosul museum as well as Hadra, Nimrud and Palmyra heritage sites (UNESCO 2015c; UN News Centre 2015; UNESCO 2015a). To Markus Hilgert (2015), Director of Berlins Museum of the Ancient Near East this pre-Islamic heritage is a historic and cultural point of reference for societal reconciliation and the (re)establishment of a cultural identity. Bokalova is also quoted on the webpage of the newly founded "Unite4Heritage" campaign (UNESCO 2015b) to support heritage threatened by extremist movements: "*Cultural sites have a universal value—they belong to all and must be protected by all. We are not just talking about stones and buildings. We are talking about values, identities and belonging*".

The growing importance of economic aspects and capital streams on the system of cultural preservation results in a competition of particular interest with global concepts. In parallel to the expanding world economy, heritage conservation became a growth industry, supported by the changing definition of heritage (Barthel-Bouchier 2012; Butina 2011). UNESCO‘s World Heritage List has somehow turned into an accreditation scheme for heritage sites, used either to attract tourists, generate revenue, or as nation building. This illustrates how transnational processes are always subject to national and local economic considerations and political agendas (Salazar 2011a, 2012). New (globalized) fashions and certain economic models are threatening heritage (Te Heuheu et al. 2012):

> Local distinctiveness and a unique heritage could well be submerged, not under the ocean, but under the overwhelming weight of world domination. Even though it might be possible to ring-fence sites of value, the values themselves will be seriously eroded if the cultural dimension, which is an integral part of the site, is lost to whatever worldwide trend happens to be fashionable at any particular moment (Te Heuheu et al. 2012: 17).

Cultural tourism has become a major source of income generation with cultural and creative industries representing an important and fast-growing sector in global economy. Cultural tourism even accounts for 40 % of the global tourism revenues (UN System Task Team on the Post-2015 UN Development Agenda 2012), relying on both tangible and intangible assets. In its publication on "Culture: a driver and an enabler of sustainable development", the UN System Task Team on the Post-2015 UN Development Agenda (2012) calls cultural heritage, cultural and creative industries, sustainable cultural tourism and cultural infrastructure strategic tools for revenue generation. They see particular potential in developing countries, often comprising of a rich cultural heritage and substantial labour force. Carman and Stig Sørensen (2009) explicitly mention the importance of cultural tourism in developing policies for African countries, while Xu et al. (2011) describe the emerging field of ecotourism in heritage sites in China.

However, conflicts between the expectations of tourists and the initial local usage or customs can arise, potentially harming the site's values.

> Tourism presents further challenges to authenticity, two of which are particularly common. First, visitors to sites often arrive with preconceptions about what they expect to see, and site managers may consciously or unconsciously attempt to ensure that these expectations are met, even if authenticity is compromised [...] Second, accommodations to meet the needs of tourists may be incompatible with preservation goals (Alberts and Hazen 2010: 68).

For example, in the case of Havana, Cuba, Mertins (2003) found that the preservation projects within the historic centre rather focused on the attractiveness of the area to tourists than on an authentic preservation of the historical fabric.

Many authors emphasize on the duality between globalization and regionalization, with heritage occupying a dual position as "*both a cause and effect*" of cultural globalization (Ireland 2012: 19). Local culture and cultural assets can be considered one means of a collective to distinguish from others. Having this in mind, Wagner (2008) asks if mankind can be one collective only, lacking the competencies of social and cultural scientists to describe such global processes of identity formation triggered by heritage. Furthermore, international governmental organizations—like the World Heritage Committee—are meanwhile fulfilling tasks that territorial states now have difficulty completing on their own. From the success of the World Heritage programme, the conclusion can be drawn that although global cultures vary enormously in their expression, content, etc., this variety can become widely available when it is coordinated by an expedient and instrumental system of management (Irr 2011).

Schemes of good governance and participatory approaches are intertwined with recent approaches of heritage conservation as well. Weise (2012: 66) states that "*living cultural heritage properties are dependent on ensuring that the community that created and maintained the heritage over the centuries continues to do so while adapting to unavoidable changing circumstances*".

International policy documents recommendations covered many social, governance, economic and broader cultural issues, suggesting appealing approaches, but still being disregarded often, especially in urban development practices. Nevertheless, they have served to sustain global cultures and heritage, what national states often had difficulties with in the past. Participatory and good governance approaches should include knowledge transfer in both directions, from communities to authorities and vice versa. However, Kulikauskas (2007) states that participation and people's involvement is still often considered as a planning instrument rather than a goal, going on to reflect that:

> Any sane minister or mayor, when confronted with conservation (and expenditure) vs. new development (and tax revenue), culture (and expenditure) vs. business (increase of tax revenue) and job creation (decrease of public expenditure and increase of tax revenue), will seldom opt for the first – unless there is a significant pressure from the community (significant enough in numbers to cost votes). To break this spell, one needs to stop confronting conservation to development, and to work together with, not against new development, helping channel it into continuous and responsible development instead of attempting to hinder the change (Kulikauskas 2007: 63).

Education on and through heritage has for a long time been hidden in other study areas or faculties; heritage studies as an own field of research emerged only in the 2000s. Since then, many universities have implemented courses in this area, again predominantly in the Euro-American region. Prott (1993) assigns educational policies a profound impact on the preservation of cultural property in the long run and equates a lack of imparting the significance of cultural values with a lack of awareness of future decision-makers concerning the needs of heritage planning. Furthermore, it requires awareness of the importance of local culture that got lost in many locations, like Suanda (2005) is stating for the case of Indonesia, claiming tools and means of bridging the gap especially in school education, to teach students to appreciate their own culture and heritage.

As a conclusion, heritage and conservation is nothing to be seen outside of current debates. Quite the contrary is true, as highlighted, most of these debates found their way into the heritage discourses, impacting on current and future focal areas of action and policies. Ireland (2012) describes twentieth and twenty-first century heritage sites as cultural anchors, touchstones, symbols of ownership, territory, belonging and identity, when stating "*What these contemporary forms of heritage have in common is their future focus, using elements of the past to represent shared values as a basis of a nation's future vision*" (Ireland 2012: 18).

Various challenges for the conservation of heritage lie ahead, requiring suitable approaches and policies. Different authors have come up with recommendations and suggestions. Barthel-Bouchier (2012) claims a need for more public support and better outreach strategies; in her opinion heritage conservation should draw more heavily on social science to build trust between experts and the public, and to give cultural heritage a voice in the new global age. Butina (2011) goes along with her when claiming an—politically as well as economically—interdisciplinary and international dialogue, with the long-term aim of making the decisive step from the globalized world towards a world community. Weise (2012: 66) concludes that with:

> The inclusion of a complex array of properties on the World Heritage List, it has become clear that a more integrated approach to conservation is required. Conservation needs to become sustainable. This requires the consideration of present day realities, economic feasibility and social benefits. Especially within properties which are still being used or are inhabited, community involvement becomes paramount for the success of any conservation efforts.

Cultural heritage and local culture have become parts of a global culture, undermining and slowly changing the Western understandings. Perception, imagination, communication and placement of heritage are changing, depending no longer on a global definition but characterized by recognition of cultural diversity and relativism (Wöhler 2008; Smith 2013). Research on local perception and identity-formation can increase the sustainability of conservation practice. A consideration of such aspects in both, theory and practice, can support realizing one so far non-fulfilled demand that Alois Riegl (1903) had raised more than a hundred years ago: considering the locally specific "*Gegenwartswert*" (present values attached) instead of focussing only on the "*Erinnerungswert*" (historic values) (Vinken 2011).

## 3.7 Cultural Heritage Conservation on National and Regional Levels

International policies on natural and especially cultural heritage have impacted strongly on the national and regional levels worldwide. This trend evolved during the late 1970s and 1980s, when the focus of international charters and policies was towards the refinement of principles at national and regional levels (Ahmad 2006). Since then, international organizations with UNESCO at the forefront, have tempted to set global standards, but with an increasing recognition of local patterns and cultures. The consideration of local culture and traditions in conservation efforts, as described before, is intimately connected with the respective political and governance system (Butina 2011). Taking a look at different world regions, the mutual influences become clear and reveal regional differences.

Two World Wars within a few years have led to tremendous losses of cultural heritage throughout the European continent. This sad fact and the ongoing losses of heritage buildings and sites during post-war years have shaped European awareness concerning the importance of conservation, on national but also supra-national level. Until, mid of the twentieth century, cultural heritage was mostly understood as the preservation of single buildings, that understanding changed and broadened after the 1964 Venice Charter towards the consideration of ensembles and quarters (Urban 2011).

The 1975 European Architectural Heritage Year was a kind of wake-up call for heritage conservation on European level. Both, the European Charter of the Architectural Heritage and the Amsterdam Declaration (Europarat 1975) discussed the broadened concept of architectural heritage and integrated conservation, and in particular the roles of authorities in managing architectural heritage. The Amsterdam Declaration explicitly addressed groups of buildings and their surroundings, old quarters, areas of towns and villages of historic or cultural interest, historic parks and gardens and contemporary buildings. It also asked for the consideration of social and economic aspects in conservation, both in urban and rural communities (Ahmad 2006). Of particular importance is the multi-level approach in conservation. Each state has its own heritage legislation, on national and/or federal state level, mostly given during the 1960s and 70s (e.g. Austria in 1959, former West-German Federal States between 1958 and end of the 70s). On top of it, the European Union has significant influence on the continent's heritage debate.

Since the 1980s, Europe has developed an increasingly positive attitude towards commemorative culture like places of remembrance. This renaissance, visible in the launch of programmes like the "European Capital of Culture" annually assigned by the European Union (EU), has to be seen in the context of increasing globalization and related tendencies of regionalization (Luger 2008). The EU strategically used the cultural (and natural) heritage to bring the joint European history to mind. Two documents of particular importance are the European Landscape Convention from 2000 and the Leipzig Charter on Sustainable European Cities (2007), indicating the two main focal areas: cultural landscapes and historic cities, both recognized as important factors for shaping the European culture. The European Landscape Convention understands places or landscapes as conceptual with associated and multi-faceted values, including understandings of association, identity, remembrance, coherence, community, sanctity and forgetting (Townend and Whittaker 2011).

The Leipzig Charter on Sustainable European Cities asks for integrated urban development strategies, linking the past heritage to future challenges. As a consequence, integrated urban (district) development has then become increasingly important in many EU Member States over the last 10 years (Ruland 2011; German Institute of Urban Affairs (Difu) and Federal Institute for Research on Building; Urban Affairs and Spatial Development within the Federal Office for Building and Regional Planning 2012). Also in terms of urban heritage, 1975 was the turning point from the post-war reconstruction phase towards more sensible conservation approaches, considering and conserving the remaining sites. Until the 1960s, much of the historic fabric left after the war years was lost then due to large-scale renovation projects following paradigms of the Athens Charter, e.g. the car-friendly city (Vinken 2010; Ruland 2011). After the fall of the Berlin Wall and the EU, entry of East-European programmes for protecting the historic cities, regarded as common heritage, were launched on national and EU level, for instance the German programme on '*Städtebaulicher Denkmalschutz*' (urban conservation). The project's aim is to protect and renovate historic urban cores in East Germany, which had suffered from poor maintenance throughout the decades before, contributing significantly to the preservation of historic urban fabric (Behr 2005; Allstedt and Metzler 2009; Ruland 2011).

Throughout Europe, today more than sixty historic city cores are listed as UNESCO World Heritage (seven out of them only in Germany, three in Austria)—an expression of an often centuries old urban culture, shaping the vision of a liveable town till today (Urban 2011). For most Europeans, heritage had changed from being a good to a product and finally to a commodity, increasingly viewed under economic criteria. Urban culture and urban heritage is considered as beneficial for urban growth and economic development. Revitalization of historic urban quarters has replaced the preservation-oriented initial phases of urban conservation, with suitable uses and vital surroundings as main challenges (Bundesministerium für Verkehr 2007; Albert 2013).

The North American continent's first approaches to protect monuments, understood as natural or archaeological sites as well as single monuments, stem from the early twentieth century, with the US 'Antiquities Act' of 1906 and the foundation of 'The Historic Sites and Monuments Board of Canada' in 1919, followed by the 1953 Historic Sites and Monuments Act. Canada and particularly the USA have a strong background in natural heritage; in fact the US system of national parks was one of the main pillars on which the UNESCO system of natural heritage was founded. Till today and against the global trend, both countries have more natural sites listed as World Heritage than cultural ones. The cultural heritage sites are either archaeological ones, mainly sites of First Nation cultures, or buildings representing important steps of the countries' history. It is probably because of the roots in natural heritage that the concept of cultural landscapes is of particular significance.

The cultural landscape concept is increasingly used to denote the overlap of natural and cultural heritage, and the values especially First Nation communities attach with certain areas. Canada is thus using the term 'aboriginal cultural landscapes' to denote these landscapes that have also been included in the associative cultural landscape category of UNESCO and the ethnographic landscape category of the US Parks Service before. Such communities have rarely been involved in the field of cultural heritage, as defined within the major academic and institutional frameworks. Heritage education on university level emerged as early as in the 1970s, initially differentiating between theory and practice, education and training. Smith (2013) states that the shift towards cultural landscapes and recognition of community-based values and related conservation approaches has bridged this gap at least partly, commenting that:

> Cultural landscape theory and practice is creating the basis for what we in Canada consider to be a new paradigm for identifying, understanding and treating cultural heritage. The most important shift is towards an ecological view of cultural resources. This view leads to an integration of natural and cultural resource management with each other and with contemporary cultural practice. Whereas previous paradigms in heritage studies might have allowed us to treat the protection of nature, or the protection of culture, as specialized and legitimate enterprises, the new paradigm requires us to accept that we are nature, and we are culture (Smith 2013: 49).

Urban heritage is still more seldom than other heritage categories, but of importance in many urban renewal projects. The conversion of centrally located lofts in US cities emerged from the end of the 1970s onwards, contributing to the upgrading of historic quarters, but subsequently also to gentrification processes (Stern and Seifert 2007b).

Australia and the Pacific States faced a denial of aboriginal heritage for a long time—heritage was considered to be either a natural one or from colonial history. Natural heritage sites then, like in Australia, were conserved because of their natural beauty and importance, but not because of intangible values the aboriginal communities ascribed to these sites. That only changed from the end of the 1970s onwards, when Australia ICOMOS drafted and adopted the Australian ICOMOS charter for the Conservation of Places of Cultural Significance in 1979 (the Burra Charter) introducing the term 'cultural significance', referring to aesthetic, historic, scientific or social value of a site. The Burra Charter was later amended several times to reflect the current concern of heritage and conservation in Australia, including the conservation of intangible values. It recognizes social and aesthetic values as part of cultural significance, as well as intangible values or intangible cultural heritage referred to by UNESCO as an integral aspect of heritage significance (Ahmad 2006).

Meanwhile, also modern heritage such as the Sydney Opera House and other urban sites are considered heritage. Also, initially exclusively natural sites were revised and subsequently considered cultural landscapes including cultural and intangible values. For example, the Australian Uluru-Kata Tjuta National Park, which was considered Natural World Heritage in 1987, was extended in 1994 to a mixed heritage and then defined as "one of the most ancient managed landscapes in the world and an outstanding illustration of successful human adaptation over many millennia to the exigencies of a hostile environment; and forming an integral part of the traditional belief system of one of the oldest human societies in the world" as described in the site justification (UNESCO 1994).

Nevertheless, although the reformulations of landscape approaches since the 1990s seeked to change this, Australia is still facing a tendency of Aboriginal cultural landscapes to be of less importance to cultural heritage managers than archaeological sites. The '*Maoritanga*', or 'Maori Renaissance' of New Zealand and related policies of biculturalism have fostered the appreciation of indigenous Maori culture while Fiji, which acquired independence from Britain in 1970, is still facing tensions between colonial and 'traditional' heritage (Ireland 2012).

The state of professional conservation in the region varies widely. While Australia and New Zealand have a quite elaborate conservation system and facilities, the majority of Pacific countries are often lacking conservators and conservation. The same is true for the educational system, with Australian universities being the focal point for higher education on the topic (Pearson 1993).

In Asia:

> The first limited conservation efforts were initiated by the ruling class and their apparatus who gained the knowledge from the collaboration with the former colonists and the 'authorized' international organization, for example through the ancient monuments conservation works such as the Angkor Wat by the French in 1920, and the Borobudur by UNESCO in 1960s, 1970s, and 1980s. The importance of the authenticity of the built fabric has also been imposed to the mind of the states' apparatus and technocrats through conservation legislations (Kwanda 2010: 7).

In India, the 1923 manual conservation from British colonial times served as a reference for legislation after independency in 1951. The Indonesian 1992 Cultural Heritage Act is similar to the 1931 Dutch's '*Monumenten Odonantie*' (cf. Sect. 5.2). Former colonial legislation was transformed and implemented in the post-colonial system. In some other countries, such as Japan and the People's Republic of China, there is a strong tradition of interest in preservation. This leads to a high degree of compliance with the goals of preservation and protection (Costin 1993; Kwanda 2010).

However, the colonial-type legislations were rooted in European understanding of heritage, often not congruent with the local one. Therefore, over time legislation was changed and adapted. Costin (1993) mentions in particular Japan, that is including intangibles like arts, manners and customs in its four defined types of cultural properties. Other nations (like India, Indonesia, Malaysia, Myanmar, Nepal and Singapore) include in their definition of a cultural property the site or land on which the item exists or is believed to exist, thereby mixing tangible and intangible assets.

One major difference lies in the source of cultural properties, as some legislation attached cultural value only to man-made materials or objects (like Malaysia, and Pakistan), while others (e.g. Japan, Papua New Guinea and the Philippines) also regard natural places such as mountains or trees as cultural heritage, if they are of cultural, historic, or scientific significance (Costin 1993).

During the 1990s, conservation in Asia was popularized by different state agencies through publications romanticizing the heritage, many of them for touristic purposes. Kwanda (2010) states that "*the past was wrapped up for commercial consumption to embrace the economic globalization*" (Kwanda 2010: 8).

Old paradigms were left behind at least partly, new theories emerging from the 1980s on were based on the idea of significance, based on the thought that not all sites can and must be preserved. Cultural significance then emerged after the Burra Charter, to be followed by participatory heritage conservation approaches, often involving newly found NGOs. Such approaches however mostly focused on single buildings.

Growing awareness of the vanishing heritage also led to efforts in academia and a number of regional or international meetings and conferences dealing with Asian or Asia-Pacific heritage. Countries like Japan and India have a well-established conservation profession, including educational facilities. In countries in Southeast Asia, countries of Brunei, Indonesia, Malaysia, the Philippines, Singapore, and Thailand, all members of the Association of Southeast Asian Nations (ASEAN), conservation facilities exist, but are mainly limited to the respective national museum (Pearson 1993).

Tangible authenticity aspects are also reflected in a number of charters, mainly based on the euro-centric background of Venice Charter, like the 2000 ASEAN Declaration on Cultural Heritage or the 2005 Hoi An Protocols For Best Conservation Practice in Asia—a guideline for assuring and preserving the authenticity of heritage sites in the context of the cultures of Asia. The 2000 ASEAN Declaration on Cultural Heritage was underlined by a mutual understanding that cultural traditions were integral to the preservation of ASEAN intangible heritage (Ahmad 2006; Engelhardt and Rumball Rogers 2009; Kwanda 2010). In terms of intangible values and authenticity, the Nara Document was of utmost importance for Asia. After the year 2000, its emphasis on the intangible heritage within Asia is evident, representing a shift from a Western focus on fabric towards intangible values. Similar to other countries, the scope of Southeast Asian heritage now covers both tangible and intangible heritage, although the broader definitions used by certain countries sometimes differ from those used by UNESCO or ICOMOS (Ahmad 2006).

Talking about urban heritage, most Asian cities are facing problems with losing their character, as they "*can hardly be distinguished except for the unique historic structures—temples, mosques, shrines, palaces and monuments—standing in the midst of confusion and dark fumes*" (Joshi 1997: 27).

In general, Asian cities are facing a process of modernization, trying to catch up with the West (Kunzmann 2008), and often following Western paradigms of urban planning, e.g. this is why China has already lost most of its historic cities and replaced it with new, often European-like, structures (Mars 2008). In the case of Mumbai, Loeckx (2009) even identified a 'mental gap' between the city's rich built heritage and the contemporary urban patterns. The latest document tackling urban issues is the Beijing Declaration concerning Urban Culture (2007), calling cities a global collective memory and an important component of cultural heritage, seeking to raise awareness for protecting urban culture and traditions.

Heritage in Latin and Middle America has to be seen in the context of the continent's past, where many pre-Hispanic cultures were erased, while new towns were erected, following Spanish or Portuguese patterns. In Spanish colonies, new settlements often strategically replaced older settlements or even bigger towns, like in the case of Mexico City, which replaced the former Aztec capital Tenochtitlan. Here, the cathedral and '*Zócalo*' (Central Square) were placed right where the Aztec Temple Mayor had been (cf. Ribbeck 2005). Thus initially, heritage in Latin America was often synonymous with remains of colonial times and archaeological sites of pre-Hispanic cultures but comparably less the heritage of today's indigenous cultures.

Rojas (2002) defines two phases in the urban heritage conservation process in Latin America and the Caribbean:

- During the first one, the conservation processes were headed by cultural elites because of concerns over the loss of specific buildings or monument areas. The scope was a limited one, partly starting already in the early twentieth century, following the interests of a societal minority, and as financing came mainly from private donors, efforts were not always sustainable.
- In the second phase, laws or regulations on heritage conservation were passed, in parallel with a comparably broader understanding of sociocultural heritage values. Additionally, different countries established specialized institutions to monitor heritage, financing and the application of legislation, mostly after the 1970s.

There is a need to head on for the next phase, in which more actors should be involved in participatory processes and partnerships (Rojas 2002).

The oldest policies on protection of cultural and archaeological heritage stem from the 1930s, like in Bolivia or Brazil, where the first law was passed in 1937 ('*Decreto-Ley sobre Protección del Patrimonio Histórico y Artístico Nacional*'). The vast majority of Latin and Middle American countries then enacted heritage legislation during the 1970s or 80s, e.g. the 1972 Mexican law on monuments and archaeological, artistic and historic sites ('*Ley Federal Sobre Monumentos y Zonas Arqueológicos, Artísticos e Históricos*'). Until only a few years ago, national politics were engaged in cultural heritage exclusively, without any public participation schemes. Thus, Lacarrieu (2013) identified cultural heritage in Latin America as part of a strategy of power, calling for conservation of fabric, aesthetics, functions and even native cultures.

Discourses on heritage are following the global paradigms, considering indigenous communities, paying attention to tangible as well as intangible values. Although intangible values might still be more often associated with indigenous communities, they are increasingly recognized as a means of shaping a local or urban culture. With regard to indigenous communities, the 1993 Mexican Declaration of Oaxaca elaborated by the Mexican National Commission for UNESCO, has be mentioned as a key document and an excellent example of adapting international convention to suit local, indigenous needs (Sullivan 1993). Emphasizing on the protection of natural resources of the Ibero-American world, it calls for respecting the relationship of indigenous people to nature.

During the last decades, Latin American cities have been undergoing a process of reconfiguration of urban space, with built heritage being a main resource. Heritage has become fashionable and an important resource, also in economic terms, indicating a change from regarding it as resource for nation building rather than an urban resource. Thus, many cities are undertaking massive renovation and renewal projects in their centres (see Chap. 4 on urban renewal); linking the built heritage with urban development and social aspects, but at the same time with massive economic interests (Hiernaux 2013; Lacarrieu 2013).

The political and administrative systems of post-independence African states were often based on concepts, philosophies and conditions rooted in colonial times, including concepts and policies of heritage conservation. Until today, the number of World Heritage sites is least on the African continent; for Birabi (2007) this can also be attributed to the prevalent culture of conservation. According to him, problems in Africa are the lack of awareness of persons in charge, e.g. government heritage ministers, as well as a lack of political will and low thresholds of awareness. On the one hand, many countries are facing a lack of financial means for an adequate conservation; on the other hand the execution of policies as well as—on World Heritage level—the very complicated nomination files are hindering a fundamental shift.

On an urban level, the situation is a comparable one. Urban development models were rooted in concepts that prevailed in more advanced economies during the mid-twentieth century. A recent UN-Habitat publication on African cities (2014) finds these approaches are of limited use to Africa, given today's very rapid urbanization, economic constraints and increasingly felt threats and impacts of environmental and climate change (UN-HABITAT 2014).

Another issue is the focus on tangible assets and fabric of many international documents while many African countries attach importance to movable artefacts, trying to avoid any obligation forcing them into a strict cadastre. As a consequence, some African states were quite late in ratifying international heritage conventions. Birabi (2007: 46) notes *that "among Africa's LDCs only Burkina Faso, Democratic Republic of Congo, Madagascar, Mali, and Senegal are State parties to the 1954 Hague Convention. Also, it was not until 1987 that Uganda ratified the 1972 UNESCO Convention*".

Only recently, the voice of African states gained importance, e.g. the 2003 Convention for the Safeguarding of Intangible Cultural Heritage was mainly initiated by African countries (Wagner 2008). As a consequence, many African states still try hard to overcome the challenging circumstances when trying to assess and protect their cultural heritage.

Arab countries have been more advanced in heritage conservation, especially concerning cultural heritage, as visible in the comparably high number of cultural and urban World Heritage sites. Unfortunately, the political circumstances have harmed many of them within the past few years, leading to many losses and efforts in safeguarding campaigns. Urbanization and related problems (e.g. in Cairo and other large cities) as well as large urban development projects (like in Dubai and

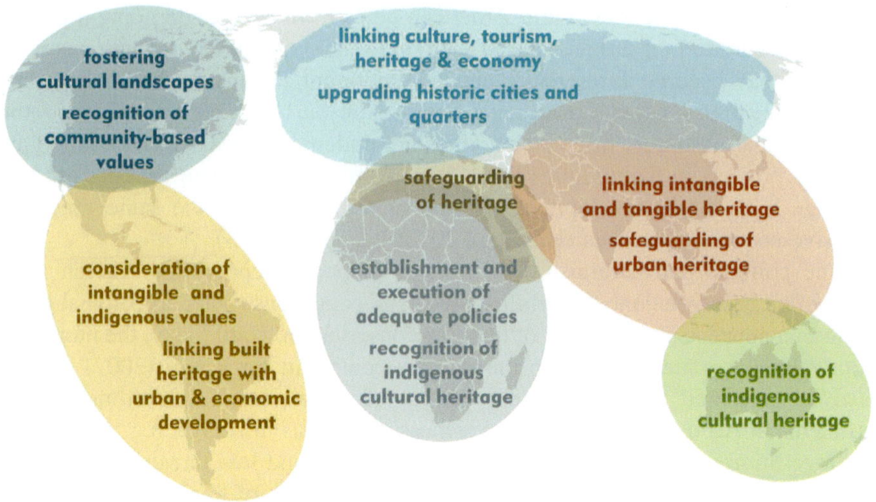


**Fig. 3.4** Current focal areas of cultural heritage conservation in different regions

other Emirates' cities) have led to risk for and even loss of urban heritage (El Mouelhi 2013; Sennett 2009).

Summing it up, the global heritage debate is not homogenous (cf. Fig. 3.4). Despite the fact that most tendencies are global ones, different world regions are focusing on certain aspects of it. Probably, one could also conclude a kind of a timeline, with many processes to emerge or rooted in the Western World, like the recent shift towards urban heritage, and then being transferred and implemented in other regions.

In this context, Birabi (2007) is speaking of European countries as 'laboratories' for the development of instruments for conservation, at the same time he warns that most international conservation charters are not imposing significant influence, especially among least developed countries (LDCs). Developing Countries in general and LDCs in particular (mostly located in Africa and Asia) are facing challenges that are either overcome or have been less crucial in the Western World. Problems include the lack of appropriate heritage legislation, a lack of conservation standards and appropriate education which is still dominated by Western nations, the illegal export of cultural property from LDCs (mainly for economic reasons and usually to Western countries), and armed conflicts that are mainly arising in poor countries rather than developed ones (cf. paragraph on Syria and other countries in Sect. 3.6). They are linked to generally weak governance systems and economic pressures for social and urban growth, thus exacerbating the challenges (Birabi 2007; Jaramillo Contreras 2012).

However, the Global South has contributed significantly to the debate, especially by recognizing heritage of indigenous communities and by putting intangible values on the global agenda. Taking a look at current trends and focal areas of cultural heritage conservation (as summed up in Fig. 3.4), such regional differences become clear. The two main topics on a global scale are the recognition of intangible or indigenous values, and the challenge of preserving urban heritage, both of them with a regional connotation. The next chapter will elaborate further on aspects of urban heritage and its regional differences.

## 3.8 Urban Heritage

Historic cities and urban heritage have become one of the focal points of heritage policies. With more than 250 inscribed sites, historic cities today constitute the largest heritage 'category' on the World Heritage List (Bandarin 2011). That reflects a growing interest in urban topics in general, but also a growing awareness of an increasing number and intensity of risks, for different reasons.

Different pioneers believed in the importance of urban fabric and heritage conservation, starting with the English Arts and Crafts movement and its representative John Ruskin (cf. Sect. 3.1) and authors like the Austrian architect and urban planner Camillo Sitte, who published his ground-breaking book "*Der Städte-Bau nach seinen künstlerischen Grundsätzen*" ("City Planning According to Artistic Principles") as early as 1889 (Sitte 1901). A few years later Patrick Geddes, a Scottish biologist, sociologist and also urban planner, emphasized on the importance of urban heritage for urban development in his book "Cities in Evolution" (1915). The term 'urban heritage' itself was invented by Gustavo Giovannoni, in his publication "*Vecchie città ed edilizia nuova*" (Old cities and new buildings) of 1913. Both Geddes and Giovannoni have integrated the concept of 'heritage management' into overarching approaches of territorial planning and urban development. Furthermore, both of them regard people as an integral part of the city. In fact, they have mainstreamed heritage management into larger policies of planning as early as one century ago—a concept still valid and reflected in cultural policy since the earliest recommendations by UNESCO in the 1960s (Veldpaus et al. 2013).

Urban development and heritage then were topics of both Athens conferences in the 1930s, but with complementary background. While the Athens Charter for the Restoration of Historic Monuments was calling for the same, trying to raise awareness of the vanishing heritage in many places, the CIAM conference was regarding heritage as virtually hindering for a modern city (cf. Mumford and Kenneth Frampton 2002, on the CIAM discourse on urbanism). Le Corbusier's '*Plan Voisin*' (1925) for example, was a product of its time when proposing the demolishment of the old Paris, leaving only few monuments like Notre Dame. He wanted to replace it with new and modern buildings which he found much more adequate for modern times:

> This is the moment when urban development and urban heritage theory really took separate paths, with one mainly concerned with the general need for expansion due to population growth and hygiene, while the other emphasised the listing of monuments (Veldpaus et al. 2013: 7).

Only from 1960s onwards have they witnessed a return towards an increasing inclusion of heritage in urban contexts in global heritage policies:

- Starting with single buildings,
- The definition of urban heritage was extended to its vicinity (ensembles), and finally
- To whole urban cores or quarters up to whole historic cities.

The Venice Charter (1964) for the first time defined principles for protecting urban ensembles, raising awareness and leading to documents and campaigns throughout the 1970s, e.g. the European Declaration of Amsterdam. For Urban (2011) the change is apparent, as the living environment worth being preserved is no longer defined by fabric only, but by social aspects alike.

Since the 1980s, international actors like the World Bank, UNDP (United Nations Development Programme), UNEP (United Nations Environment Programme) und UNHCS (UN Centre for Human Settlements (HABITAT) developed policies with and for national or regional governments. Meanwhile, all of them are important global players in urban debates, influencing on topics by a large number of relevant publications but also by economic means. World Bank for example, is one of the main donor agencies for the urban sector (Pugh 1996). All of them have included the topic of urban heritage in their portfolio in the recent years. In parallel, environmental concerns grew and found their way into urban politics and policies.

A broadening of the perspective towards urban heritage then occurred with the growing recognition of intangible heritage, also in the urban context:

> Cities are among 'societies' most precious cultural artefacts. This can be seen in the visual and decorative arts, music and dance, theatre and literature that develop there and in the variety and diversity of street life evident in most cities. In most cities, there are buildings, streets, layouts and neighbourhoods that form a central part of the history and culture of that society (UNCHS 1999: 58).

In 1991, the First International Symposium of World Heritage Cities took place. Two-years later the 'Organization of World Heritage Cities (OWHC)' was founded with the aim to create a global network of member and "*to thereby contribute to the global debate on urban heritage management and sustainable development*" (OWHC, 2011).

In a content analysis of 62 cultural heritage policy documents between 1950 and 2008 (taken from a list of the Getty Conservation Institute, excluding such specifically referring to natural heritage, movable heritage, and underwater heritage), Veldpaus et al. (2013: 10f) found:

> [...] a clear increase in the use and number of words related to the urban scale from the 1960s onwards. [...] When the urban scale-related terminology is further analysed, it is shown to be slowly evolving from being about explicitly appointed sites such as historic towns and settlements (1970s and 1980s) towards more general and less defined names, e.g. (historic) urban areas in the 1980s and places and landscapes in the 1990s. At the turn of this century the concern for landscapes as a cultural heritage re-emerged with a bigger role for the European Landscape Convention. Later, this was confirmed by the different documents on HULs issued by UNESCO, ICOMOS, and the International Committee on Historic Towns and Villages, which had been discussing the need for an updated or new charter for historic cities since 2005, to replace the 1987 Washington Charter. These documents show the use of a wider range of urban-related terms, as well as a shift in the type of terms towards a more general and inclusive terminology. This comes at the same time that the documents clearly start to mention, and distinguish between, the words 'tangible' and 'intangible'; especially after the establishment of the Nara document on authenticity.

Ongoing debates on urban development projects, like in the historic centre of Vienna, and around the Cologne Cathedral, or projects of urban verticalization around Westminster and the Tower of London, and urban changes and losses in cities like Kathmandu, raised a worldwide awareness for related policies (Turner 2013). The Vienna Memorandum on 'World Heritage and Contemporary Architecture—Managing the Historic Urban Landscape' (UNESCO 2005) then acknowledged the importance of linking urban heritage and urban development when stating that:

> The central challenge of contemporary architecture in the historic urban landscape is to respond to development dynamics in order to facilitate socio-economic changes and growth on the one hand, while simultaneously respecting the inherited townscape and its landscape setting on the other (Paragraph C/14).

Since 2005, the 'World Heritage Cities Programme' is one out of six thematic programmes under the umbrella of the World Heritage Committee. Its aim to develop a theoretical framework for conserving urban heritage and to assist the state parties in protecting and managing their sites then led to the formulation of the 'Recommendation on the Historic Urban Landscape' (UNESCO 2011). In paragraph I/8 the Recommendation defines Historic Urban Landscapes (HUL) as: "*The urban area understood as the result of a historic layering of cultural and natural values and attributes, extending beyond the notion of 'historic centre' or 'ensemble' to include the broader urban context and its geographical setting*".

> This wider context includes the site's topography, geomorphology and natural features, built environment –both historic and contemporary–, open spaces, land use patterns and spatial organization, as well as all other elements of the urban structure, next to social and cultural practices and values, economic processes and the intangible dimensions of heritage. All these contribute to the city's singularity, its genius loci, and the specificity of the urban experience, which should guide any decision as regards the planning and design of interventions (WHITRAP 2014).

In an address given on the Recommendation, the Deputy Director of UNESCO's World Heritage Centre, Rössler (2014), explained the Historic Urban Landscape approach as:

> [...] an attempt to refocus urban development into a more inclusive process and to increase the long-term sustainability of planning and design interventions by taking into account the existing built environment, intangible heritage, cultural diversity, socio-economic and environmental factors and local community values. We consider these aspects the "software of urban development", while the productive support for infrastructure, with attention to mobility, water and energy supply, and waste treatment, would constitute the hardware. Both are critical to sustainable urban and national development.

Building upon the HUL approach, ICOMOS adopted "The Valletta Principles for the Safeguarding and Management of Historic Cities, Towns and Urban Areas" (2011) for the safeguarding of historic towns and urban areas and their settings. Here, tangible and intangible elements are both mentioned as elements of historic towns and urban areas. They are defined as:

> The tangible elements include, in addition to the urban structure, architectural elements the landscapes within and around the town, archaeological remains, panoramas, skylines, view-lines and landmark sites. Intangible elements include activities, symbolic and historic functions, cultural practices, traditions, memories, and cultural references that constitute the substance of their historic value (Paragraph 1/a).

Furthermore, the principles recognize historic towns as 'living organisms' and thus subject to continuous change, which is now no longer regarded as a threat only but as potentially improving the area's quality.

Nevertheless, according to Turner (2013) heritage and development are still seen as an oxymoron, with mutually exclusive components. The formal conservation language (established by the 1972 World Heritage) definitions of 'monuments, groups of buildings and sites', following international and national documents is still rooted very much in 'object' thinking, while it is moving to living cities and urban landscapes. Also, it incorporated aspects of the Nara document's thoughts on cultural diversity, not only between peoples and place but also through time. Past approaches, based on the physical assets and tangible dimension, were replaced by more integrative ones, changing towards inclusive and landscape-based views, including intangible and community-based values thus also moving away from the traditional Western fabric-oriented paradigms. The HUL concept is allowing for (manageable) change, accepting and permitting a more flexible approach.

Figure 3.5 visualizes the shift that has occurred in the focal areas of major heritage documents since the 1960s. It is based on the analysis of those documents that are mentioned as 'standard-setting' in the 2005 Vienna Memorandum and the 2011 HUL Recommendation. The analysis reveals a threefold derivation of the Historic Urban Landscape approach. Its major roots stem from discourses on defining heritage. Since the 1960s, this has changed from natural landscapes and monument sites, then including surroundings of heritage sites, inclusion of historic (urban) gardens and whole towns and quarters throughout the 1980s, then finally to the recent turn towards urban landscapes. In parallel to the 'fabric', another discourse emerged on the intangible or cultural aspects of heritage, starting with the 1964 Venice Charter and the appearance of 'authenticity'.

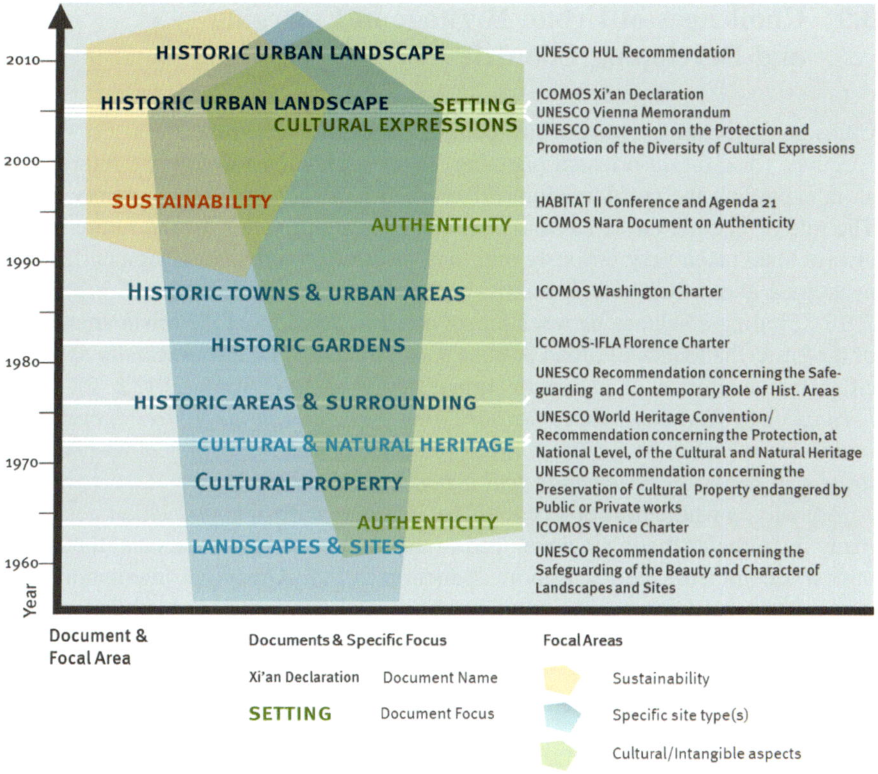


**Fig. 3.5** Roots and background of the historic urban landscape approach (*Sources* UNESCO 2005, 2011 recommendation on the historic urban landscape)

The next decades' debates on this concept took place on a more regional level, entering back on the global stage in the 1990s, and finally mainstreamed into the HUL approach. In parallel, sustainability emerged as a global topic, with sustainable planning and development becoming the guiding principles also on the urban scale. Subsequently, sustainability became one of the objectives also in heritage conservation, first focussing more on natural heritage, later also mainstreamed in cultural heritage and urban heritage as a component of sustainable urban development (Albert 2013; NSW Heritage Office 2004). As a consequence, the HUL approach tempts to be very ample, inclusive and promising, seeking to learn from past mistakes, but it has not yet proven its suitability. Summing up their analysis on the Historic Urban Landscape Approach, Veldpaus et al. (2013: 15) stated "*In conclusion, the century old ideas still inspire but the challenge lies in practice*".

## 3.9 Challenges of Urban Heritage in Emerging and Developing Countries

Cities are structurally woven into complex systems, requiring cross-sectoral perspectives for a future-oriented planning. Such orientation towards the future is not at all excluding the consideration of the urban heritage—quite the contrary is true. The role of historic city areas and centres in sustaining cities for the future has not always been taken very seriously until now in strategic urban development. Greater awareness about their role at political, planning and civic levels is of need (Ripp 2013). Finding a balance between preserving and developing the urban areas is one of the basic challenges of urban politics. For Urban (2011), this especially applies to cities that comprise of a historic urban core. To UNESCO (2013a: 5) "*urban heritage is of vital importance for our cities—now and in the future. Tangible and intangible urban heritage are sources of social cohesion, factors of diversity and drivers of creativity, innovation and urban regeneration*".

Actually, a whole portfolio of tools for integrated urban conservation does exist; many instruments that were developed in different cultural contexts and throughout almost a complete century by now (Bandarin 2011). Almost all international recommendations on urban conservation tackle social, economic, and cultural issues—unfortunately all of them are often disregarded in practice (Kulikauskas 2007). Latest approaches are considering fabric, intangible values and community expectations at the same time, conserving heritage while allowing future economic as well as social development. Such a wish list sounds promising but at the same time hard to implement, due to its very holistic approach, requiring the integration of a whole variety of actors. In an article on the Historic Urban Landscape approach (2011), Francesco Bandarin, former Director of the UNESCO World Heritage programme, stated that:

> The planning and regulatory tools put in place are not always adequate to address the new challenges. Urban conservators are increasingly aware of the gap existing between the ideal world of the 'Charters' and the practical realities, especially in emerging societies, and are advocating that new principles, approaches and tools have to be identified to cope with the new challenges (Bandarin 2011: 179).

Current challenges that historic cities or historic urban landscapes are facing include:

- Tensions between globalization and local development,
- Population growth and urbanization,
- incompatible new developments, including increasing pressures for land conversion, inside and outside the historic precincts,
- market exploitation and economy-driven urban policies, and
- Unsustainable and/or mass tourism demanding for specific historic 'packages' (Alberts and Hazen 2010; Bandarin 2010, 2011; UNESCO 2013a).

Global processes have a direct impact on the identity and visual integrity of historic cities, on their broader setting and on the people who live in them (Bandarin 2011). Therefore, Yeo and Han (2012) claim a need to rethink heritage conservation which should consider the regeneration of social values as well, instead of focusing on economic aspects only. Heritage discourses should therefore be more socially inclusive to bring more opportunities to the local community. In the words of Alberts and Hazen (2010: 65):

> Problems associated with individual structures become magnified when extended to protecting historic cities and cultural landscapes. One of the goals of designating whole cities, or parts of cities, as World Heritage sites is to try to maintain integrity, but the expanded scale presents a challenge when management authorities struggle to administer and fund the preservation of large areas. The fact that sizable populations inhabit most historic cities and cultural landscapes exacerbates this challenge and leads to controversy over proposed developments that may put preservationists' goals and inhabitants' needs at odds.

Bandarin (2010), calls the capacity of historic cities to accommodate and benefit from such radical and rapid changes that accompany urban growth, while maintaining heritage values, an increasingly critical issue. This is especially true for cities in emerging and developing countries that are facing growth and change at a much higher rate.

Figure 3.6 reveals a major problem of the World Heritage List: the predominance of European sites, with more than 50 % of the overall listed inhabited urban centres and cities that are World Heritage Sites. When taking a closer look at the average year of inscription (calculated per world region as defined by UNESCO),

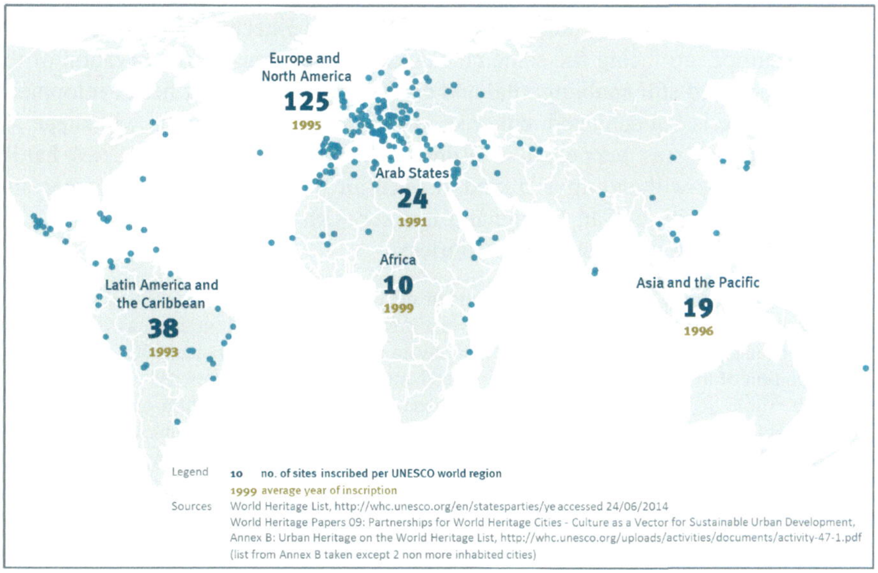


**Fig. 3.6** Historic urban centres and cities inscribed on the UNESCO List of Cultural World Heritage

the recent efforts in establishing a more balanced list are revealed, as the latest inscriptions come from Africa and Asia. However, recent trends in Europe towards integrated approaches of urban conservation within the HUL framework and under the Leipzig Charter, have led to a shift towards consideration of larger areas like landscapes and cities instead of single buildings. Therefore, a large share of the European sites also only got inscribed quite recently. Furthermore, heritage in developing countries is facing more difficulties compared to the Euro-American states, as described previously. This is especially true in the case of urban heritage, exacerbated by the integrated approach the HUL is emphasizing.

As Fig. 3.6 clearly shows, the vast majority of urban heritage sites are still European, which is also indicating the importance urban heritage has for the European states. Roots of European cities go back to the fifteenth century, when a classical canon of forms and design was developed, e.g. urban open spaces in core areas and their multiple uses. Such open spaces like the Piazza Navona in Rome, the Piazza del Campo in Siena, or the Place de la Concorde in Paris then became global icons of architectural history. However, urban planning shifts in early twentieth century led to losses of such open spaces in central urban locations, including losses of functions. Since the 1970s, another shift in paradigm then re-invented such spaces and surrounding areas, which before had served as parking or traffic zones. A restoration of urban shapes and functions were accompanied by a growing awareness of the inherent values, also for the quarters as a whole (Knirsch 2004).

However, UNESCO has announced to eventually remove different European cities from the World Heritage List—after having done so with the Dresden Elbe Valley in 2009—because of new urban development plans, which might harm the historical context of their protected parts (Alberts and Hazen 2010). Urban historic cores in Europe are facing the same challenges—balancing the preservation of the historic fabric and still enabling vital and creative powers and further development. If such balance is not achieved, either short-term economic interests will shape the urban appearance (as happened in Vilnius, where conservation policies hardly exist), or the city will end up as a kind of open-air museum, serving as scenery for tourists, like the city of San Gimignano in Italy (Urban 2011).

In historic cities of developing countries:

> City administrations are now facing a stage of development where the conservation of the urban heritage - and its integration into wider development opportunities - has become a major challenge. It is a challenge that is likely to increase in importance, with a growing recognition of the need to preserve and, indeed, strengthen the structures and edifices upon which whole societies and lifestyles have been built. There can be little or no socially sustainable development without preserving cultural continuity. The cultural identity of cities and nations is an essential element in helping present and future generations retain their natural and built patrimony, as well as helping to build a better and sustainable, people-centred culture in the future (UN-HABITAT 2002: 75).

In contrast, much of this might be recognized, but is hardly implemented. While a growing number of cities do have strategies to protect their heritage, by far not all of them incorporate social appropriation. The topic of 'societal use' has found its way into urban heritage strategies, unfortunately not (yet) reflected in urban governance schemes and policies—here one could talk of 'business as usual'. According to Lacarrieu (2013), this is in particular the case when housing in areas that are of historic value but no single monuments are concerned—here administrations have shortcomings. Birabi (2007) as well, identified a lack of political will to apportion or mobilize finances for urban heritage conservation in developing countries, as well as a lack of academic institutions addressing conservation and management of the urban cultural heritage.

Tweed and Sutherland (2007) recognize a growing awareness of governments concerning the contribution of cultural heritage to the wellbeing of different urban actor groups. However, they see a clear deficiency in dealing with less tangible or intangible urban features like street patterns that conventional policies—usually the listing of individual monuments and buildings or the designation of conservation areas—are not able to deal with. Unfortunately, it is particularly these features shaping the urban and cultural identity.

The rehabilitation and regeneration of historic centres has been recognized as an efficient tool for urban development for some decades now, synthesizing cultural values with economic opportunities and benefits. In their article on "Why development needs culture", Bandarin et al. (2011) argue that development is strongly pushed by culture, creative industries and cultural heritage, not only in terms of economic growth but also of qualitative standards of equity and wellbeing. But on the other side of the coin, conservation threats to appropriate urban development were also identified by the many proponents of the economic viability of the city (Turner 2013).

Urban heritage areas generate comparably higher economic returns than areas lacking such assets (UN System Task Team on the Post-2015 UN Development Agenda 2012). Income is mainly generated by touristic uses; simultaneously, urban upgrading is attracting higher income groups to live in the area and business to settle there. To UNESCO (2013a), the historic cities included in the World Heritage List deliver significant socioeconomic benefits at both levels—local and national—through tourism, related goods and services, as well as through other functions. But they also admit drawbacks, e.g. the displacement of former resident groups in the course of socioeconomic urban upgrading (UN-HABITAT 2002). There is a need for research on how heritage conservation can tackle negative gentrification effects and contribute to the process of social inclusion demanded in local context. Turner (2013: 83) claims that:

> Most cities have layered values reflecting the processes that created them, as in Bruges or Prague. One of the attributes might be in the building concept and form. At the next level, urban attributes might, in addition to plot size, include building lines, roofscape, materials, vegetation, ceremonies and events. […] The amount of layering and approach to the policies of conservation will be determined by the documentation with its authentication and its integrity in determining the extent for conservation.

The historic urban landscape in its actual understanding is composed of and linking a variety of assets, such as a high quality of the architectural and physical

**Fig. 3.7** Layers of the city, adapted from UNESCO (2013a). *Photo* Praça Dom Vital (Recife) by author

environment, the persistence of a sense of place, and forms of historic and artistic expression forming the basis for the people's identity. The key to understanding and managing any historic urban environment is the recognition that the city is not something static, but dynamic, with economic, social and cultural forces that shaped it and keep shaping it (Bandarin 2011; UNESCO 2013a). The way forward demands urban tools respecting such variety; to Turner (2013: 79) "*settings and context, spirit and feeling are all part of the elusive genus loci of the city*".

Figure 3.7 gives an overview about the different layers of a city, composed of natural as well as cultural values, of past and present ones, tangible and intangible, in the urban centre and its surrounding. Their interplay and intertwining then add up to a historic urban landscape (Turner 2013; UNESCO 2013a). Simultaneously, it depicts the complexity of the approach and the way historic urban areas are regarded in heritage discourses by now. Urban properties cannot be sheltered against every impact or change, as they are living cities (Ringbeck and Rössler 2011).

Quite some challenges lie ahead—the promising HUL approach is still a very newly constructed one, with not many success stories yet. The holistic background from which it emerged is both, promising and hindering, for emerging and developing societies. On the one hand intangible values, aspects of context and setting as well as community values are now for the first time forming integral parts of the urban heritage. This broadened perspective, however would require new forms of governance, policies and also recognition of community values on the local scale, scaled down from the global vision. Change and future development have also been accepted in the HUL approach, but the archetype a certain city is emulating will have significant impacts on all strategies and the treatment, and maybe even the definition of its heritage. "*Cities are dynamic organisms. There is not a single 'historic' city in the world that has retained its 'original' character: the concept is a moving target, destined to change with society itself*" (UNESCO 2013a: 24).

## 3.10 Chapter Summary

**Global heritage paradigm shifts**

This chapter has clearly shown the origins of global cultural heritage debates. Rooted in a predominantly Euro-American understanding, with material, original state and authenticity as major justifications for being considered as cultural heritage, such understanding has led to a biased understanding of what is heritage for decades. As elaborated on the previous pages only since the 1990s two major shifts have occurred:

- A shift from single buildings towards (urban) landscape approach, and
- The growing consideration of intangible heritage and holistic approaches combining both, tangible and intangible heritage.

In both cases the process was triggered outside the Euro-American sphere, with the aim to adapt the global heritage understanding to other cultural environments and to introduce heritage concepts which suit the cultural understandings of the non-Western world.

**Urban heritage**

The growing awareness of the vanishing heritage in urban areas has led to international efforts. Historic cities and urban heritage have become one of the focal points of heritage policies. Key concerns are:

- Acknowledging the importance of linking urban heritage and urban development, and
- Understanding urban areas as a result of historic layering of cultural and natural values and attributes.

Global processes have a direct impact on historic cities, due to urban change, economy-driven new urban developments, and non-suitable urban policies. Cities in the Global South face comparably more challenges in the protection of their urban heritage, due to

- Population growth,
- Administrative shortcomings, and
- A lack of suitable policies and their implementation.

## References

Ahmad Y (2006) The scope and definitions of heritage: from tangible to intangible. Int J Heritage Stud 12(3):292–300

Albert M-T (2013) Heritage studies—paradigmatic reflections. In: Albert M-T, Bernecker R, Rudolff B (eds) Understanding heritage: perspectives in heritage studies. de Gruyter, Berlin/Boston, pp 9–17

Alberts HC, Hazen HD (2010) Maintaining authenticity and integrity at cultural world heritage sites. Geogr Rev 100(1):56–73

Allstedt T, Metzler A (2009) Auferstanden in alter Pracht. Stadtansichten vor der Wende und heute, Zeitgeist Media, Düsseldorf

Amen M, Toly N, McCarney P, Segbers K (2011) Sighting or slighting cities in international relations. In: Amen M, Toly NJ, McCarney PL, Segbers K (eds) Cities and global governance: new sites for international relations. Ashgate, Farnham, pp 13–32

Bandarin F (2010) Revising the approach to urban conservation. World heritage world heritage cities (No. 55, February 2010):6–15

Bandarin F (2011) A new international instrument: the proposed UNESCO recommendation for the conservation of historic urban landscapes. Informationen zur Raumentwicklung 3/4 (Denkmalschutz und Stadtentwicklung):179–182

Barthel-Bouchier D (2012) Heritage and social change: anticipating future trends. In: Negussie E (ed) Changing world, changing views of heritage: heritage and social change. Proceedings of the ICOMOS Scientific Symposium 2010. ICOMOS, Dublin, Ireland, pp 82–86

Behr A (2005) Das Städtebauförderprogramm Städtebaulicher Denkmalschutz. Ein Instrument zur Erhaltung historischer Stadtzentren. Informationen zur Raumentwicklung 06/2005, Strategien für historische Stadtzentren:365–378

Birabi AK (2007) International urban conservation charters: catalytic or passive tools of urban conservation practices among developing countries. City & Time 3(2) (4):39–53

Bundesministerium für Verkehr BuS (ed) (2007) Baukultur als Wachstumsimpuls: Gute Beispiele für europäische Städte. Ausgewählte Fallstudien. Bundesministerium für Verkehr, Bau und Stadtentwicklung, Berlin

Butina B (2011) From the seven wonders of the ancient world to the UNESCO world heritage: political and economic aspects of institutionalised cultural preservation. Hrvatski Geografski Glasnik 73(2):141–154

Cameron C, Rössler M (2013) The shift towards conservation: early history of the 1972 world heritage convention and global heritage conservation. In: Albert M-T, Bernecker R, Rudolff B (eds) Understanding heritage: perspectives in heritage studies. de Gruyter, Berlin/Boston, pp 69–76

Carman J, Stig Sørensen ML (2009) Heritage studies. an outline. In: Stig Sørensen ML, Carman J (eds) Heritage studies. methods and approaches. Routledge, London, New York, pp 11–28

Conradin K, Engesser M, Wiesmann U (2015) Four decades of world natural heritage—how changing protected area values influence the UNESCO label. Die Erde 146(1):34–46

Costin CL (1993) Legal and policy issues in the protection of cultural heritage in South Asia and the Pacific. In: Mac Lean MGH (ed) Cultural heritage in Asia and the Pacific: conservation and policy. proceedings of a symposium held in Honolulu, Hawaii. The Getty Conservation Institute, Los Angeles, pp 27–62, 8–13 Sept 1991

Dehio G (1905) Denkmalschutz und Denkmalpflege im Neunzehnten Jahrhundert. Rede zur Feier des Geburtstags Sr. Majestät des Kaisers gehalten in der Aula der Kaiser-Wilhelms-Universität Straßburg am 27. Januar 1905. Straßburg

El Mouelhi H (2013) Culture and identity in cairo urban development. In: Seifert A (ed) Mkuki na Nyota, Dar es Salaam

Engelhardt RA, Rumball Rogers P (2009) Hoi an protocols for best conservation practice in Asia: professional guidelines for assuring and preserving the authenticity of heritage sites in the context of the cultures of Asia. Bangkok

Europarat (1975) Europäische Denkmalschutz Charta. http://www.callnrw.de/broschuerenservice/download/1350/online_denkmalschutz.pdf. Accessed 05 May 2015

Falser MS (2010) From the venice charter (1964) to the Nara Conference (1994)—changing concepts of authenticity? In: Falser MS, Lipp W, Tomaszewski A (eds) Theory and practice in conservation and preservation—interactions between theory and practice. Proceedings of the international conference of the ICOMOS international scientific committee for the theory and philosophy of conservation and restoration, 23–27 April 2008 (Vienna). Edizioni Polistampa, Florenz, pp 115–132

First International Congress of Architects and Technicians of Historic Monuments (1931) Athens charter for the restoration of historic monuments. http://www.icomos.org/en/charters-and-texts/179-articles-en-francais/ressources/charters-and-standards/167-the-athens-charter-for-the-restoration-of-historic-monuments. Accessed 28 July 2014

Francioni F (2012) Forty years on: world heritage and international law. World Heritage, vol 63. UNESCO, Paris

German Institute of Urban Affairs (Difu), Federal Institute for Research on Building; Urban Affairs and Spatial Development within the Federal Office for Building and Regional Planning (eds) (2012) 5 Years after the LEIPZIG CHARTER—integrated urban development as a prerequisite for a sustainable city. Federal Ministry of Transport, Building and Urban Development, Berlin

Gfeller AE (2013) Negotiating the meaning of global heritage: 'cultural landscapes' in the UNESCO world heritage convention, 1972–92. J Glob Hist 8(03):483–503. doi:10.1017/S1740022813000387

Hiernaux D (2013) Die historischen Stadtzentren Lateinamerikas: Auf dem Weg zu einer kreolischen Gentrifizierung? In: Huffschmid A, Wildner K (eds) Stadtforschung aus Lateinamerika: Neue urbane Szenarien: Öffentlichkeit—Territorialität—Imaginarios. Transcript, Bielefeld, pp 377–395

Hilgert M (2015) Es sind auch unsere Wurzeln FAZ

ICOMOS (2008) Charter for the interpretation and presentation of cultural heritage sites ICOMOS. http://www.international.icomos.org/charters/interpretation_e.pdf. Accessed 21 Aug 2014

ICOMOS (2011) The Valletta principles for the safeguarding and management of historic cities, towns and urban areas. ICOMOS. http://www.international.icomos.org/Paris2011/GA2011_CIVVIH_text_EN_FR_final_20120110.pdf. Accessed 28 Aug 2014

Ireland T (2012) Excavating globalisation from the ruins of colonialism: archaeological heritage management responses to cultural change. In: Negussie E (ed) Changing world, changing views of heritage: heritage and social change. Proceedings of the ICOMOS scientific symposium 2010. ICOMOS, Paris, pp 18–26

Irr C (2011) World heritage sites and the culture of the commons globalization and autonomy online compendium:2

Jaramillo Contreras M (2012) Beyond the protection of material cultural heritage in times of conflict. Protecting Cultural Heritage in Times of Conflict ICCROM, Rome

Joshi J (1997) Planning for sustainable development. Urban Management in Nepal and South Asia, Lajmina Joshi, Kathmandu

Knirsch J (2004) Stadtplätze. Architektur und Freiraumplanung. Alexander Koch, Leinfelden-Echterdingen

Koch G (2013) Studying heritage in the digital era. In: Albert M-T, Bernecker R, Rudolff B (eds) Understanding heritage: perspectives in heritage studies. de Gruyter, Berlin/Boston, pp 169–181

Kulikauskas P (2007) International charters on conservation: the lost C(l)auses. City Time 3 (3):61–68

Kunzmann KR (2008) Herausforderungen des asiatischen Jahrhunderts für die Raumentwicklungspolitik in Deutschland. Informationen zur Raumentwicklung 8.2008 (Raum- und Stadtentwicklung in Asien):523–533

Kwanda T (2010) Tradition of conservation: redefining authenticity in Javanese architectural conservation, 22–26 June 2010. Paper presented at the Heritage 2010—international conference on heritage and sustainable development. Evora, Portugal

Lacarrieu M (2013) Kulturerbe, Materialität und die Frage nach dem Städtischen. In: Huffschmid A, Wildner K (eds) Stadtforschung aus Lateinamerika: Neue urbane Szenarien: Öffentlichkeit—Territorialität—Imaginarios. Transcript, Bielefeld, pp 419–439

Leipzig Charter on Sustainable European Cities (2007) EU. http://ec.europa.eu/regional_policy/archive/themes/urban/leipzig_charter.pdf. Accessed 23 July 2014

Locke WJ (1904) Recommendations of the Madrid Conference (1904). The architectural journal being the journal of the royal institute of british architects (RIBA) XI (Third Series):343–346

Loeckx A (2009) Towards an urban debate. Questioning Mumbai's voices & issues. In: Shannon KG, Jannina (ed) Reclaiming (the Urbanism of) Mumbai. Explorations in/of Urbanism. Sun Publishers, Amsterdam, pp 179–193

Luger K (2008) Welterbe-Tourismus. Ökonomie, Ökologie und Kultur in weltgesellschaftlicher Verantwortung. In: Luger K, Wöhler K (eds) Welterbe und Tourismus: Schützen und Nützen aus einer Perspektive der Nachhaltigkeit. StudienVerlag, Innsbruck, pp 17–42

Mars N (2008) Cities without History. In: Wolfrum S, Nerdinger W (eds) Multiple City. Stadtkonzepte 1908 | 2008. Jovis, Berlin, pp 274–278

Mertins G (2003) Städtetourismus in Havanna (Kuba). Geographische Rundschau 55(3):20–25

Meskell L (2013) UNESCO's world heritage convention at 40: challenging the economic and political order of international heritage conservation. Curr Anthropol 54(4):483–494

Mumford E, Frampton Kenneth (2002) The CIAM discourse on urbanism, 1928–1960. The MIT Press, Cambridge

Negussie E (2012) Conserving the rock-hewn churches of Lalibela as a World Heritage site: a case for international support and local participation. In: Negussie E (ed) Changing world, changing views of heritage: heritage and social change. Proceedings of the ICOMOS scientific symposium 2010. ICOMOS, Paris, pp 76–81

NSW Heritage Office (2004) Heritage & sustainability. A Discussion Paper.23

Organization of World Heritage Cities (OWHC) (2011) Organization of world heritage cities. OWHC. http://www.ovpm.org/en. Accessed 31 May 2011

Pearson C (1993) Issues that affect cultural property, specifically objects, in South Asia and the Pacific. In: Mac Lean MGH (ed) Cultural heritage in Asia and the Pacific: conservation and policy. Proceedings of a symposium held in Honolulu, Hawaii. The Getty Conservation Institute, Los Angeles, pp 63–76, 8–13 Sept 1991

Peña Moreno S (2013) Global to local: how world heritage sites can bring international conservation goals to ground level. World Heritage, vol 70. UNESCO, Paris

Perry J, Falzon C (2014) Climate change adaptation for natural world heritage sites. A practical guide, vol 37. World Heritage Paper Series UNESCO, Paris

Prott LV (1993) The impact of policy on cultural heritage protection. In: Mac Lean MGH (ed) Cultural heritage in Asia and the Pacific: conservation and policy. Proceedings of a symposium held in Honolulu, Hawaii. The Getty Conservation Institute, Los Angeles, pp 1–14, 8–13 Sept 1991

Pugh C (1996) Sustainability and sustainable cities. In: Pugh C (ed) Sustainability, the environment and urbanisation, vol Neudruck 2002. Earthscan, London, pp 135–177

Ribbeck E (2005) Die Welt wird Stadt: Stadtbilder aus Asien. Afrika und Lateinamerika, Jovis, Berlin

Riegl A (1903) Der moderne Denkmalkultus. Sein Wesen und seine Entstehung. W, Braumüller, Wien und Leipzig

Rigg J (1997) Southeast Asia—the human landscape of modernization and development. Routledge, London and New York

Ringbeck B, Rössler M (2011) Between international obligations and local politics: the case of the Dresden Elbe Valley under the 1972 World Heritage Convention. Informationen zur Raumentwicklung 3/4 (Denkmalschutz und Stadtentwicklung):205–212

Ripp M (2013) Krisen: Chancen für die Altstadt? Zur Rolle des gebauten Kulturerbes als urbaner Resilienzfaktor. Forum Stadt Vierteljahreszeitschrift für Stadtgeschichte, Stadtsoziologie, Denkmalpflege und Stadtentwicklung 2(2013):149–161

Rojas E (2002) Urban heritage conservation in latin America and the carribean. A task for all social actors. Sustainable development department technical papers series: 25

Rössler M (2014) The UNESCO recommendation on the historic urban landscape. Address, European heritage heads forum, 22 May 2014, Conference Address edn., Leuven

Rudorff E (1897) Heimatschutz. Reichl, St. Goar

Ruland R (2011) Die Rolle des historische Erbes in der Stadtentwicklung in Deutschland. Informationen zur Raumentwicklung 3/4 (Denkmalschutz und Stadtentwicklung):183–191

Salazar NB (2011) Imagineering cultural heritage for local-to-global audiences. In: Halbertsma M, van Stipriaan A, van Ulzen P (eds) The heritage theatre: globalisation and cultural heritage. Cambridge Scholars, Newcastle upon Tyne, pp 49–72

Salazar NB (2012) Shifting values and meanings of heritage from cultural appropriation to tourism interpretation and back. In: Lyon S, Wells EC (eds) Global tourism. cultural heritage and economic encounters, vol 30. Society for Economic Anthropology Monographs. Altamira, Lanham, pp 21–41

Schmitt TM (2009) Global cultural governance. decision-making concerning world heritage between politics and science. Erdkunde 63(2):103–121

Sennett R (2009) Istanbul within a Europe of cities. Istanbul—City of Intersections:12–13

Silberman NA (2013) Heritage interpretation as public discourse: towards a new paradigm. In: Albert M-T, Bernecker R, Rudolff B (eds) Understanding heritage: perspectives in heritage studies. de Gruyter, Berlin/Boston, pp 21–33

Sitte C (1901) Der Städte-Bau nach seinen künstlerischen Grundsätzen. Carl Graeser & Co., Wien

Smith J (2013) Cultural landscape theory and practice: moving from observation to experience. In: Albert M-T, Bernecker R, Rudolff B (eds) Understanding heritage: perspectives in heritage studies. de Gruyter, Berlin/Boston, pp 49–66

Stern MJ, Seifert SC (2007) Culture and urban revitalization: a harvest document. Creativity and neighborhood development: strategies for community investment. University of Pennsylvania, Philadelphia

Stovel H (1998) Risk preparedness: a management manual for world cultural heritage. ICCROM, Rome

Suanda E (2005) Art education in Indonesia. Towards an integrated curriculum. In: UNESCO (ed) Educating for creativity: bringing the arts and culture into Asian education. Measuring the impact of arts in education; Hong Kong SAR, China, 9–11 January 2004 and transmissions and transformations: Learning through the arts in Asia; New Delhi, India, 21-24 March 2005. United Nations Educational, Scientific and Cultural Organization (UNESCO), Bangkok, p 85

Sullivan S (1993) Conservation policy delivery. In: Mac Lean MGH (ed) Cultural heritage in Asia and the Pacific: conservation and policy. Proceedings of a symposium held in Honolulu, Hawaii. The Getty Conservation Institute, Los Angeles, pp 15–26, 8–13 Sept 1991

Taylor WM, Levine MP (2011) Prospects for an ethics of architecture. Routledge, London and New York

Te Heuheu T, Kawharu M, Tuheiava RA (2012) World heritage and indigeneity. World heritage, vol 62. UNESCO, Paris

The Getty Conservation Institute (2015) Cultural heritage policy documents. http://getty.edu/conservation/publications_resources/research_resources/charters.html. Accessed 01 Oct 2015

Townend S, Whittaker K (2011) Being accounted for: qualitative data analysis in assessing ‘place’ and ‘value’. In: Schofield J, Szymanski R (eds) Local heritage, global context: cultural perspectives on sense of place. Heritage, Culture and Identity. Ashgate, Farnham, pp 65–77

Turner M (2012) World heritage and sustainable development. World Heritage Rev Sustain Develop 65:8–15

Turner M (2013) UNESCO recommendation on the historic urban landscape. In: Albert M-T, Bernecker R, Rudolff B (eds) Understanding heritage: perspectives in heritage studies. de Gruyter, Berlin/Boston, pp 77–87

Tweed C, Sutherland M (2007) Built cultural heritage and sustainable urban development. Landscape Urban Plan 83:62–69
UN News Centre (2015) UN condemns destruction of Nimrud heritage site in northern Iraq. United Nations. http://www.un.org/apps/news/story.asp?NewsID=50256#.VPrixOH-UhQ. Accessed 07 March 2015
UN Security Council (2015) Security council press statement on ISIL's destruction of religious and cultural artefacts in Mosul (SC/11804-IK/700). UN security council. http://www.un.org/press/en/2015/sc11804.doc.htm. Accessed 07 March 2015
UN System Task Team on the Post-2015 UN Development Agenda (2012) Culture: a driver and an enabler of sustainable development. UNESCO, Paris
UNCHS (1999) Cities as solutions in an urbanizing world. In: Satterthwaite D (ed) The earthscan reader in sustainable cities. Earthscan, London, pp 55–61
UNESCO (1972) Convention concerning the protection of the world cultural and natural heritage. UNESCO. http://whc.unesco.org/en/conventiontext/. Accessed 25 July 2014
UNESCO (1994) World heritage committee eighteenth session. WHC-94/Conf.003/7. Rev. Phuket, Thailand
UNESCO (2003) Convention for the safeguarding of intangible cultural heritage. UNESCO. http://www.unesco.org/culture/ich/index.php?pg=00006. Accessed 21 Aug 2014
UNESCO (2005) Vienna memorandum on "world heritage and contemporary architecture—managing the historic urban landscape" UNESCO. http://whc.unesco.org/archive/2005/whc05–15ga-inf7e.pdf
UNESCO (2007) Strategy for reducing risks from disasters at world heritage properties UNESCO. http://whc.unesco.org/archive/2007/whc07-31com-72e.pdf. Accessed 21 Aug 2014
UNESCO (2011) Recommendation on the historic urban landscape. UNESCO. http://www.historicurbanlandscape.com/themes/196/userfiles/download/2014/3/31/3ptdwdsom3eihfb.pdf. Accessed 28 Aug 2014
UNESCO (2013a) New life for historic cities: the historic urban landscape approach explained. UNESCO, Paris
UNESCO (2013b) Operational guidelines for the implementation of the world heritage convention. UNESCO world heritage centre. http://whc.unesco.org/archive/opguide13-en.pdf . Accessed 25 July 2014
UNESCO (2014a) Brazil. http://whc.unesco.org/en/statesparties/br. Accessed 20 July 2014
UNESCO (2014b) Indonesia. http://whc.unesco.org/en/statesparties/id. Accessed 20 July 2014
UNESCO (2014c) Introducing UNESCO. UNESCO. https://en.unesco.org/about-us/introducing-unesco. Accessed 29 July 2014
UNESCO (2014d) Lists of intangible cultural heritage and Register of best safeguarding practices. UNESCO. http://www.unesco.org/culture/ich/index.php?lg=en&%20pg=00011#tabs. Accessed 20 July 2014
UNESCO (2014e) Nepal. http://whc.unesco.org/en/statesparties/id. Accessed 20 July 2014
UNESCO (2015a) Director-General Irina Bokova firmly condemns the destruction of Palmyra's ancient temple of Baalshamin, Syria. UNESCO. http://en.unesco.org/news/director-general-irina-bokova-firmly-condemns-destruction-palmyra-s-ancient-temple-baalshamin. Accessed 24 Aug 2015
UNESCO (2015b) Number of World Heritage Properties by region. http://whc.unesco.org/en/list/stat#d1. Accessed 09 Oct 2015
UNESCO (2015c) UNESCO calls for mobilization to stop "cultural cleansing" in Iraq. UNESCO. http://whc.unesco.org/en/news/1242/. Accessed 07 March 2015
UNESCO (2015d) Unite4Heritage. UNESCO. http://www.unite4heritage.org/index.php. Accessed 01 June 2015
UNESCO (2016) Lists of intangible cultural heritage and the register of best safeguarding practices. http://www.unesco.org/culture/ich/en/lists?multinational=3&type=00002&display1=regionIDs. Accessed 10 May 2016
UN-HABITAT (2002) The state of the world's cities 2001. United Nations, Nairobi

UN-HABITAT (2014) The state of African cities 2014. Re-imagining sustainable urban transitions. UN-Habitat, Nairobi

Urban F (2011) Denkmalschutz und Stadtentwicklung—Einführung. Informationen zur Raumentwicklung 3/4 (Denkmalschutz und Stadtentwicklung):I–III

Veldpaus L, Pereira Roders AR, Colenbrander BJF (2013) Urban heritage: putting the past into the future. Hist Env 4(1):3–18

Vinken G (2010) Zone Heimat. Deutscher Kunstverlag, Berlin, München, Altstadt im modernen Städtebau

Vinken G (2011) Lokale Sinnstiftung—Die Bedeutung der Denkmale. In: Löw M, Terizakis G (eds) Städte und ihre Eigenlogik. Ein Handbuch für Stadtplanung und Stadtentwicklung, vol Interdisziplinäre Stadtforschung, Band 11. Interdisziplinäre Stadtforschung. Campus Verlag, Frankfurt am Main, pp 73–82

von Droste B (2012) Empowerment of the Convention. From theory to practical application (1981–2000). World Heritage, vol 64. UNESCO, Paris

Wagner G (2008) Kulturelles Gedächtnis versus Verhandlung. Das Welterbe und die Konstruktion glokaler Identität. In: Luger K, Wöhler K (eds) Welterbe und Tourismus: Schützen und Nützen aus einer Perspektive der Nachhaltigkeit. StudienVerlag, Innsbruck, pp 87–102

Weise K (2012) Kathmandu valley world heritage property: conservation and community. Involving communities in world heritage conservation—concepts and actions in Asia. In: International conference in celebration of the 40th anniversary of the world heritage convention. UNESCO, Buyeo, Korea

Wöhler K (2008) Heritagefication: Zur Vergegenwärtigung des Kulturerbes. In: Luger K, Wöhler K (eds) Welterbe und Tourismus: Schützen und Nützen aus einer Perspektive der Nachhaltigkeit. StudienVerlag, Innsbruck, pp 43–58

World Heritage Institute of Training and Research for the Asia and the Pacific Region (WHITRAP) (2014) The historic urban landscape. http://www.historicurbanlandscape.com/. Accessed 28 Aug 2014

Xu J, Yao L, Mo L (2011) Simulation of low-carbon tourism in world natural and cultural heritage areas: an application to Shizhong District of Leshan City in China. Energy Policy 39(7):4298–4307. doi:10.1016/j.enpol.2011.04.047

Yeo HK, Han SM (2012) Heritage conservation as urban regeneration policy in globalizing cities: social exclusion and gentrification at the vicinity of the Jongmyo world heritage site in Seoul, Korea. In: Negussie E (ed) Changing world, changing views of heritage: heritage and social change. Proceedings of the ICOMOS scientific symposium 2010. ICOMOS, Paris, pp 56–62

# Chapter 4
# Urban Regeneration

**Abstract** The focal areas of urban regeneration have changed profoundly over the last century. Current regeneration schemes are mostly dealing with inserting landmark buildings in historic surroundings which are upgraded and renovated in the course of the regeneration project, e.g. in popular waterfront developments. Today, there is a whole variety of urban regeneration scales, ranging from small-scale projects to upgrade historic urban cores to large scale and more economy-driven projects. Urban regeneration projects originated in the Euro-American region, but became popular on a global scale. Urban regeneration in emerging and developing countries underwent three phases, from a renovation and beautification phase to a preference of modernization which with a focus on traffic infrastructure and verticalization and only little consideration of the centres. Currently urban centres are reconsidered and redefined as important components of an often fragmented city. Urban heritage and its contribution to urban sustainability is of growing importance in urban development. Regeneration does not necessarily consider historic buildings or intangible heritage, but increasingly their potential, importance and value for the urban population is recognized, including developing and emerging countries.

**Keywords** Urban renewal · Regeneration schemes

Cities are a very complex phenomenon—they have developed throughout history, fulfilling various economic, political and cultural functions, and are still undergoing rapid changes (Camagni 2000). Cities are located in different countries, in different climates and with different cultural backgrounds. Today, almost any current publication on any aspect of urban planning begins with the statement that more than half of the global population is living in cities now, indicating the importance urban planning and urban management are having these days. Throughout the past century, cities have always been a focal area of planners, following changing principles:

 
S. Sandholz, *Urban Centres in Asia and Latin America*,
The Urban Book Series, DOI 10.1007/978-3-319-43735-4_4

> Shaping the form of cities has been a central preoccupation of city building from the dawn of civilization. The purposes and values of settlements have varied considerably over the centuries, as have theories of good urban form, but the importance of designing settlements has never flagged (Hack 2012: 33).

Cities can potentially decline for different reasons, like the deterioration of their physical environment, or a sinking economic competitiveness. Thus urban regeneration is needed, with different approaches in different stages of the urban development and in different times. Urban regeneration has become a worldwide fashion during the last decades, what can be seen as a response to the rapid and fundamental social, economic and institutional changes societies and cities are undergoing (Chien-Yuan Lin 2007). For a successful urban regeneration, it is crucial to understand that communities are geographically and socially constructed places and with key identifying features (i.e. 'genius loci'). People live and work in the place and shape it, individually and collectively, adding human, social and cultural values (Colantonio and Dixon 2011).

This chapter begins with an overview on key moments of urban regeneration since the 1850s. After elaborating on the inclusion of urban heritage (in particular historic centres) in regeneration processes, regeneration in Europe and in emerging and developing countries of the Global South will be highlighted. The chapter then deals with the role of urban heritage and regeneration in sustainable urban planning.

Before providing a summary of what purposes regeneration followed in the past decades, there is a need to clarify the different meanings behind what is called 'regeneration' or 'renewal' in this research work, before elaborating on current paradigms and major approaches on a global scale and in the different world regions. There is a whole variety of aims and approaches that are subsumed under 'renewal', 'regeneration', 'redevelopment', 'rehabilitation' and so on. There are even different definitions for the different terms; they vary in goal, scale, over time and in different languages. Different authors use different words, e.g. the Council of Europe (2004) talks about 'urban rehabilitation' since the end of the 1970s to describe the upgrading of single buildings; meanwhile, they use the word also for comparable approaches on an urban scale.

Authors that use the phrase 'regeneration' in the context of historic environments include: Yeo and Han (2012) in the case of heritage conservation as urban regeneration policy in Seoul; Birabi (2007) in the context of urban conservation in developing countries; or Gonay and Dokmeci in a study on culture-led regeneration of Istanbul waterfront. In a study on the regeneration of London, Butler and Hamnett (2009: 53) states that "*Regeneration involves an interaction between the built environment and social policy*". Following the definition used by the UK government, Colantoinio and Dixon (2011: 7f) define regeneration as a "*set of activities that reverse economic, social and physical decline in areas where the market will not resolve this without government support*". They conclude that any definition of regeneration overlaps in terms of a 'social dimension' which they have in common, however they vary in terms of the extent to which these social aspects are included (Colantonio and Dixon 2011).

Urban renewal as well is a phrase used in a number of publications. In a critical review on trends of urban renewal in the Netherlands, Musterd (2008) revealed a shift over time from urban to city to social renewal. Bervoets and Loopmans (2013) use both renewal and regeneration, in the same sense in their research on neoliberal urban renewal in Ouagadougou, Burkina Faso. In a study of overall 81 papers on urban renewal (and similar words), Zheng et al. (2014) found a whole number of different aspects subsumed under this phrase, e.g. the promotion of land values, improving environmental quality and socio-economic aspects like the consideration and inclusion of vulnerable groups. They conclude, however, that urban renewal, urban regeneration, urban redevelopment and urban rehabilitation share similar meanings in the field of urban planning but are significantly different in terms of scale (Zheng et al. 2014). In their study, they found renewal and regeneration to have almost similar meanings and applied on a comparably large scale. Redevelopment is more holistic and applied on comparably smaller scales, aiming at the improvement of economic, physical, social and environmental conditions (as elaborated by Ercan 2011, in her study on historic neighbourhoods in Istanbul).

In summary, both urban renewal and regeneration aim at an improvement of the physical, social, economic and ecological aspects of urban areas (Ercan 2011; Zheng et al. 2014). Among the actions to achieve such improvement are redevelopment, rehabilitation and heritage preservation. Therefore this research follows the approach of Zheng et al. (2014), who use the phrases of urban renewal and urban regeneration interchangeably.

## 4.1 Key Moments and Focal Points of Urban Regeneration

Urban regeneration had its first appearance on a global scale in the 1850s. Since then, different waves and trends have influenced on global paradigms. Apparently, almost all of them emerged in Europe and North America and then spilled over to other continents. What is striking is the time such trends need to become famous in other countries. While it took quite a while until the Paris renovation, under Baron Georges-Eugène Haussmann, became popular around the globe and especially in Latin America (but then kept up for decades), recent fashions like waterfront developments became fashion on a global scale within a few years only.

The probably best known and most influential urban renewal project ever undertaken was the Paris' urban reform by Haussmann between 1853 and 1870. He was appointed by Emperor Napoleon III to come up with an urban reform plan for central Paris. At that time, the Paris' population had grown rapidly within a few years without any extension of built-up areas; thus the inner-city housing situation was deteriorating and population density extremely high, resulting in severe health and other problems. His urban regeneration project is still known as the 'Haussmann renovation' of Paris, and included the construction of new boulevards and parks as well as improvement of urban infrastructures. The most apparent undertaking was the construction of large multi-lane boulevards flanked by representative new housing blocks. To clear the space, Haussmann dismantled whole blocks in the inner-city area, meaning the displacement of thousands of families, with thousands of workers busy for almost two decades reshaping Paris (Twickel 2010).

In fact, the Paris image known today is the Haussmann Paris, with the long boulevards, leading to central open spaces. Haussmann also designed rules for the buildings along these boulevards, in terms of height, number of storeys, style and materials. Transportation and accessibility were also main aspects of the renovation measures, as the plan included the construction of new railroad stations, the Gare de Lyon and the Gare du Nord. Furthermore, water supply and other infrastructure like central marketplace (Les Halles) or hospital facilities were improved. Major urban parks in central spots were created, as well as promenades and smaller plantations all around the city. In the end, today's Paris is the Haussmann Paris. His vision and outline of the modern Paris, with its new street plan including boulevards meeting in central and representative places as well as the building types he delineated became fashion and have shaped urban renovation projects at least until the early twentieth century. In fact, parts of his plans were completed only in the 1920s. At that time, his vision had already been implemented in many other cities, especially in Latin America, where colonial cities were reformed with European capital and templates. 'Frenchified' Latin American urban elites were in favour of adapting the Paris' model, then claiming their city to be the 'Paris of South America'—a title still associated with Buenos Aires, Argentina. Mexico City's '*Paseo de la Reforma*' is probably the first copy of a Parisian boulevard in the Americas, designed in the 1860s (Almandoz 2002). Rio de Janeiro's urban reform under famous engineer Francisco Perreira Passos after 1903 also had Haussmann roots. Even non-capital cities like Recife in Brazil followed (cf. Sect. 5.3), leading to the name 'Pernambucan Paris', after the urban renovation in the end of nineteenth and beginning of twentieth century (Proença Leite 2006).

The rapid urban growth during industrialization in European cities led to high urban densities combined with precarious housing situations of low-income groups. As a consequence, the beginning of twentieth century witnessed growing concerns about the need for improved planning. Healthier living environments with improved sanitary facilities, modern construction methods and in general, the model of the functional city became fashionable, manifested in the CIAM Athens Charter 1931 and shaping the global debate until the 1960s. Heritage was hardly considered, if not regarded as outdated.

After the end of World War II many of Europe's cities were destroyed—in the urban reconstruction phase after 1945, functionality continued to be an important planning paradigm. After the war, two different tendencies could be perceived, with reconstruction on the one hand (cf. Tokya-Seid 2003; or Vinken 2010, on the reconstruction of Cologne after the war, where the historic part around the cathedral was reconstructed, while other parts were completely rebuilt, following modern planning paradigms) and the replacement by a less compact and car-friendly planning following urban planning trends at that time on the other (von Beyme et al. 1992).

By the late 1960s, most Western European countries substituted this approach with one of large-scale redevelopment and renewal—the car-friendly city was the most popular planning trend (also spreading to other continents, e.g. to Brazil and the planning of its new capital city Brasília). In parallel, historic centres were 'renovated' in the sense that they were demolished and rebuilt. Urban spaces were organized in a functional way, with cultural, sports and leisure facilities distributed all around the city. As a consequence, traditional urban spaces dissolved and changed in the post-industrial city (Bittner 2008). Throughout the 1970s and 1980s, processes of public participation emerged, as well as organized protests against large-scale redevelopment processes, e.g. in Berlin-Kreuzberg (Ruland 2011; Winston 2009). Still, destruction of much of the older inner-city building stock continued; instead suburban (high-rise) estates emerged to satisfy the growing housing demands (as e.g. Lichtenberger 1990, describes for the city of Vienna), in particular social housing. Such debates on demolition versus renovation are still ongoing in many countries, with many urban authorities preferring the demolition option (Winston 2009).

Urban renewal in the US throughout the 1950s and 1960s evolved to counteract the many losses of urban centres that occurred in the post-war phase. Manufacturing as well as the functional diversity declined and cities lost their urban nucleus—both residents and employers moved to the suburbs. The (white) middle class population moved out, while low-income classes—often ethnic minorities—moved in or stayed in the centres. The subsequent ethnic and social polarization, exacerbated by political fragmentation, supported the so-called urban crisis of the 1970s (Eeckhout and Jacobs 2008; Mallach and Brachman 2013). One result of this suburbanisation was the spatial segregation of cities—inner-city industrial areas were abandoned, and instead mono-functional office buildings emerged. As a consequence, urban centres were vivid during daytime but almost empty at night (Eeckhout and Jacobs 2008). Jacobs has criticized such modernist planning in her famous book, 'The Life and Death of Great American cities' (1961). In her book, she emphasizes the connections she observed between the physical environment of neighbourhoods, and the social interactions between the citizens that inhabited them (cf. Sect. 2.1). Without a primary focus on urban heritage she however created an awareness of the potentials and challenges of (historic) urban centres.

Out of these urban preservation movements, neo-traditionalism as a form of postmodern urbanism evolved in the United States in the 1970s. During recent decades, this neo-traditional development has gained a global recognition, 'Death and Life' is well cited among architects and urban planners (Salesses et al. 2013). So-called new urbanism is probably the major planning approach that can be categorized under this neo-traditionalism. New urbanism movements arose in the US in the early 1980s—it calls for a revival and recombination of several traditional urban elements (Sharifi and Murayama 2013). While trends of decentralization continued, at the same time the number of inner-city revitalisation projects increased, in Euro-American cities and beyond. For instance, in 1984 the path-breaking regeneration project in the centre of Rio de Janeiro was initiated, being the first of its kind in Brazil. It aimed at a mixed use for the area, combining housing, trade and conservation (Ribbeck 2005; Kamarid and Leupert 2009).

Inner-urban activities were promoted, e.g. by the promotion of open spaces for market of festival events. One unwelcome effect was growing gentrification on the one hand, and trends to turn historic centres or landmarks to open-air-museums for tourists on the other

> This museum character, reinforced through 'shopertainment' strategies has strengthened the fragmentation process in its own way, since only isolated, contrasting fragments of static tableaux are used to suggest a shared past (Eeckhout and Jacobs 2008).

Another trend that emerged in the US after the mid-1950s is waterfront development. Due to urban growth within the past 150 years, harbour areas in most coastal and riverine cities are now located in very central urban areas. As a consequence, urban administrations rediscovered these areas. They aimed at a conversion to value by means of rehabilitation of the physical environment, as well as the preservation and development of the historical buildings and cultural heritage with measures to promote entrepreneurship. In parallel, cities seek to improve their image and develop their touristic potential to also have a comparative advantage in global location competition (Stiftung Lebendige Stadt 2005; Krieger 2008). Major projects in the USA are Union Wharf in Boston, Charles Center in Baltimore, Fisherman's Wharf in San Francisco, Manhattan's Battery Park City or the South Boston Waterfront (cf. Heeg 2008, on property-led development in Boston). This trend then spread to other continents, e.g. to Minato Mirai 21 in Yokohama (Japan), Singapore's Marina Bay (cf. Fig. 4.1 left), Sydney's Darling Harbour in Australia or Kowloon and Victoria Island in Hong Kong (Kraas and Nitschke 2008; Kamarid and Leupert 2009). Other renowned examples are the Barcelona waterfront that was developed in the course of the 1992 Olympic Games, the Imperial War Museum North at the Manchester ship Canal, Chicago's Navy Pier or the 'Hafencity' in Hamburg (Dziomba 2007; Krieger 2008; UNECE 2009). In the latter case, the original idea of converting the city's historic warehouse district ('*Speicherstadt*') was explicitly based on the London dockland model (Daase 1995; Twickel 2010).

**Fig. 4.1** (*Left*) Singapore's Marina Bay; (*right*) Golden Horn and Galata Bridge, Istanbul, Turkey

Istanbul is following the trend with its ambitious—and like others not conflict-free—approach to create the Golden Horn Cultural Valley Project (cf. Fig. 4.1 right), composed of almost 20 sub-projects and including three World Heritage sites (Gunay and Dokmeci 2012).

Waterfront development, as one type of reurbanization (cf. Gerhard 2012, on reurbanisation in a global city perspective) is also fashionable in the Global South (Kraas 2010), e.g. Victoria and Alfred Waterfront in Cape Town (South Africa), or the waterfront in Jakarta (Indonesia). The latter undertaken is only one example for social problems that may arise, as here vulnerable groups were evicted (Simone 2010). Other Asian examples are the creation of Zhongshan Shipyard Park in Pearl River Delta, China, a conversion of 10 ha formerly industrial harbour area, or the conversion of Mumbai's Eastern Docklands, a large scale undertaken covering the immense amount of 738 ha. This large-scale undertaking is supposed to contribute to Mumbai's authorities' wish to become an actor on the global (urban) stage

> Mumbai's approach to the redefinition of its physical and projected image is a monumental challenge that could effectively change its reputation from a city of dirty industrial voids and contaminated natural resources into a model city for comprehensive [re]urbanisation and innovative urban design (Tiranishti and Gjoklaj 2009: 163).

In Latin America, the Puerto Madero in Buenos Aires (cf. Pütz and Rehner 2007) or the ongoing waterfront project Porto Maravilha for the Rio de Janeiro Olympic Games are among the major waterfront projects. Recently also, the city of Recife has planned to develop a similar project, including high-rise apartment blocks which are partly replacing a former quayside (Zancheti 2005; Furtado et al. 2014; Truffi 2014). Urban heritage in this context is often rather the decorative scenery for newly constructed buildings and new uses.

One of the most famous waterfront projects are the London Docklands. To counteract the ongoing physical decay of the inner city the London Docklands Development Corporation (LDDC) was established in 1981 as a head organization for the regeneration of depressed docklands in the east of London. The London dockland programme became a 'blueprint' for urban renaissance programmes around the UK and subsequently on a global scale. LDDC operated until 1994, subsequently other programmes followed, e.g. at Canary Wharf. All these projects aim at the regeneration of former deprived harbour areas along the River Thames, by establishing office locations (in particular for financial services) and riverside residential areas, combining modern architecture with parts of the traditional buildings (Klotzhuber 1995; Twickel 2010). Among others, the fundamental urban development projects in London led to a major privatization of former public spaces, e.g. as tackled in a recent newspaper article that found London to privatize itself 'to death' (Martin 2015). In fact, today Canary Wharf belongs 60 % to US financial group Morgan Stanley (Twickel 2010). Such privatization of public spaces can be witnessed in many cities, e.g. at Potsdamer Platz in Berlin, a place with a long history that was rebuilt by famous architects after reunification. Today it is partly privatized, e.g. Sony Center courtyard, and controlled by a private security (Twickel 2010). This is somehow indicating the latest fashion: landmark buildings in combination with 'eventization', e.g. during mega-sport events (Poytner 2009; Steinbrink et al. 2011; Aragão and Maennig 2013; Rubens de Menezes and Figueira de Souza 2014) and privatization of public spaces (Oliva i Casas 2007; Carmona and Wunderlich 2012; UN-HABITAT 2013).

Cultural facilities in icons of modern architecture designed by famous architects, often in deprived inner-city locations—so-called flagship projects—have been attracting attention since Renzo Piano's and Richard Rogers' Centre Pompidou in Paris in the 1970s. The so-called 'Bilbao-Effect' (intimately linked to the eventization trend) was named after the 1997 opening of the Guggenheim Museum in Bilbao, Spain, a landmark designed by Frank Gehry. An article published in The Economist (2013) on the Bilbao effect was asking "If you build it, will they come?". This illustrates very well the reason behind such undertakings. An ostensibly cultural project is meant to serve for economic upgrading of a quarter, a city or a whole region. For example, the UK Imperial War Museum North was opened in 2002 in Greater Manchester, UK. The museum, designed by Daniel Libeskind, was erected in a run-down former industrial site overlooking the Manchester Ship canal, an area heavily bombed during WWII. The Guggenheim Bilbao was constructed in Northern Spain, one of the poorest areas of the country, with the clear aim to bring visitors to the city that would else not come (Polinna 2008; Plazaab et al. 2009), something that Plaza et al. (2015: 179, 195) call a "*cultural re-imaging through iconic art museums*" or symbolic capital that benefits the local economies.

Spin-offs of international museums like the Guggenheim have become popular 'urban renewal projects', and cities are competing against each other to host such offshoot: Abu Dhabi's Saadiyat Island will be home to local branches of the Guggenheim and the Louvre; and the West Kowloon Cultural District in Hong Kong will house the M+ museum of Chinese contemporary art, supposed to be Hong Kong's answer to London's Tate Modern. One effect of such 'wow factor architecture' is a kind of global urban appearance, which differs in the landmark building of a famous contemporary architect, but that does not reveal much of its urban past. Cities somehow have become global actors, polarizing local processes of socio-spatial redistribution within the urban area (Bittner 2008; Butler and Hamnett 2009). However, there is a growing consent that "*successful regeneration is not merely about signature buildings or megaprojects*" (Mallach and Brachman 2013: 3), but going beyond, including physical, economic and social conditions of the surrounding area.

Such large-scale projects are usually realized in form of public-private partnerships (PPPs). While until the mid-1970s, private investors were included in renewal projects mainly on a contracting base after defining the project itself, things then changed towards cooperation schemes between public and private actors. Today tax incentives and entrepreneurial approaches in urban politics are common tools in urban upgrading schemes. In the course of the shift towards an entrepreneurial city, an increasing inclusion of civil societal actors in urban governance schemes became apparent (Heeg 2008). The latest model of governance on a sub-municipal level are business improvement districts (BIDs), that aim at securing private capital for enhancing the attractiveness of urban central areas (Peyroux et al. 2012; Silomon-Pflug et al. 2013).

Regeneration approaches have gone global (Bodenschatz and Laible 2007). While regeneration schemes very clearly emerged in a European or US-American context and then were exported to other countries, things have changed latest with waterfronts developments that became an international fashion within a few years only. The cities now erecting big museums or other landmarks, designed by international celebrity architects and following the Bilbao—and 'eventization'—trends, can be found as much or even more often in the Global South. Actually, a (post-industrial) shift is occurring from the event city towards a creative city, resulting in changing concepts of 'urban' making the dynamic and vital city itself an 'experience'. It seems that the city has become again the space for societal interaction, innovation and creativity. The 'creative' became the target group—they are treated with preference resulting in a disregard of other important urban topics and actor groups (Bittner 2008). As a conclusion, taking a look at urban regeneration during the past century, a shift of approaches and of underlying paradigms becomes visible. Table 4.1 sums up the different guiding principles of urban regeneration, its respective focal areas, social aspects and related urban heritage policies over the past decades from the 1940s until today.

Throughout the twentieth century, the attitude towards urban heritage and the historic centre as well as its consideration in regeneration projects has changed quite a few times, depending on different guiding principles in force at that time (cf. Ruland 2011, in the case of Germany). European cities paid comparably more attention to preservation of historic building stock, while in the US and also in countries like China business districts are often constructed within the metropolises' centres replacing original fabric (Sassen 2000; Twickel 2010). In this context, Allen (2014) uses the phrase of 'Shanghai-ization' to describe the concept of using modern architecture for branding a world leading city—urban heritage is not part of this concept.

On a global scale, Ronneberger (2000) noticed a growing 'mallification' of urban centres, as central areas are upgraded or modified to urban landscapes with different uses and events. Such trends are influencing on urban open and also green spaces (cf. Secretariat of the Convention on Biological Diversity 2012; Trzyna 2014, on co-benefits of mixed-use urban green spaces) as it is excluding certain uses and thus users. Shaftoe (2008) claims that despite Europe comprising of the longest history and considerable experience with planned and public urban open space, there are examples of such spaces anywhere around the world. This is why to him, open spaces seem to be a fundamental need of human nature and requests a strategic management of such spaces to make them more attractive, liveable and vital. However, "*there will never be a one-size-fits-all utopian model of public space [...]*" (Carmona and Wunderlich 2012: 286).

Aside the creative city, today's planning and regeneration role models are heading towards more integrated approaches, comprising economic, ecological and social aspects. In contrast to the more functionalist view on public open spaces in the 1960s and 1970s, today, social and heritage aspects are considered important in

**Table 4.1** The evolution of urban regeneration and its focal areas (*Sources* Chien-Yuan Lin, 2007; Roberts and Sykes 2000; Council of Europe, 2004; Colantonio and Dixon 2011; Zheng et al. 2014)

| Decade and focal area | Major strategy | Key actors & stakeholders | Economic focus | Social aspects | Physical emphasis | Urban heritage policies |
|---|---|---|---|---|---|---|
| 1940s & 50s physical redevelopment | Reconstruction and extension of older urban areas, often based on 'Master plan' | Government, growing inclusion of private sector developers | Mainly public sector | Improvement of housing and living standards | Replacement of inner-city areas and peripheral development | Preservation and reconstruction of single buildings |
| 1960s social welfare | Continuation of 1950s theme, suburban and peripheral growth | Move towards public and private sector balance | Growing influence of private sector | Social and welfare improvement | Continuation of 1950s theme in parallel rehabilitation of existing areas | Ensure integrated conservation of sites and groups of buildings in old city centers |
| 1970s economic prosperity | Focus on 'in situ' renewal and neighborhood schemes, still peripheral development | Growing role of private sector, local government decentralisation | Ongoing growth of private investment | Community-based action and empowerment | More extensive renewal of older urban areas, large-scale rehabilitation policies | Continuation of 1960s approach, restrictions for demolishing old buildings |
| 1980s prosperity-led regeneration | Flagship & out of town projects, various schemes of (re)development | Emphasis on private sector, growth of partnerships | Dominance of private sector with selective public funds, development of urban tourism in old neighborhoods | Community self-help with very selective state support | Major schemes of replacement and new development, 'flagship themes' | Urban rehabilitation increasingly treated as fundamental component of urban and local development |
| 1990s community partnership | Emphasis on integrated approaches, move towards more comprehensive form of policy and practice | Partnerships stay dominant, community-based approaches & private sector ones in parallel | Greater balance between public, private and voluntary funds | Emphasis on role of community, participatory processes | New focus on heritage on community scale, more 'modest' than in 1980s, ongoing flagship projects | Application of sustainable development principles in spatial planning strategies & urban heritage |
| since 2000s Creative & sustainable city | Large-scale urban projects including heritage areas, promotion of dense historic centres as environmentally beneficial | Private investors & public-private-partnerships | Trend towards private funding | Increasing promotion of cultural diversity and human well-being | Ongoing large-scale flagship projects, reconstruction | cultural diversity, sustainable development, & climate change adaptation guiding principles in spatial planning & heritage conservation |

urban development and regeneration. Public spaces become places of individual and societal social identification, shaped by their 'Genius Loci' and affecting minds and spirits of its users (Knirsch 2004; Shaftoe 2008).

Among the main problems of urban areas is environmental degradation, deficiencies in urban services and adequate housing, deterioration of existing infrastructure, a lack of access to key resources and violence—problems even worse in poor and highly segregated countries with a high level of social exclusion (IHDP 2005).

Therefore, one of today's most urgent needs is to prepare for the sustainable growth and expansion of cities in rapidly urbanizing countries. To do so, it is crucial to gain "*a deeper understanding of the key spatial dimensions of these cities and their changes over time*" (Angel et al. 2011: 2).

Social capital, including culture as well as cultural participation, is becoming a new paradigm in urban renewal (UNECE 2009, for the case of Europe; UNESCO Institute for Statistics 2012). Winston (2009) is emphasizing the need to include social aspects with community and neighbourhood regeneration, in any physical regeneration scheme to make it successful. Stern and Seifert (2007) agree with Winston (2009), as to them culture is the right tool for urban revival in the twenty-first century, although being aware of potential negative consequences of culture-based development—gentrification and the expansion of economic inequality.

Also, recent publications have stressed the potential linkage between culture and risk reduction, e.g. the World Disasters Report (2014) that emphasizes on the importance to understand how and which values people and societal groups on different aspects of their lives

> People choose to follow certain doctrines because these match their cultural heritage, priorities, values and economic interests. There are similarities with people's response to the idea of climate change, which may provoke tremendous emotional reactions and denial that it exists (p. 41).

In this sense, the 'cultural globalization' (Mucke 2014a) can potentially exacerbate risk levels. Probably the biggest urban challenge is adapting cities to impacts of climate change (Bigio 2003; Pelling 2011; Mucke 2014b). In this context, reducing urban sprawl and returning to planning more compact cities, the need to plan more open and green spaces (Beck 2012), and the implications of urbanisation on risk and vulnerability of different societal groups (Garschagen 2014), have gained momentum. This is particularly the case in the rapidly urbanizing cities of the Global South (Bulkeley 2010; Angel et al. 2011), where the addition of high levels of informality and lack of budgets (Mucke 2014b; Schauber 2014), and long-term planning (cf. Petrella 2010, in the case of African cities; Abbate 2010, on the lack of proper planning, hindering innovation and freezing urban potentials) is potentially impeding regeneration (cf. Chap. 5 on the case studies).

In Europe within the past years, a 'renaissance' of the historic urban centre has been going on (cf. Dziomba and Matuschewski 2007, on urban renaissance; or Rautenberg 2011 on the growing importance of heritage in urban regeneration). To Tokya-Seid (2003), it is however questionable whether this is more than a short-term reanimation of lost things. While the original medieval city centre has vanished to large extents and for various reasons, the interest in preserving its 'left-overs' seems to grow. In this process, questions of 'authenticity' of the historic fabric are of comparably low importance. Therefore, even reconstructions are now popular again, e.g. in the cases of the Berlin castle or Frankfurt's historic centre (Rodenstein 2010). Inner-city projects are focussing comparably more on modern urban marketing, branding an urban image and cultural tourism than on housing, trade and urban life. In this sense, the historic centre is not more than a means to an end (c.f. Vinken 2010, in the cases of Cologne and Bristol; Tokya-Seid 2003).

## 4.2 Urban Regeneration and Related Policies in Different World Regions

Area-based conservation and later renewal of the historic inner city occurred in most European countries around the same time: the 1961 Monument Act in the Netherlands was the first; followed by the '*Loi Malraux*' in France, the 1976 Civic Amenities Act in the United Kingdom; the Urban Planning Act in the same year in Italy; and the 1973 Monument and Historic Buildings Act in Turkey (Tiesdell et al. 1996). It is only since the 1960s that historic areas and quarters of cities have had a significant re-evaluation of their positive qualities. It came about as a reaction to the evident social, cultural and physical disruption of lives and thus loss of place attachment of the local actors. Integrated heritage conservation within Europe as defined by European Union (Council of Europe and Directorate of Culture and Cultural and Natural Heritage 2004), has the aim of ensuring the perpetuation of the cultural heritage and of seeing that it is not only maintained as part of an appropriate built or natural human setting, but also suitably used and adapted to society's needs. It has an essential objective of integration of heritage—including historic inner-city areas—that constitutes the cultural environment into the human environment of present-day society.

The urban renewal concept and paradigms have changed profoundly around Europe throughout the past decades, mainly due to the inclusion of the population's needs (Lichtenberger 1990; Atkinson 2008). The 'ecological' dimension was included in renewal processes in the 1990s as a consequence of the worldwide sustainability debate (Council of Europe and Directorate of Culture and Cultural and Natural Heritage 2004). However, the base remained the same: the definition of the 'European City', its assets and values.

Europe possesses a long tradition in mixed urban functions, comprising residential functions as well as workplaces. This paradigm was followed also in renewal processes that mainly combined both functions with touristic activities. European cities comprise of compact structures, an urban society that is to a wide extent coherent, and an urban culture focusing on the centre. These attributes are important factors for the locational advantage of European cities in global competition (Bundesministerium für Verkehr 2007). European concepts ('*Leitbild*') are focusing on the compact (European) city model, by means of preservation, renewal and upgrading of historic city centres (Behr 2005).

Until 1994, the European Union's involvement in specific urban interventions was relatively limited; explicit urban interventions had not been a feature of regional policy. In 1994, the URBAN Community Initiative was launched at EU policy level in response to the growing awareness of the challenges facing Europe's towns and cities. It was extended until 2006, then replaced by follow-up programmes. These programmes were crucial drivers for the regeneration of EU cities, and initiated a new planning paradigm on EU level: integrated area-based regeneration combining economic, social, cultural and environmental aspects and civic involvement, also encouraging PPPs in urban regeneration (Colantonio and Dixon 2011). During these years, the recognition of the high importance of European cities, but also an awareness concerning growing urban disparities, grew. Urban regeneration projects were found to be an adequate tool, and have been on the agenda of Member states since the mid-1990s onwards.

Since the 1990s, Carpenter (2011: 100f) sees an increasing recognition of the importance of addressing urban challenges at the EU level, finally resulting in the most recent development on the EU urban policy scene, the 2007 Leipzig Charter on Sustainable European Cities

> Ministers for the first time have recognised collectively the importance of integrated urban development and the role of partnerships, which in theory lay the foundation for greater emphasis on urban policy in the future. Given these policy priorities, it is likely that the social sustainability agenda will come to dominate urban policy at the EU level in the years to come.

The Leipzig Charter (2007) adopted during the Informal Ministerial Meeting on Urban Development and Territorial Cohesion in Leipzig, May 2007, during the German EU Presidency, has set an agenda for a joint policy on European cities, emphasizing on the common urban concept that joins Europe. The charter is based on the belief that the focus of urban planning is the city centres and that revitalizing old residential areas will be crucial parts of urban planning. It has influenced the growing importance of integrated urban (district) development within Europe (German Institute of Urban Affairs (Difu) and Federal Institute for Research on Building; Urban Affairs and Spatial Development within the Federal Office for Building and Regional Planning 2012). According to the Council of Europe (2004), the urban challenge in Europe is upholding the cultural heritage's role in the face of globalization, reinforcing local identities and sustainable development. That signifies a progress from linear urban development processes (e.g. building towns from new or greenfield sites) to more complex processes of urban recycling or rehabilitation (working with existing cities) to improve the surrounding and well-being of residents and users. Also, 'Community' has become a focal point of sustainable urban development within the EU and European policy: key objectives at EU level are economic prosperity, social equity and cohesion, environmental protection (Colantonio and Dixon 2011).

Since the 1980s and 1990s, a return of habitants of the city centre could be perceived in European and also US cities, especially in newly renewed areas. One reason for moving (back) to the centre was historic surrounding. The residents' activities are a crucial contribution to the vitality of an urban quarter, creating greater demand for facilities in the city centre, and, thereby, increasing the number and mix of uses within the quarter—potentially resulting in gentrification processes. Urban areas are a vital part of Europe: complex demographic, social and economic forces have (re-)shaped the structures of European cities, requiring continuous adaptation and regeneration. Thus, to revitalize historic urban quarters, European cities are attempting to attract residential uses—a crucial point in any revitalization plan (Tiesdell et al. 1996; Council of Europe and Directorate of Culture and Cultural and Natural Heritage 2004). The concept of the 'European City', understood as a densely built core area with mixed uses and functions, is turning to be the general principle for planners and policy makers again (among others because of their comparable advantages in climate change adaptation, cf. Satterthwaite and Dodman 2009). This is why Gaines and Jäger (2009) assume that one could right now experience this concept being used as an archetype for other cities worldwide.

Urban regeneration is not a clearly defined issue; it ranges from large-scale projects for economic growth to neighbourhood interventions to improve quality of life (Colantonio and Dixon 2011). However, many inner-city renewal measures inevitably lead to changes in the functional and social character of the surroundings—resulting in gentrification. These problems of gentrification, and even segregation also, could be perceived to an even larger extent when urban renewal projects were implemented in Latin America and other countries of the Global South.

Urban regeneration in emerging and developing countries can be divided into three main phases

- Until the mid-twentieth century, city centres were renovated and beautified with boulevards and representative buildings, following the example of Paris and its transformation planned by Haussmann.
- The phase up until mid-1980s can be defined as period of 'modernization', geared to the principles of the Athens Charta: inner cities were adapted to the needs of modern transportation by cutting lanes through the centre, service functions replaced residential functions and excessive 'verticalisation' dominated the increasingly specialized urban centres (Coy 2007). The inner cities very often were not in the scope of urban planning, as one consequence degradation and dislocation processes emerged in that period, resulting in a 'loss of the centre' (Gaebe 2004, in Mexico City; Coy 2007; Luger 2008, in Bhaktapur/Kathmandu; Kraas 2010; Coy and Töpfer 2014, in São Paulo).
- For the past 30 years, the number of private and public conservation and renewal measures has been rising, not only in industrial countries, but also in emerging and developing countries (Gaebe 2004). Within the past 20 years, the position of urban centres within the fragmented city had to be redefined almost everywhere. This phase can be defined by the loss of functions and at the same time by the bid for renewal (Coy 2007).

Urban challenges of emerging and developing countries are different from Europe and America: urbanisation processes are (still) much more dynamic, socio-spatial fragmentation is higher, infrastructure is often lacking significantly, the informal urban sector plays a significant role and often urban planning and governance is poorly developed (Choguill and Choguill 1996; Drakakis-Smith 2000; Herrle et al. 2006; Oliva i Casas 2007; Kraas 2010; Angel et al. 2011). Urban development in Europe and Latin America is differing more in terms of implementing planning, e.g. compression or development of urban centres or limiting urban sprawl, than in the goals themselves. Achieving those goals is further hampered by different interests, differenced in political systems, lacking controls, corruption and frequent modification of goals (Gaebe 2004). The same finding seems to become also true for Asian cities (e.g. in Singapore and Hanoi: Böhme et al. 2003; Vorlaufer 2009).

Ashworth and Tunbridge (2000) call it dangerous to allow the assumptions and experiences of the conservation 'front runners' in Western Europe and North America to determine urban conservation practice in societies with quite different attitudes towards the built environment. It can be assumed that successful inner-city revitalization projects in developing or emerging countries, with different urban patterns and urban history, make demands going beyond European concepts.

Due to economic reasons, preservation of cultural heritage and revitalization of inner cities was a concern of richer countries for decades. Poorer ones may possess an immense cultural heritage; however, the ability to develop adequate preservation measures is obviously more limited. That results in the problem that lots of preservation measures or financing instruments stem from supra-national organizations or other countries—that has practical implications for what is or is not preserved, and how and to whom it is interpreted (Ashworth and Tunbridge 2000; Smith 2013).

Latin American, African and Asian cities often lack financial means and instruments for urban upgrading; the multitude of regeneration projects thus is planned with the private sector or enterprises. As a consequence, the urban sector is under high pressure as supply functions disappear (Gaebe 2004). Revitalization is putting a high pressure on less prominent and competitive land uses than social housing (Ashworth and Tunbridge 2000). Thus, gentrification that was known from European cities became an even bigger issue in inner-city renewal in emerging and developing countries.

So far, only limited research has been carried out on the link between the concepts of inner-city renewal in these countries, and the effects not only on the physical environment and provision of adequate housing, but also on the inclusion of beliefs and place attachments of local population in planning.

Despite their longing for being acknowledged as global cities following global trends (e.g. Johannesburg, cf. Mayr 2011), African cities are less popular for urban renewal processes, except very few key projects like the Victoria and Alfred Waterfront of Cape Town. Urban centres in many African countries are lacking policies defining their role and setting up a long-term planning and management (Petrella 2010); existing prestigious urban centre renewal schemes are mainly private-sector led, following neoliberal tendencies, e.g. in Ouagadougou, the capital of Burkina Faso (Bervoets and Loopmans 2013). However, urban culture has recently been identified as a key topic of African urbanism (UN-HABITAT 2014), going hand in hand with a raising awareness of conservation charters and heritage resources throughout the continent (Birabi 2007). Also heritage tourism is becoming more popular, thus preservation of cultural heritage in urban areas and

**Fig. 4.2** (*Left*) Doha skyline (*Source* W. Lange); (*right*) inner-city reconstruction of traditional architecture in Souk Waqif, a major tourist destination (both pictures taken in 2006)

cultural landscapes, as Keitumetse (2009) has researched for Botswana, is being increasingly notified also as an economic asset. This becomes visible in African and also Arab states, e.g. in cities like Marrakesh and Fes in Morocco or other Northern African cities that have followed large-scale renewal and projects including restoration schemes, in particular in the historic centres and central market areas. Unfortunately, recent political incidents have impacted negatively on some historic centres, e.g. the Old Town of Timbuktu which had even been out on the List of World Heritage in Danger, accompanied by Arab cities from Yemen or Syria. In the Gulf States, many of the historic settlement cores have been replaced in today's major cities in their quest for becoming a 'global city' with a 'skyline'. As a consequence, some of them are even rebuilding a souk or other traditional areas (e.g. in Doha, Qatar, see Fig. 4.2 left and right), also and foremost for tourist purposes (Goldman 2011; Haines 2011; Ong 2011a).

Throughout the past 20 years, numerous historic urban centres in Latin America and the Caribbean have been transformed, boosted by discourses of political and technocratic power that emphasized on the centres as focal points of their urban policies (Hiernaux 2013). During colonial times, pre-Hispanic cities, their buildings, cultures and ways of life, were destroyed and replaced by new settlements or cities, constructed after the conquerors' role model (García Canclini 2013). Urban centres were planned according to Spanish (and to a lesser extent Portuguese) models with the main functions and buildings located around the central 'plaza' (Bähr and Mertins 1995). These colonial cities witnessed a fast demographic and spatial expansion at the end of the nineteenth and early twentieth century, when European and US-American models of industrialization appeared. As a consequence, urban planning followed European role models of 'modernization', and many centres were transformed in the Paris' Haussmann style with large boulevards and representative facades, e.g. Rio de Janeiro or Recife in Brazil (cf. Rojas 2002; Proença Leite 2007; Kamarid and Leupert 2009; Cadena de Melo Filho 2012) (cf. Sects. 4.1 and 5.3). During the 1920s and 1930s, the centres were transformed: financial sector moved in and housing and trade moved out. Some decades later, office uses and the financial sector moved out of the centre to newly constructed high-rise areas, going hand in hand with urban expansion long main arterial roads. Subsequently, urban centres started to degrade: richer classes migrated out; the central areas became inhabited by middle and lower classes, businesses that initially were located in the centre also left and moved to the periphery. Such kinds of processes can be observed in the vast majority of Latin American cities, and has led to the emergence of satellite towns, boosted by road constructions. The consequence was a huge urban expansion—like decades before in European cities like Madrid, Paris, London or Berlin—but in a comparably much accelerated way, shaping Latin American cities. The more this politically supported suburbanisation advanced, the less social, cultural and political activities moved out of the historic centres, which lost importance and centrality. As a consequence, urban agglomerations became highly fragmented (Bähr and Mertins 1995; Coy 2006, 2010). There the majority of historic urban centres in Latin America were and remain still in a process of decay, comprising of the destruction of historic fabric, lack of traffic control, a strong presence of informal economy (especially street vendors), and a comparably low quality of goods for sale (Almandoz 2002; García Canclini 2013; Hiernaux 2013).

Another challenge the centres are facing is the decline of resident population in the central area, resulting in increasing decay and insecurity, which in the end has led to a shift of urban policies (cf. Coy 2007; Coy and Töpfer 2014, in the case of Brazil and particularly São Paulo). Hiernaux (2013) dates the beginning of this shift back to the early 1980s, when different buildings were declared historical heritage, e.g. Mexico City's Plaza Mayor. Today, the heritage dimension is a key element of any regeneration project in the historic centre areas, correlating with a trend of nostalgia concerning the past. One consequence is the term 'historic centre' itself, which is indicating the historic values a geographical space has for the urban or national identity.

In this context, Hiernaux (2013) talks about an '*imaginario*' that was created in cities in Europe and the USA, and then spread to Latin America. As a result, the creative class has returned back to these centres, where living, working and finally tourism were re-established. The synergies between such positive '*imaginarios*' result in the upgrading of historic centres, their heritage and living qualities, an expanding urban tourism and the expanding real estate market, as can be perceived in Latin America. On the other hand, an ongoing gentrification of urban centres, induced by upgrading and increasing prices, is leaving aside the fact that the desire for what is regarded as traditional urban life was shaped to a large extent by the groups that are now forced out—e.g. the poor or informal trade. In the end, gentrification, as can be found in Latin America, is composed of two contradictory images: the built heritage left from ruling classes, and a lifestyle largely shaped by subalterns (Nobre 2002; Proença Leite 2007; Rothfuß 2007). Such kind of gentrification is somehow new and locally specific to the Global South. Hiernaux (2013) calls it a '*Creole Gentrification*', in other words: a process shaped by imported models of regeneration but adapted to local realities and happening on a much faster speed than in the North.

In reality, this would mean a cautious return to the centre where a growing number of businesses meeting international standards are coinciding with traditional and often informal economies which itself attract tourists. The role of both tangible and intangible heritage in this process can and should not be underestimated (Hiernaux 2013). Latin American cities are characterized by colonial heritage and some—if at all—minor elements of pre-Hispanic era, amalgamated with modern buildings and spatial patterns that itself often follow(ed) European metropolises or role models of US-American modernity. The concept of colonial heritage itself is a postcolonial one (García Canclini 2013), increasingly seen intertwined with urban regeneration approaches.

Renewal projects in historic Latin American cities logically deal with the colonial past of the continent and the majority of its cities (García Canclini 2013). They also have to be seen as authorities' reaction to poor living quality in these areas (Vega and Braig 2013). Since the 1980s, many central urban areas were renewed, such as the historic centre of Havana in Cuba (Mertins 2003), the centre and '*Corredor Cultural*' of Rio de Janeiro (Kamarid and Leupert 2009), São Paulo (Coy 2007), as well as the renewal of the Bairro do Recife (de Albuquerque Lapa and Almeida de Melo 2007; Pontual 2007) in Brazil or Mexico City (Hiernaux 2013). Waterfront development projects in central urban areas in Buenos Aires or Rio de Janeiro have attracted even global attention (cf. Sect. 4.1). Especially throughout the earlier renewal phases, low-income groups and informal sectors were disregarded or even forced out (Vega and Braig 2013). Nevertheless, also upgrading with a focus on low-income neighbourhoods was done, including housing and infrastructure improvements, e.g. in Ribeira Azul in Salvador, Brazil, that was renewed under the '*Viver Melhor*' programme scheme (Baker 2006).

In Asia, neoliberal urban policies gear towards obtaining an economy-driven status of 'Global City' or 'World City'. Shatkin (2011) claims that even research of urban Asia so far mostly dealt with meaning and agencies of development of global urban culture, with a focus on multinational developers and architecture, and planning firms as the primary actors in urban mega-projects. Projects like Malaysia's Multimedia Super Corridor, Hong Kong's Cyberport or the spectacular skylines of Chinese Metropolises showcase this understanding that Michel (2010) describes as especially attractive for cities of East and Southeast Asia after a long economic boom phase. Obtaining a 'world-class' status or in other words some degree of global significance, is definitely an objective for many Asian cities. Ong (2011b: 13) sees such tendency "*is no longer simply to turn to Western prototypes, but rather to develop from homegrown solutions to Asian metropolitan challenges, distinctive urban profiles, political styles, and aesthetic forms*".

Meanwhile, one Asian city is copying the other, e.g. from established Chinese cities. Asian cities are experimenting with architectural forms: the goal is to position and brand their cities as symbol of progress and development on a global scale to attract foreign capital, e.g. Singapore's recent development is another example for this mega-project trend, which is at the same time paying attention to brand the city as iconic and unique (Goldman 2011; Shatkin 2011). Such projects include urban upgrading, in this case in particular by cultural flagship projects, often in combination with waterfront development. This understanding of urban upgrading resulted and still results in large-scale losses of historic fabric and urban centres. As a consequence, throughout the past years a discussion on the globalization of gentrification and urban renaissance arose (Michel 2010). In an interview (Tweeddale 2012) Pritzker Price laureate 2012, Chinese architect Wang Shu, complained about the loss of urban traditions in contemporary China which to him is either borrowing from the Western world or from cities like Hong Kong or Singapore. In his opinion Chinese want change and that simply implies new things for which they have to go to the cities: "*In fact China had a great tradition of cities—many, many beautiful cities—but in the past 30 years we have demolished more than 90 per cent of traditional cities. We have almost totally rebuilt a new country. Is this country's name China? It's just a new country. But the people also lost memory, tradition, lost history.*"

The rise of modernism as a model of urban design and architecture emerged during a phase of state-driven modernisation in the post-independence phase throughout the 1950s and 1960s. In practice, planning has often failed because of not considering the social practices of the urban population—not to talk about participation in planning processes. In reality however, what has resulted very often is a fragmented urban landscape, including areas for business, shopping, housing (Shatkin 2011) and historic fabric for tourists and nostalgic locals.

In fact, it is difficult to draw such conclusions for the whole of Asia, as this is a very heterogeneous continent, in cultural, history, as well as economic contexts (cf. Brillantes and Flores 2012, in the context of the Philippines, Indonesia and Thailand; Dahiya 2012). This kind of urban development is well underway for the major cities; in particular those in emerging countries; while others, mainly the ones in poorer countries and regions, are struggling with lack of budget and governance, and often very high urbanisation rates. Urban policies straddle deep class divisions, particularly in cities with a weak state presence (Ong 2011a, b).

Today, there is a whole variety of urban regeneration scales, ranging from small-scale—often tourism-centred—projects to upgrade historic urban cores to large scale—more economy-driven—projects involving public and private financial means. For example, Mumbai is losing much of its historic buildings and patterns, being replaced by modern architecture. To Loeckx (2009), this signifies a mental gap between the urban built heritage and contemporary urban forms that shape the metropolis. Mehotra (2009) finds this and any other Indian city to be divided into two parts: a static city, with permanent buildings; and a kinetic, temporary and dynamic one, allowing for change. Policies and legislations protect only the static parts, often leaving cultural assets and values of a lot of societal groups and (informal) uses aside (Roy 2011).

Former public open spaces are increasingly privatized and only accessible for selected parts of its population (Gosseye 2009). Kathmandu in Nepal is lacking concerted strategies linking heritage conservation with social and economic development (Government of Nepal; Department of Archaeology 2007). Hong Kong on the other hand is increasingly cherishing old buildings that embody its recent past as a preeminent harbour city to distinguish the city from other Asian shopping destinations (Ong 2011b).

Large-scale renewal projects of historic urban areas can be found in Chinese cities like Beijing or Gouangzhou, where many of the traditional former agricultural village structures were replaced by high-rise buildings (Kraas and Nitschke 2008; Mars 2008), but also outside China, e.g. in Mumbai, Bangkok, Hanoi, Ho Chi Minh City (Kraas and Nitschke 2008). Also, initiatives to preserve and modernize historic quarters can be found, e.g. in Shanghai old town or the colonial part around Colaba in Mumbai. Simultaneously, new quarters are constructed in central locations (Dahiya 2012), partly in form of waterfront development projects like in Kowloon and Victoria Island, Hong Kong (Kraas and Nitschke 2008). Siu (2011) comments about an urban renaissance that Hong Kong is currently undergoing, including city reconstruction and identity politics. 'Identity' in this city also includes a debate on what Hong Kong's history is, a negotiation for cultural and social identity between colonial past and Chinese present

> In the name of making Hong Kong into a leading world city in China and Asia, many of the accumulated details of the lives of generations are lost in policy blueprints. Old neighborhoods together with their colonial pasts will be erased, replaced by essentialized 'Chinese' cultural icons or signature landmarks on the grandest scales. They are planned for the city to signify super (post-)modernity and global engagement (Siu 2011: 139).

Here one can witness that 'collective memory' or 'identity' can potentially mean totally different things within the same urban boundaries, depending on a person's or group's background and not necessarily considering historic buildings. In the words of Michel Rautenberg (2011: 522f) "*there is a fundamental incompatibility between local popular representations and entrepreneurial or cultural regeneration, even when the latter is supported by local authorities*".

## 4.3 The Role of Urban Heritage in Regeneration and Sustainable Urban Planning

Urban heritage plays a significant role in sustainable urban planning for two main—and potentially contradictory—reasons. Most cities comprise of at least some historic urban fabric, therefore, any regeneration project has to deal with that fabric, intentional or not, because of its mere existence in some regeneration project area. So in this case, the historic fabric could even be regarded as unfavourable for the regeneration, resulting in the decision to replace it. Another—and more recent—regeneration approach is regarding this urban heritage as particularly valuable, for economic, social and historical reasons. As a consequence, its preservation is an essential project component. Use or reuse concepts for the historic buildings are one essential part of the planning—often combined with cultural flagship projects.

The terms of 'sustainable communities' and 'sustainable cities' have already existed since the late 1960s; 'sustainability' then became fashion with the

Brundtland report (1987). That report included a chapter on cities, focusing in particular on challenges in the developing world. Almost at the same time, publications from thinkers on eco-cities or ecological planning came up. Today, many cities have developed comprehensive green or sustainability plans, e.g. the cities of Copenhagen, London, Chicago or New York (Beatley 2012). Literature on sustainable urban development and cities amplified during the 1990s, including some guidelines and policies of international significance (Pugh 1996). A multitude of authors and institutions have worked on different aspects of sustainable urban development and have tried to come up with frameworks (c.f. Robrecht 2000), due to the fact that an agreed definition as to what the terms 'sustainable cities' and 'sustainable human settlements' mean was lacking (Satterthwaite 1999). Satterthwaite (1999) argued that in this debate it is important to also ensure the provision of facilities that make city environments valued more by their inhabitants: "*It includes a concern to preserve the city's cultural heritage*" (1999: 88). Beatley (2012: 94) recently identified a remarkable global emergence of new commitments to sustainability since the turn of the millennium

> Today, sustainability is understood as also being about cities and the built environment, about the social and the economic as much as the ecological; and it is now permeating and penetrating cultural consciousness in a way earlier versions did not. Much of the activity and activism is bubbling up from the bottom, and sustainability is no longer confined to the realm of experts in state and federal offices. The language of our society and culture has changed.

Concepts of sustainable livelihood are based on some core principles: they all are holistic, focusing on multiple actors with multiple influences, multiple livelihood outcomes and rely on participatory approaches (Jones 2004). Urban regeneration has to involve public participation processes to be sustainable; simultaneously, for the development of a more sustainable environment, effective governance as well as more knowledge to deal with urban regeneration are needed. Many policy instruments and incentive programmes on urban sustainability have been proposed and experienced in different countries. (Chien-Yuan Lin 2007). On a European level the concept of sustainable development has been linked with 'sustainable cities' since the 1990s

> In recent years, the sustainable urban development agenda has been broadened and incorporated into planning practices and governments' policies for urban-regeneration projects. In the 1980s, regeneration projects focused mainly on the physical and economic renewal of degraded areas. However, since the 1990s, especially in Britain, regeneration programmes have combined the stimulation of economic activities and environmental improvements with social and cultural vitality (Colantonio and Dixon 2011: 31).

Concepts of community and neighbourhood are essential components of such regeneration. Thus, since 2005, 'local community' has been regarded as a focal point for the delivery of sustainable development in EU urban policy. Gaines and Jäger (2009) mention the chance to make Europe's cities the test laboratories for sustainable and long-term urban concepts, that could be exported abroad, as the urban development of the future will mostly take place in the Global South, e.g.

urban cores in most European cities are densely built and therefore ideally suited for pedestrians and bicyclists (Satterthwaite and Dodman 2009).

Meanwhile, sustainability implies cultural diversity and heterogeneity (Gaines and Jäger 2009), including the need to acknowledge a variety of values, perceptions and attitudes. Among the three guiding urban principles that Kraas (2005) proposes to achieve urban sustainability is the one of a 'cultural city' that can be achieved by caring about the cultural elements of a city, by preserving cultural heritage and cultural identity as well as by supporting cultural diversity. To Hardoy et al. (1992), it is obvious that development necessarily comprises a respect for cultural patrimony

> Culture implies knowledge and a vast wealth of traditional knowledge of relevance to sustainable natural resource use (and to development) is ignored or given scant attention in development plans. But the term 'cultural sustainability' seems rather imprecise for the need to recognize the importance of culture and respect it within development. Culture is never static; to argue that it should be sustained is to deny an important aspect, changing and developing nature (Hardoy et al. 1992).

Today, visions of a sustainable city are becoming more abundant: by broadening the concept to a city 'worth living in', including aspects of open spaces and their uses, about feeling at home and being proud about someone's city, as in a study carried out by Baur et al. (2010). Urban zoning and especially 'naturally grown' land use patterns that emerged without regulations, are certainly influencing the vision of a sustainable city (Aoki 2006). Also public spaces, shaped by different people in different times, are a basic hardware of any sustainable city and are increasingly researched (Hassenpflug 2000; Gaines and Jäger 2009). Furthermore

> Urban regeneration contributes to sustainable development by solving economic declining problems in the inner city, making contribution to the economic development and social equity, and shifting the urban sprawl pressure in suburban greenfields and thus facilitating the development of more compact cities (Chien-Yuan Lin 2007: 1).

One of the challenges of urban regeneration lies in the fact that only a redevelopment or renovation of the physical environment will not guarantee an economically successful project (Chien-Yuan Lin 2007). A key challenge for urban regeneration is to link neighbourhoods with their surroundings, and the city's overall vision and development plans. From a monitoring perspective, it is difficult to measure the 'softer' aspects of social sustainability like well-being, happiness and neighbourhood satisfaction arising from urban regeneration (Colantonio and Dixon 2011).

One of the four mega trends determining the future of the world that the 2014 World Risk Report mentions is cultural globalization (Mucke 2014a), fostering international urban trends and finally uniformity. On the contrary, the sustainability approach requires an individual analysis of urban development potentials, going beyond the universally accepted basic principles. In an article on sustainable urban development, Kopatz (2000) claims that there are no standardized solutions or panacea to achieve the goals of the same. Spaces vary and thus requirements have to be developed on an individual basis, considering spatial characteristics and the inhabitants' potentials (Koch 2001)—cities are human artefacts that reflect, as well as influence, their inhabitants (IHDP 2005). Gaines and Jäger (2009) claim that cities should learn from their past in the way that the urban building history forms part of a unique structure; cities are built witnesses to history depicting human nature and habits throughout times. Therefore, it is crucial to find a way to link urban past to present, as buildings are not just objects, but transformations of space through objects (Hillier and Hanson 1984).

It is widely agreed that "*old buildings carry traditional culture and historic memory and thus they have become important cultural capital for local development. Historic places and landscapes tell the story of collective memory*" (Chien-Yuan Lin 2007: 9). The social dimension in urban sustainability debate includes cultural diversity and social integration, while acknowledging the importance of physical environment, e.g. urban design and public spaces, thus also comprising cultural heritage (Polese and Stren 2000; Colantonio and Dixon 2011; Pelling 2011). Cultural heritage is a largely non-renewable resource. Therefore different authors include it in the sustainability debate

> Although new items can and will be added, they cannot replace existing treasures. Neither replicas nor reproductions can take the place of lost or irreversibly damaged property. Cultural property is threatened by dangers of both natural and human origin (Costin 1993: 27).

Like 'heritage', also 'sustainability' is a global concept that has to be scaled down to local context and adapted to cultural circumstances. Without any doubt, achieving 'sustainable cities' are of utmost importance (Gaines and Jäger 2009). However, 'sustainable cities' may have a whole variety of definitions—there is hardly any ongoing debate on urban planning that does not focus on 'sustainability'. Any planning approach is claimed to be sustainable, whether it is focusing on climate efficiency, green spaces, ecological buildings and many others. Increasingly, the recognition of built fabric and its preservation is winning recognition; the understanding of sustainable urban planning is no more directed towards only 'new' aspects. This somehow is a second shift in the sustainability debate, which for a long time was a non-urban one.

On a global scale, in September 2015, the Millennium Development Goals (MDGs) have been replaced by the Sustainable Development Goals (SDGs). Whereas urban issues were hardly included in the MDGs, with the exception of the aim to reduce the number of global slum dwellers, there is an urban SDG, Goal 11, 'Make Cities and Human Settlements inclusive, safe, resilient and sustainable': its targets include '11.4: strengthen efforts to protect and safeguard the world's cultural and natural heritage', and also '11.3: by 2030 enhance inclusive and sustainable urbanisation and capacities for participatory, integrated and sustainable human settlement planning and management in all countries' (Open Working Group on Sustainable Development Goals 2014). It remains to be seen how this promising approach is enforced.

## 4.4 Chapter Summary

Urban Regeneration
Since its appearance on a global scale in the 1850s the focal areas of urban regeneration have changed profoundly. Current trends of urban regeneration are mostly dealing with inserting landmark buildings in historic surroundings which are upgraded and renovated in the course of the regeneration project, e.g. in waterfront developments. Today, there is a whole variety of urban regeneration scales, ranging from small-scale projects to upgrade historic urban cores to large-scale and more economy-driven projects.

Urban regeneration in emerging and developing countries can be divided into three main phases

- Renovation and beautification until the mid-twentieth century,
- 'Modernization', until the mid-1980s, with a focus on traffic infrastructure and verticalisation and only little consideration of the centres, and a
- Redefinition and reconsideration of urban centres within the fragmented city since the 1990s

Regeneration does not necessarily consider historic buildings or intangible heritage, but increasingly their importance is recognized, including developing and emerging countries. Urban heritage and its contribution to urban sustainability are increasingly recognized.

## References

Abbate C (2010) A vicious circle. In: Plunz R, Sutto MP (eds) Urban climate change crossroads. Ashgate, Farnham, pp 129–134

Allen M (2014) The paris debate: must preservation inhibit urban renewal? ArchDaily. http://www.archdaily.com/524597/the-paris-debate-must-preservation-inhibit-urban-renewal/. Accessed 07 July 2014

Almandoz A (ed) (2002) Planning latin America's capital cities 1850–1950. Routledge, London

Angel S, Parent J, Civco DL, Blei AM (2011) Making room for a planet of cities. Lincoln Institute of Land Policy, Cambridge

Aoki Y (2006) Formulating sustainable systems. In: Tamagawa H (ed) Sustainable cities—Japanese perspectives on physical and social structures. United Nations University Press, Tokio, pp 50–69

Aragão T, Maennig W (2013) Mega sporting events, real estate, and urban social economics—the case of Brazil 2014/2016, vol 47. Hamburg Contemporary Economic Discussions, Hamburg

Ashworth GJ, Tunbridge JE (2000) The tourist-historic city. Retrospect and prospect of managing the heritage city. Adv Tourism Res Ser. Elsevier, Oxford

Atkinson R (2008) Introduction to this special issue on the future of European cities. Urban Res Pract (Special issue: the future of european cities) 1(3):217–221

Bähr J, Mertins G (1995) Die lateinamerikanische Groß-Stadt: Verstädterungsprozesse und Stadtstrukturen. Wissenschaftliche Buchgesellschaft, Darmstadt

Baker JL (2006) Integrated urban ugrading for the poor: the experience of Ribeira Azul, Brazil. Working paper number 3861. World Bank, New York. doi:10.1596/1813-9450-3861

Baur R, Feuz M, Derungs CG, Gmünder A, Hausheer T, Jann M, Krass P, Mv Lupin, Maag T, Tgetgel U, Zwissler M (2010) meine, deine schönste stadt der welt.Merkmale urbaner Lebensqualität. Lars Müller Publishers, Baden

Beatley T (2012) Sustainability in planning: the arc and trajectory of a movement, and new directions for the twenty-first-century city. In: Sanyal B, Vale LJ, Rosan CD (eds) Planning ideas that matter: livability, territoriality, governance, and reflective practice. MIT Press, Cambridge, pp 91–124

Beck H (2012) Understanding the impact of urban green space on health and wellbeing. In: Atkinson S, Fuller S, Painter J (eds) Wellbeing and place, vol Farnham. Ashgate, pp 35–52

Behr A (2005) Das Städtebauförderprogramm „Städtebaulicher Denkmalschutz“. Ein Instrument zur Erhaltung historischer Stadtzentren. Informationen zur Raumentwicklung 06/2005, Strategien für historische Stadtzentren:365–378

Bervoets W, Loopmans M (2013) The divisive nature of neoliberal urban renewal in Ouagadougou, Burkina Faso. In: Samara TR, He S, Chen G (eds) Locating right to the city in the global south. Routledge, Oxon, pp 60–80

Bigio AG (2003) Cities and climate change. In: Kreimer A, Arnold M, Carlin A (eds) Building safer cities: the future of disaster risk. The World Bank, Washington, D.C., pp 91–99

Birabi AK (2007) International urban conservation charters: catalytic or passive tools of urban conservation practices among developing countries. City Time 3(2)(4):39–53

Bittner R (2008) Cities of Perfection: Stadtfluchten oder Stadtversprechen. In: Wolfrum S, Nerdinger W (eds) Multiple city. Stadtkonzepte 1908|2008. Jovis, Berlin, pp 234–238

Bodenschatz H, Laible U (eds) (2007) Großstädte von morgen. Internationale Strategien des Stadtumbaus. Braun, Berlin

Böhme H, Körte A, Tokya-Seid M (eds) (2003) WOHNEN—BAUEN—PLANEN. Erneuerung historischer Kernstädte in Südostasien und Europa im historisch-architektonischen Vergleich, Wissenschaftliche Buchgesellschaft, Darmstadt

Brillantes ABJ, Flores HP (2012) Decentralization and good urban governance in Southeast Asia: focus on the Philippines, Indonesia, and Thailand. In: Yap KS, Thuzzar M (eds) Urbanisation in Southeast Asia: issues and impacts. Institute of Southeast Asian Studies, Singapore, pp 293–319

Bulkeley H (2010) Cities and governance. In: Plunz R, Sutto MP (eds) Urban climate change crossroads. Ashgate, Farnham, pp 29–38

Bundesministerium für Verkehr BuS (ed) (2007) Baukultur als Wachstumsimpuls: Gute Beispiele für europäische Städte. Ausgewählte Fallstudien. Bundesministerium für Verkehr, Bau und Stadtentwicklung, Berlin

Butler T, Hamnett C (2009) Regenerating a global city. In: Imrie R, Lees L, Raco M (eds) Regenerating London. Governance, sustainability and community in a global city, vol Reprint 2010. Routledge, Milton Park, pp 40–57

Camagni R (2000) Das urbane Milieu: Voraussetzung für Innovation und wirtschaftlichen Erfolg. In: Matejovski D (ed) Metropolen: Laboratorien der Moderne, vol Schriftenreihe., des Wissenschaftszentrums Nordrhein-Westfalen, Band 5. Campus Verlag, Frankfurt/New York, pp 292–307

Carmona M, Wunderlich FM (2012) Capital spaces: the multiple complex public spaces of a global city. Routledge, London and New York

Carpenter J (2011) Integrated Urban Regeneration and sustainability: approaches from the European Union. In: Colantonio A, Dixon T (eds) Urban regeneration & social sustainability. Best practice from European cities. Wiley-Blackwell, Chichester, West Sussex, pp 83–101

Chien-Yuan Lin J (2007) Urban regeneration and sustainable development. Nature-conservation-led approach for Green Island in Taiwan. Lincoln Institute of Land Policy, Cambridge

Choguill CL, Choguill MBG (1996) sustainable infrastructure for low-income communities. In: Pugh C (ed) Sustainability, the environment and urbanisation, vol Neudruck 2002. Earthscan, London, pp 84–102

Colantonio A, Dixon T (2011) Urban regeneration & social sustainability. Best practice from European cities. Real Estate Issues. Wiley-Blackwell, Chichester, West Sussex

Costin CL (1993) Legal and policy issues in the protection of cultural heritage in South Asia and the Pacific. In: Mac Lean MGH (ed) Cultural heritage in Asia and the Pacific: conservation and policy. Proceedings of a symposium held in Honolulu, Hawaii, September 8-I3, I99I. The Getty Conservation Institute, Los Angeles, pp 27–62

Council of Europe, Directorate of Culture and Cultural and Natural Heritage (2004) Guidance on urban rehabilitation. Council of Europe Publishing, Strasbourg

Coy M (2006) Gated communities and urban fragmentation in Latin America: the Brazilian experience. GeoJournal 66:121–132

Coy M (2007) Innenstadtentwicklung und Innenstadterneuerung in São Paulo—Akteure, Wahrnehmungen, Interessenkonflikte. In: Rothfuß E, Gamerith W (eds) Stadtwelten in den Americas., vol Passauer Schriften zur Geographie 23. Passavia Universitätsverlag, Passau, pp 57–69

Coy M (2010) Stadtentwicklung und Stadtpolitik. Sozioökonomische Fragmentierung und Beispiele zukunftsorientierter Planung. In: Costa S, Kohlhepp G, Nitschack H, Sangmeister H (eds) Brasilien heute: Geographischer Raum—Politik—Wirtschaft—Kultur. Vervuert, Frankfurt am Main, pp 51–73

Coy M, Töpfer T (2014) Inner-City development in megacities between degradation and renewal: the case of São Paulo. In: Kraas F, Aggarwal S, Coy M, Mertins G (eds) Megacities: our global urban future. Springer, Dordrecht Heidelberg New York London, pp 101–120

Daase M (1995) Prozesse der Stadterneuerung in innenstadtnahen Wohngebieten am Beispiel von Hamburg-Ottensen. In: Nagel N (ed) Stadtentwicklung und Stadterneuerung. Hamburg—London—Singapur. vol Mitteilungen der Geographischen Gesellschaft in Hamburg, Band 85. Franz Steiner, Stuttgart, pp 3–142

Dahiya B (2012) Cities in Asia, 2012: Demographics, economics, poverty, environment and governance. Cities Volume 29. Supplement 2:S44–S61

de Albuquerque Lapa T, Almeida de Melo RF (2007) Interventions in historic areas, mobility and urban conservation: the case study of Bairro do Recife District. City Time 2(3–4):37–46

de Melo Cadena, Filho DR (2012) Para além da fisionomia—identifição da paisagem cultural do centro do Recife Universidade Federal de Pernambuco. UFPE, Recife

Drakakis-Smith D (2000) Third world cities. Routledge introductions to development series, 2nd edn. Routledge, New York

Dziomba M (2007) Städtebauliche Grossprojekte der Urbanen Renaissance. disP 171(4):12–24

Dziomba M, Matuschewski A (2007) Grossprojekte in der Stadtentwicklung—Konfliktbereiche und Erfolgsfaktoren. disP 171 (4):5–11

Eeckhout B, Jacobs S (2008) (Dys)Functionalism in a post(sub)urban landscape. In: Wolfrum S, Nerdinger W (eds) Multiple city. Stadtkonzepte 1908| 2008. Jovis, Berlin, pp 34–37

Ercan MA (2011) Challenges and conflicts in achieving sustainable communities in historic neighbourhoods of Istanbul. Habitat Int 35(2):295–306. doi:10.1016/j.habitatint.2010.10.001

Furtado F, Alcântara E, Bezerra O (2014) Is urban sustainability possible in the face of accelerated property development and major public works? In: Edgerton E, Romice O, Thwaites K (eds) Bridging the boundaries: human experience in the natural and built environment and implications for research, policy, and practice, vol 5., Advances in people-environment studiesHogrefe, Boston, pp 245–260

Gaebe W (2004) Urbane Räume. Eugen Ulmer GmbH & co., Stuttgart

Gaines J, Jäger S (2009) Albert speer & partner. Ein Manifest für nachhaltige Stadtplanung. Think Local, Act Global. Prestel, München

García Canclini N (2013) Zur Metamorphose der lateinamerikanischen Stadtanthropologie. In: Huffschmid A, Wildner K (eds) Stadtforschung aus Lateinamerika: Neue urbane Szenarien: Öffentlichkeit—Territorialität—Imaginarios. Transcript, Bielefeld, pp 33–43

Garschagen M (2014) Urbanisation and risk—challenges and opportunities In: Bündnis Entwicklung Hilft (Alliance Development Works) and United Nations University—Institute for Environment and Human Security (UNU-EHS) (ed) world risk report 2014—focus: the city as a risk area. Bündnis Entwicklung Hilft (Alliance Development Works) and United Nations University—Institute for Environment and Human Security (UNU-EHS), Berlin/Bonn, pp 12–18

Gerhard U (2012) Reurbanisierung—städtische Aufwertungsprozesse in der Global city-Perspektive. In: Brake K, Herfert G (eds) Reurbanisierung: Materialität und Diskurs in Deutschland. Springer, Wiesbaden, pp 52–68

German Institute of Urban Affairs (Difu), Federal Institute for Research on Building; Urban Affairs and Spatial Development within the Federal Office for Building and Regional Planning (eds) (2012) 5 Years after the LEIPZIG CHARTER—Integrated urban development as a prerequisite for a sustainable city. Federal Ministry of Transport, Building and Urban Development, Berlin

Goldman M (2011) Speculating on the next world city. In: Roy A, Ong A (eds) Worlding cities: Asian experiments and the art of being global. Studies in urban and social change, Wiley-Blackwell, Chichester, pp 229–258

Gosseye J (2009) Mumbai's maidans. From fields of fire to non-places. In: Shannon KG, Jannina (ed) Reclaiming (the Urbanism of) Mumbai. Explorations in/of Urbanism. Sun Publishers, Amsterdam, pp 122–134

Government of Nepal; Department of Archaeology (2007) Kathmandu world heritage site: integrated management framework. Government of Nepal, ministry of culture, tourism and civil aviation, department of archaeology in close collaboration with the world heritage centre and UNESCO-Kathmandu Office, Kathmandu

Gunay Z, Dokmeci V (2012) Culture-led regeneration of Istanbul waterfront: golden horn cultural valley project. Cities 29(4):213–222. doi:10.1016/j.cities.2011.08.010

Hack G (2012) Shaping urban form. In: Sanyal B, Vale LJ, Rosan CD (eds) Planning ideas that matter: livability, territoriality, governance, and reflective practice. MIT Press, Cambridge, pp 34–63

Haines C (2011) Cracks in the Façade: landscapes of hope and desire in Dubai. In: Roy A, Ong A (eds) Worlding cities: Asian experiments and the art of being global. Studies in urban and social change. Wiley-Blackwell, Chichester, pp 160–181

Hardoy JE, Mitlin D, Satterthwaite D (1992) Environmental problems in third world cities. Earthscan, London

Hassenpflug D (2000) Citytainment oder die Zukunft des öffentlichen Raums. In: Matejovski D (ed) Metropolen: Laboratorien der Moderne, vol Schriftenreihe., des Wissenschaftszentrums Nordrhein-Westfalen, Band 5. Campus Verlag, Frankfurt/New York, pp 308–323

Heeg S (2008) Von Stadtplanung und Immobilienwirtschaft. Die 'South Boston Waterfront' als Beispiel für eine neue Strategie städtischer Baupolitik. transcript, Bielefeld

Herrle P, Jachnow A, Ley A (2006) Die Metropolen des Südens: Labor für Innovationen? Mit neuen Allianzen zu besserem Stadtmanagement. Stiftung Entwicklung und Frieden, vol Policy Paper 25. Bonn. doi:ISSN 1437–2800

Hiernaux D (2013) Die historischen Stadtzentren Lateinamerikas: Auf dem Weg zu einer kreolischen Gentrifizierung? In: Huffschmid A, Wildner K (eds) Stadtforschung aus Lateinamerika: Neue urbane Szenarien: Öffentlichkeit—Territorialität—Imaginarios. Transcript, Bielefeld, pp 377–395

Hillier B, Hanson J (1984) The social logic of space. Cambridge University Press, Cambridge

IHDP (2005) Science plan—urbanisation and global environmental change. IHDP, Bonn

International Federation of Red Cross and Red Crescent Societies (2014) World Disasters Report 2014—Focus on Culture and Risk. International Federation of Red Cross and Red Crescent Societies, Geneva. doi:ISBN 978-92-9139-214-8

Jacobs J (1961) The death and life of great American Cities. Random House, New York

Jones S, Carswell G (2004) The earthscan reader in environment, development & rural livelihood. Earthscan, London

Kamarid A, Leupert D (2009) Stadterneuerung am Praça XV de Novembro—Entwicklungskonzept für einen Stadtplatz in Rio de Janeiro. VDM Verlag Dr, Müller, Saarbrücken

Keitumetse S (2009) Methods for investigating locals' perceptions of a cultural heritage product for tourism. Lessons from Botswana. In: Stig Sørensen ML, Carman J (eds) Heritage studies. Methods and Approaches. Routledge, London, New York, pp 201–216

Klotzhuber I (1995) Die Isle of Dogs in den Londoner Docklands. Management und Zukunft eines derelikten innenstadtnahen Hafengebietes. In: Nagel N (ed) Stadtentwicklung und Stadterneuerung. Hamburg—London—Singapur. vol Mitteilungen der Geographischen Gesellschaft in Hamburg, Band 85. Franz Steiner, Stuttgart, pp 143–290

Knirsch J (2004) Stadtplätze. Architektur und Freiraumplanung, Alexander Koch, Leinfelden-Echterdingen

Koch M (2001) Ökologische Stadtentwicklung: Innovative Konzepte für Städtebau, Verkehr und Infrastruktur. Kohlhammer, Stuttgart; Berlin; Köln

Kopatz M, Hennicke P (2000) Zukunftsfähige Stadtentwicklung: Realistisches Leitziel oder Beschönigungsformel? In: Matejovski D (ed) Metropolen: Laboratorien der Moderne, vol Schriftenreihe., des Wissenschaftszentrums Nordrhein-Westfalen, Band 5. Campus Verlag, Frankfurt/New York, pp 223–247

Kraas F (2005) Urban sustainability—sustainability in urban areas: basic considerations. In: Gaese H, F. Kraas, Mi Mi Kyi (eds.) Sustainability in rural and urban environments. Proceedings of the first German-Myanmar workshop in Yangon/Myanmar, 17–21 November 2003. Cologne, pp 31–42

Kraas F (2010) Urbanisierung als weltweite Herausforderung. In: Debiel T, Messner D, Nuscheler F, Roth M, Ulbert C (eds) Globale Trends 2010. Frieden, Entwicklung, Umwelt. Stiftung Entwicklung und Frieden, Frankfurt, pp 181–199

Kraas F, Nitschke U (2008) Megaurbanisierung in Asien: Entwicklungsprozesse und Konsequenzen stadträumlicher Reorganisation. Informationen zur Raumentwicklung Themenheft „Raum- und Stadtentwicklung in Asien" (8-2008):447-456

Krieger A (2008) Dynamik und Management von Waterfront Development. In: „Waterfront Development"—Dokumentation der internationalen Städtekonferenz im Juni 2007 in Wien, Urania/TechGate Wien, 2008 2007. Stadtentwicklung Wien, pp 27–32

Leipzig Charter on Sustainable European Cities (2007) EU. http://ec.europa.eu/regional_policy/archive/themes/urban/leipzig_charter.pdf. Accessed 23 July 2014

Lichtenberger E (1990) Stadtverfall und Stadterneuerung. VÖAW, Wien

Loeckx A (2009) Towards an urban debate. Questioning mumbai's voices & issues. In: Shannon KG, Jannina (ed) Reclaiming (the Urbanism of) Mumbai. Explorations in/of Urbanism. Sun Publishers, Amsterdam, pp 179–193

Luger K (2008) Welterbe-Tourismus. Ökonomie, Ökologie und Kultur in weltgesellschaftlicher Verantwortung. In: Luger K, Wöhler K (eds) Welterbe und Tourismus: Schützen und Nützen aus einer Perspektive der Nachhaltigkeit. StudienVerlag, Innsbruck, pp 17–42

Mallach A, Brachman L (2013) Regenerating America's legacy cities. Policy Focus Report Series, Lincoln Institute of Land Policy, Cambridge

Mars N (2008) Cities without history. In: Wolfrum S, Nerdinger W (eds) Multiple city. Stadtkonzepte 1908| 2008. Jovis, Berlin, pp 274–278

Martin I (2015) The city that privatised itself to death: 'London is now a set of improbable sex toys poking gormlessly into the air' The Guardian (24022015)

Mayr A (2011) Auf dem Weg zur Weltstadt: Johannesburg und Dehli zwischen globaler Anschlussfähigkeit und Selbstfindung. In: Löw M, Terizakis G (eds) Städte und ihre Eigenlogik. Ein Handbuch für Stadtplanung und Stadtentwicklung, vol Interdisziplinäre Stadtforschung, Band 11. Interdisziplinäre Stadtforschung. Campus Verlag, Frankfurt am Main, pp 189–201

Mehotra R (2009) Kinetic city. Issues for urban design in South Asia. In: Shannon KG, Jannina (ed) Reclaiming (the Urbanism of) Mumbai. Explorations in/of Urbanism. Sun Publishers, Amsterdam, pp 142–152

Mertins G (2003) Städtetourismus in Havanna (Kuba). Geographische Rundschau 55(3):20–25

Michel B (2010) Global City als Projekt: Neoliberale Urbanisierung und Politiken der Exklusion in Metro Manila. Transcript, Bielefeld

Mucke P (2014a) Political challenges and perspectives. In: Bündnis Entwicklung Hilft (Alliance Development Works) and United Nations University—Institute for Environment and Human Security (UNU-EHS) (ed) World Risk Report 2014—Focus: The city as a risk area. Bündnis Entwicklung Hilft (Alliance Development Works) and United Nations University—Institute for Environment and Human Security (UNU-EHS), Berlin/Bonn, pp 53–62

Mucke P (2014b) Urbanisation—trends and risk assessment. In: Bündnis Entwicklung Hilft (Alliance Development Works) and United Nations University—Institute for Environment and Human Security (UNU-EHS) (ed) World Risk Report 2014—Focus: The city as a risk area. Bündnis Entwicklung Hilft (Alliance Development Works) and United Nations University—Institute for Environment and Human Security (UNU-EHS), Berlin/Bonn, pp 5–10

Musterd S, Ostendorf W (2008) Integrated urban renewal in The Netherlands: a critical appraisal. Urban Research & Practice 1(1):78–92. doi:10.1080/17535060701795389

Nobre EAC (2002) Urban regeneration experiences in Brazil: Historical preservation, tourism development and gentrification in Salvador da Bahia. Urban Design International:109–124. doi:10.1057/palgrave.udi.9000066

Oliva i Casas J (2007) Confusion in Urban Design: the public city versus the domestic city. Techne Press, Amsterdam

Ong A (2011a) Hyperbuilding: Spectacle, Speculation, and the Hyperspace of Sovereignity. In: Roy A, Ong A (eds) Worlding Cities: Asian experiments and the art of being global. Studies in urban and social change. Wiley-Blackwell, Chichester, pp 205–226

Ong A (2011b) Worlding cities, or the art of being global. In: Roy A, Ong A (eds) Worlding Cities: Asian experiments and the art of being global. Studies in urban and social change, Wiley-Blackwell, Chichester, pp 1–26

Open Working Group on Sustainable Development Goals (2014) Outcome Document—Open Working Group on Sustainable Development Goals. United Nations Department of Economic and Social Affairs, Division for Sustainable Development. http://sustainabledevelopment.un.org/focussdgs.html. Accessed 02 Sept 2014

Pelling M (2011) Adaptation to Climate Change: From resilience to transformation. Routledge, London and New York

Petrella L (2010) Urban Africa—challenges and opportunities for planning at a time of climate change. In: Gossop C (ed) Sustainable city/developing world, vol Isocarp Review 06. Routledge, pp 52–76

Peyroux E, Pütz R, Glasze G (2012) Business Improvement Districts (BIDs): the internationalization and contextualization of a 'travelling concept'. European Urban and Regional Studies 19(2):111–120. doi:10.1177/0969776411420788

Plaza B, González-Casimiro P, Moral-Zuazo P, Waldron C (2015) Culture-led city brands as economic engines: theory and empirics. Ann Reg Sci 43(1):179–196. doi:10.1007/s00168-014-0650-0

Plazaab B, Tironic M, Haarichd SN (2009) Bilbao's art scene and the "Guggenheim effect" Revisited. Eur Plan Stud (17:11):1711–1729. doi:10.1080/09654310903230806

Polese M, Stren R (eds) (2000) The social sustainability of cities: diversity and the management of change. scholarly publishing division; 1, edition edn. University of Toronto Press, Toronto

Polinna C (2008) Learning from London. In: Wolfrum S, Nerdinger W (eds) Multiple city. Stadtkonzepte 1908|2008. Jovis, Berlin, pp 94–97

Pontual V (2007) Prácticas urbanísticas em áreas históricas: o bairro de Recife. Revista Bibliográfica de Geografía y Ciencias Sociales, vol XII, nº 752. Universidad de Barcelona, Barcelona

Poytner G (2009) The 2012 Olympic games and the reshaping of East London. In: Imrie R, Lees L, Raco M (eds) Regenerating London. Governance, sustainability and community in a global city, vol Reprint 2010. Routledge, Milton Park, pp 132–148

Proença Leite R (2006) Patrimônio e enobrecimento no Bairro do Recife. Revista CPC 2:17–30. doi:10.1590/S1980-44662006000200003+

Proença Leite R (2007) Contra-usos da cidade: lugares e espaço público na experiência urbana contemporânea. vol 2. ed., ver. e ampl. edn. Ed. da UNICAMP/ Ed. UFS, Campinas/Aracaju

Pugh C (1996) Sustainability and sustainable cities. In: Pugh C (ed) Sustainability, the environment and urbanisation, vol Neudruck 2002. Earthscan, London, pp 135–177

Pütz M, Rehner J (2007) Macht in konfliktreichen Grossprojekten der Stadtentwicklung: Revitalisierung des Hafens Puerto Madero in Buenos Aires. disP 171 (4):36–49

Rautenberg M (2011) Industrial heritage, regeneration of cities and public policies in the 1990s: elements of a French/British comparison. International Journal of Heritage Studies 18(5):513–525. doi:10.1080/13527258.2011.637945

Ribbeck E (2005) Die Welt wird Stadt: Stadtbilder aus Asien. Afrika und Lateinamerika, Jovis, Berlin

Roberts P, Sykes H (eds) (2000) Urban regeneration. A Handbook, Sage, London

Robrecht H (2000) Nachhaltige Stadtentwicklung—ein systematischer Rahmen. In: Matejovski D (ed) Metropolen: Laboratorien der Moderne, vol Schriftenreihe., des Wissenschaftszentrums Nordrhein-Westfalen, Band 5. Campus Verlag, Frankfurt/New York, pp 260–277

Rodenstein M (2010) Forgetting and Remembering: Frankfurt's *Altstadt* after the Second World War. In: Fenster T, Yacobi H (eds) Remembering, forgetting and city builders re-materialising cultural geography. Ashgate, Farnham, UK, pp 159–174

Rojas E (2002) Urban heritage conservation in Latin America and the Carribean. A task for all social actors. Sustainable development department technical papers series:25

Ronneberger K (2000) Container des 21. Jahrhunderts. In: Matejovski D (ed) Metropolen: Laboratorien der Moderne, vol Schriftenreihe des Wissenschaftszentrums Nordrhein-Westfalen, vol 5. Campus Verlag, Frankfurt/New York, pp 324–342

Rothfuß E (2007) Tourismus-Gentrification im Pelourinho. Urbane Deformation des historischen Stadtzentrums von Salvador da Bahia (Brasilien). Passauer Schriften zur Geographie 23:41–56

Roy A (2011) The blockade of the world-class city: dialectical images of Indian urbanism. In: Roy A, Ong A (eds) Worlding cities: Asian experiments and the art of being global. Studies in urban and social change. Wiley-Blackwell, Chichester, pp 259–278

Rubens de Menezes T, Figueira de Souza J (2014) Transportation and urban mobility in mega-events: the case of Recife. Procedia—Social and Behavioral Sciences 162:218–227. doi:10.1016/j.sbspro.2014.12.202

Ruland R (2011) Die Rolle des historische Erbes in der Stadtentwicklung in Deutschland. Informationen zur Raumentwicklung 3/4 (Denkmalschutz und Stadtentwicklung):183–191

Salesses P, Schechtner K, Hidalgo CA (2013) The collaborative image of the city: mapping the inequality of urban perception. PLoS ONE 8(7). doi:10.1371/journal.pone.0068400

Sassen S (2000) Über die Auswirkungen der neuen Technologien und der Globalisierung auf die Städte. In: Matejovski D (ed) Metropolen: Laboratorien der Moderne, vol Schriftenreihe., des Wissenschaftszentrums Nordrhein-Westfalen, Band 5. Campus Verlag, Frankfurt/New York, pp 29–50

Satterthwaite D (1999) Sustainable cities or cities that contribute to sustainable development? In: Satterthwaite D (ed) The Earthscan Reader in Sustainable Cities. Earthscan, London, pp 80–106

Satterthwaite D, Dodman D (2009) The role of cities in climate change. In: Institute W (ed) State of the world 2009: into a warming world. Norton Washington DC, pp 75–77

Schauber A (2014) Light and dark—citizens and invisible city-dwellers. In: Bündnis Entwicklung Hilft (Alliance Development Works) and United Nations University—Institute for Environment and Human Security (UNU-EHS) (ed) World Risk Report 2014—focus: the city as a risk area. Bündnis Entwicklung Hilft (Alliance Development Works) and United Nations University—Institute for Environment and Human Security (UNU-EHS), Berlin/Bonn, pp 18–24

Secretariat of the Convention on Biological Diversity (2012) Cities and Biodiversity Outlook. Secretariat of the Convention on Biological Diversity, Montreal

Shaftoe H (2008) Convivial urban spaces. Earthscan, London
Sharifi A, Murayama A (2013) Changes in the traditional urban form and the social sustainability of contemporary cities: A case study of Iranian cities. Habitat Int 38:126–134
Shatkin G (2011) Planning privatopolis: representation and contestation in the development of urban integrated mega-projects. In: Roy A, Ong A (eds) Worlding cities: Asian experiments and the art of being global. Studies in urban and social change. Wiley-Blackwell, Chichester, pp 77–97
Silomon-Pflug F, Stein C, Heeg S, Pütz R (2013) Urban policy mobility studies and the entrepreneurial city. Geographische Zeitschrift 101:201–217
Simone A (2010) City Life from Jakarta to Dakar. Movements at the Crossroads. Routledge, New York
Siu HF (2011) Retuning a Provincialized middle class in asia's urban postmodern: the case of Hong Kong. In: Roy A, Ong A (eds) Worlding cities: Asian experiments and the art of being global. Studies in urban and social change. Wiley-Blackwell, Chichester, pp 129–159
Smith J (2013) Cultural landscape theory and practice: moving from observation to experience. In: Albert M-T, Bernecker R, Rudolff B (eds) Understanding heritage: perspectives in heritage studies. de Gruyter, Berlin/Boston, pp 49–66
Steinbrink M, Haferburg C, Ley A (2011) Festivalisation and urban renewal in the Global South: socio-spatial consequences of the 2010 FIFA World Cup. S Afr Geogr J 93(1):15–28. doi:10.1080/03736245.2011.567827
Stern MJ, Seifert SC (2007) Cultivating "natural" cultural districts. Creativity and neighborhood development: strategies for community investment. University of Pennsylvania, Philadelphia
Stadt Stiftung Lebendige (ed) (2005) Stadt am Wasser: Neue Chancen für Kommunen und Tourismus, vol 4. Schriftenreihe Lebendige Stadt, Frankfurter Societäts-Druckerei GmbH, Frankfurt/Main
Tiesdell S, Oc T, Heath T (1996) Revitalizing historic urban quarters. Architectural Press, Oxford
Tiranishti J, Gjoklaj E (2009) Potential in Mumbai's post-industrial waste landscapes. In: Shannon KG, Jannina (ed) Reclaiming (the Urbanism of) Mumbai. Explorations in/of Urbanism. Sun Publishers, Amsterdam, pp 160–178
Tokya-Seid M (2003) Bristol-Köln. In: Böhme H, Körte A, Tokya-Seid M (eds) WOHNEN—BAUEN—PLANEN. Erneuerung historischer Kernstädte in Südostasien und Europa im historisch-architektonischen Vergleich, Wissenschaftliche Buchgesellschaft, Darmstadt, pp 19–74
Truffi R (2014) A batalha pelo Cais José Estelita. Carta Capital, vol 2014. Recife
Trzyna T (2014) Urban protected areas: profiles and best practice guidelines. Best practice protected area guidelines series No. 22. IUCN, Gland
Tweeddale A (2012) Wang Shu: "Memory is deeper than symbols". Architectural Review Asia Pacific (127)
Twickel C (2010) GENTRIFIDINGSBUMS oder eine Stadt für Alle. Nautilus, Hamburg
UN-HABITAT (2013) Streets as Public Spaces and Drivers of Urban Prosperity. UN-Habitat, Nairobi
UN-HABITAT (2014) The State of African Cities 2014. Re-imagining sustainable urban transitions. UN-Habitat, Nairobi
UNECE (2009) Self-made Cities. In search of sustainable solutions for informal settlements in the United Nations Economic Commission for Europe region. United Nations, Geneva
UNESCO Institute for Statistics (2012) Measuring cultural participation. 2009 Framework for Cultural Statistics Handbook No. 2. UNESCO-UIS, Montreal
Vega CA, Braig M (2013) Wem gehört die Altstadt? Straßenhandel, Stadtpolitik und 'Globalisierung des Lokalen' in Mexiko-Stadt. In: Huffschmid A, Wildner K (eds) Stadtforschung aus Lateinamerika: Neue urbane Szenarien: Öffentlichkeit—Territorialität—Imaginarios. Transcript, Bielefeld, pp 397–417
Vinken G (2010) Zone Heimat. Deutscher Kunstverlag, Berlin, München, Altstadt im modernen Städtebau

von Beyme K, Durth W, Gutschow N, Nerdinger W (eds) (1992) Neue Städte aus Ruinen: Deutscher Städtebau der Nachkriegszeit. Prestel, München

Vorlaufer K (2009) Südostasien. WBG, Darmstadt

Winston N (2009) Urban regeneration for sustainable development: The Role of Sustainable Housing? European Planning Studies 17 (12 December 2009):1781–1796. doi:10.1080/09654310903322306

Yeo HK, Han SM (2012) Heritage conservation as urban regeneration policy in globalizing cities: social exclusion and gentrification at the vicinity of the Jongmyo World Heritage site in Seoul, Korea. In: Negussie E (ed) Changing World, Changing Views of Heritage: heritage and social change. Proceedings of the ICOMOS Scientific Symposium 2010. ICOMOS, Paris, pp 56–62

Zancheti SM (2005) Development versus urban conservation in Recife a problem of governance and public management. City Time 1(3)(2):15–27

Zheng HW, Shen GQ, Wang H (2014) A review of recent studies on sustainable urban renewal. Habitat Int 41:272–279. doi:10.1016/j.habitatint.2013.08.006

# Chapter 5
# Heritage and Identities in Selected Urban Centres

**Abstract** The cities of Kathmandu, Nepal, Yogyakarta, Indonesia and Recife, Brazil were analyzed based up on their urban development, drivers of change, urban and heritage policies, as well as the phases of urban regeneration. In all three cities the urban outline is changing, induced by rapid urban change. The centres, however, still comprise of heritage buildings and places, often attached to intangible values still appreciated by the urban population. The comparative analysis of the case studies reveals major impact clusters influencing on the urban centre. Common problems of conservation legal frameworks are overlaps or contradictions between documents issued on different administrative levels or by different authorities and outdated contents. In addition, the execution of the legal framework in force is often poor and hardly reflecting interlinkages of tangible and intangible heritage. The attachment to the historic urban core is still strong in all cases. The centres are of importance, the same applies to distinct places of remembrance and particularly to different forms of intangible heritage. However, the overall awareness and recognition of intangible values seems comparably fuzzier than in the case of tangible heritage and may result in potential losses. To develop suitable strategies for regenerating the historic centres the triggers of urban change as well as their interlinkages have to be taken into account.

**Keywords** Kathmandu · Nepal · Yogyakarta · Indonesia · Recife · Brazil · Urban heritage · Historic centre · Place attachment · Comparative analysis · Urban regeneration · Heritage legislation

In this section the results of the empirical research will be presented. As elaborated in the introductory chapter overall three cities or, more explicitly, their historic centres, were analyzed: Kathmandu in Nepal, Yogyakarta in Indonesia and Recife in Brazil.

The three cities have been chosen purposefully for different reasons. Table 5.1 gives an overview on some key facts. Located in the Global South, all three still comprise of historic centres with fabric built in past epochs. However, in all cases decay processes were or are still visible in the centres and urban authorities tried and are trying to counteract this by means of urban regeneration projects.



S. Sandholz, *Urban Centres in Asia and Latin America*,
The Urban Book Series, DOI 10.1007/978-3-319-43735-4_5

**Table 5.1** Key facts of the Case Study Cities

| | Kathmandu, Nepal | Yogyakarta, Indonesia | Recife, Brazil |
|---|---|---|---|
| |  |  |  |
| Population and administration | • 1.18 million inhabitants (Metropolitan Area)<br>• 5 municipalities<br>• Largest urban agglomeration in country<br>• Capital City of Nepal,<br>• Initially 3 urban centres | • 3.7 million inhabitants (Special Region), City of Yogyakarta 511.000<br>• 4th largest urban agglomeration in country<br>• 3 cities in "Kartamantul" metropolitan area, Yogyakarta 14 municipalities<br>• Capital City of Special Province, Capital of Indonesia 1945-49 | • 3.8 million inhabitants (Metropolitan Area)<br>• 5th largest urban agglomeration in country<br>• Metropolitan area with 17 municipalities<br>• Capital City of Pernambuco State<br>• „twin-city „Recife-Olinda |
| Climate and location | • humid subtropical/subtropical highland climate, monsoon June – August<br>• City area 50.67 km², average elevation 1,350 m above sea level<br>• Located in fertile valley along main trade route from China to Tibet<br>• Today center of Newari/Nepalese culture and university town | • Tropical monsoon climate, rainy period October – June<br>• City area 32.5 km²<br>• Located at prominent place between Merapi Volcano and sea<br>• Today center of Javanese culture and university town | • Tropical climate, rainy period March – July<br>• City area 218.4 km²<br>• Located at strategically important place near westernmost point of Brazil and sugarcane plantations<br>• Center of Nordestino Culture, university hub |
| History | • Independent country, never colony but British legate in Kathmandu<br>• Hindu kingdom until 2008, 1990s-2008 civil war, since them parliamentary democracy | • Officially founded 1755 as Sultanate, under Dutch Colony<br>• Still Sultanate until today with Sultan als head of province | • Portuguese Colony from 1537 on, 1630-54 under Dutch forces<br>• Traditional link between Brazil and Europe |
| Urban structure/ monuments | • Hindu-Buddhist urban layout(s),<br>• Initially 3 urban centers & kingdoms in the Kathmandu Valley, now merged to one agglomeration<br>• distinctive buildings/urban layout from 14-15th century | • Hindu-Buddhist urban layout<br>• Historic center with palace, preserved ensembles and quarters from Dutch colonial times, Javanese and Dutch style<br>• Main monuments: palace area, Dutch fort, historic quarters | • Twin-city Recife and Olinda, with Recife having strong Dutch influences visible until today,<br>• Many historic churches (baroque) and buildings (partly younger, from 19th century in Bairro do Recife), fortifications |
| Legislation | • Different legislations for UNESCO zones and others, ensembles and cultural landscapes difficult to preserve<br>• Lacking: coordination among authoritities to stop illegal activities | • Legislation for single buildings, ensembles difficult to preserve,<br>• Lacking: legislation for „cultural landscapes" and laws adapted to Javanese context | • Legislation for single buildings, ensembles and heritage zones,<br>• Law to protect intangible values<br>• Lacking: political will to execute laws and allocate budgets |

All three case studies were investigated based on the same methodology and are described by following the same structure in the chapters below. Each case will go from large-scale—national and urban level—to medium- and small-scale—the centre and its different parts and attached values. At first, a short introduction into the national and urban development history of the respective country will be given, which is followed by an introduction to the case study city itself, focussing on a detailed description of its 'historic urban landscape' assets. In this context the phrase 'historic urban landscape' refers to the UNESCO definition and seeks to provide a holistic overview of culture, planning history and the current state of urban heritage, tangible and intangible one. While this overview is done based on a comprehensive literature research the subsequent sections include results of the empirical research, starting with an analysis of the urban drivers of change which is based on the expert interview outcomes as well as on own surveys.

Subsequently urban and heritage policies as well as the phases of urban regeneration in the case study cities are assessed based on the findings of literature research and expert interviews. Processes, patterns and dynamics of the inner city are set out in detail on the basis of an own survey, expert interviews and the questionnaire results. In the light of this analysis, then the importance of tangible and intangible assets to the peer group is assessed and compared to the experts' statements before concluding on the role of heritage in shaping urban identity in the case study city.

In the last chapter, the results obtained in Kathmandu, Yogyakarta and Recife are compared with one another to draw conclusions on similarities and differences, in the legal framework, in urban regeneration approaches, and in the importance of the historic centre for the development of place attachment and an urban identity.

## 5.1 Case Study I: Kathmandu, Nepal[1]

The empirical analysis investigated the perception of Kathmandu in general and the historic city centre in particular. Methods used for this research include own field surveys as well as literature and documents survey in relation to urban planning in Kathmandu and its heritage, such as legislations, international surveys carried out by organizations like UNESCO, and various reports.

In Kathmandu overall twelve expert interviews were carried out, with experts from urban planning, conservation, academia, and administration. The in-depth interviews took between 30 min and 1.5 h and were recorded, transcribed and afterwards evaluated with MAXQDA software, using the same coding for all interviews and case studies. Interview partners do not appear with their full names but with codes from K01 to K12 in the text. Additionally a list is given in Annex II. Interviewee K01 is working with the Kathmandu Valley Preservation Trust, one of the major conservation NGOs of the country. K02 is program manager for another NGO, the National Society for Earthquake Technology which is concerned with urban development as well. Overall four interviewees work for different authorities: K02 and K12 are responsible for different urban development aspects in the Department of Urban Development and Building Construction, K04 is employed at Kathmandu Valley Development Authority while K05 is a high-ranking official of the Department of Archaeology. Four more interviewees add the view of academia and private sector; all four are working as assistant or senior professors for Tribhuvan University—the largest university of the country—in the fields of geography, engineering, architecture and urban planning or conservation. At the same time all are working or have worked as private consultant in their sectors

[1]This chapter was completed before Kathmandu was hit by two major earthquakes in April/May 2015. It was then decided to leave the chapter as it is, but to add a reflection on the earthquakes' impact on Kathmandu's historic centre in Chap. 8.

**Table 5.2** Period the questionnaire respondents are already residing in Kathmandu ($n$ = 80)

| Grown up in Kathmandu | For more than 5 years | Between 2 and 5 years | Between 6 months and 2 years | Less than 6 months | No answer |
|---|---|---|---|---|---|
| 35.0 % | 45.0 % | 13.8 % | 0.0 % | 0.0 % | 6.3 % |

as well. Interviewee is a shop owner and businessman in Kathmandu while K11 is a renowned journalist publishing on urban and social concerns on regular intervals.

In addition, questionnaires served as data sources for this case study. They were handed to postgraduate students enrolled in Master Programs on various topics with a link to sustainability, either from a social, a planning or an ecological perspective; all of them enrolled at Tribhuvan University. Overall 80 responses were collected. All students were of Nepali citizenship, around one third of them being from Kathmandu originally and another 45 % already living there for more than 5 years' time (see Table 5.2). All respondents were residing in Kathmandu when filling the questionnaire. Age distribution ranged from 22 to 45, with an average age of 28.2 years.

In the questionnaire some questions focussed on the 'centre area', which in this case was defined as the Kathmandu Durbar Square heritage area and its vicinity (cf. Fig. 5.5: Map of Kathmandu centre area). Sections 5.1.3–5.1.7 will provide a more detailed description of the area itself, after giving an overview about the urban history, planning and conservation policies as well as urban renewal projects.

## 5.1.1 A Brief Country and Planning History of Nepal

Nepal, mainly known for the Himalayan Mountains, is a multiethnic country rich of history, arts and architecture; all three closely connected to each other. Today Nepal belongs to the least developed countries of the world, being ranked at place 145 out of 187 in the Human Development Index (UNDP 2014). Nepal's urbanization level is comparably low with only 18.6 % in 2015. Nevertheless this means a distinct urban growth, e.g. compared to the year 2000 when the rate was still 13.4 % (United Nations—Department of Economic and Social Affairs 2014). Until the mid of the 1990s the major reasons for rural–urban migration were the classic 'push' and 'pull' factors: loss of arable lands and income sources due to unsustainable land uses and a shortage of inherited land, and better health care, schooling and employment opportunities in the cities. Since the mid-1990s the migration numbers accelerated due to the conflict between the Maoists and the establishment that took place in the rural areas in particular (UN-HABITAT 2010a). Nepal currently is the fastest urbanizing country in South Asia (Muzzini and Aparicio 2013), currently counting 29 million inhabitants (UN-HABITAT 2010b)—with the peculiarity that in Nepal most of the urban growth was and is concentrating solely in the capital of Kathmandu, which is growing annually by 4.10 % (calculated for 2010–2015,

Population Division of the Department of Economic and Social Affairs of the United Nations Secretariat 2011). Around 2.5 million dwellers are settling in the almost built-up Kathmandu Valley (Muzzini and Aparicio 2013).

Nepal opened its borders to the outside world only after 1950. Until then, it was difficult to get access as a foreigner. Additionally, travelling abroad was not seen as adequate for high caste people for a long time, even or in particular for high-ranking leaders. As a consequence, major political, societal and cultural changes occurred only after this turning point. At that time the country was already looking back on centuries of history. Early history ranges back more than 2,500 years ago. Since then different dynasties have ruled the country or parts of it, some of them coming from Indian territories. Among the most influential ones was the Malla dynasty that came into power in the twelfth century. It is of particular importance for Kathmandu, as under their reign Central Nepal was split in three independent kingdoms, each of with the capital city located within the Kathmandu Valley. Instead of fighting against each other, the three kings rather tried to outdo each other by arts and architecture, resulting in the construction of the three Durbar Squares (central areas with temples, religious buildings, open spaces and palace area) of Kathmandu, Patan (also called Lalitpur) and Bhaktapur. Therefore the phase between the fifteenth and the eighteenth century is often called Nepal's 'Golden Age'.

The country was then reunified under the Shah dynasty which ruled Nepal from 1769 to 2008. Recognizing the threat to be colonized by the British like India, Nepal remained in self-imposed isolation for more than a century during this time. The country itself never fell under British Colony; however, the British were influential as for some decades a British Envoy was installed in Kathmandu. From mid of the nineteenth century then the influential Rana family came to power, occupying the Prime Minister position, and relegating the still formally ruling Shah Kings. Rana and Shah Families started to build palaces inside and outside the then urban boundaries, adapting European architectural elements. New access roads were constructed to access the new palaces and middle-class families started to follow the new building style that mixed traditional Nepalese with foreign neo-classical elements.

The political situation changed only after 1950 when late King Tribhuvan was reinstalled as head of state. The next 50 years were then marked by different democratic movements, reforms and counter-reforms, including an attempt to foster self-governance in the Panchayat system which was introduced in 1960 after dissolving the parliament (Bista 2000). Despite massive development attempts, e.g. in terms of education and infrastructure, the country somewhat failed to improve, partly induced by its rapidly growing population (Donner 1971). A civil war (1996–2006) resulted in the abolishment of monarchy in 2008, since then governmental coalitions keep changing on regular intervals (The Asia Foundation 2012), an obstacle for sustainable development and long-term strategies. Obviously, architectural and cultural heritage was not at top of the agendas during these times, resulting in poorly executed legislation and the deterioration of many buildings throughout the country.

### 5.1.2 Introduction to Kathmandu

Kathmandu is one out of five municipalities located in the 665 $km^2$ large Kathmandu Valley (see Fig. 5.1) and the centre of the Nepal Central Development Region. The word 'Kathmandu' itself is used for both, the municipality and the agglomeration. In the following text it will be used in the context of the whole agglomeration. Situated in the central hills, the surrounding mountains form a natural barrier for the further urban expansion (cf. Fig. 5.1). Having an average elevation of 1,350 m above sea level the surrounding mountains reach almost 3,000 m, resulting in limited access only. Kathmandu has a humid subtropical/subtropical-highland climate, with a monsoon period from June to August.

The Valley is prone to multiple natural disasters; in particular earthquakes, annual flooding, and landslides. Such extreme natural hazards form an obstacle for a sustainable urban, social and economic development and are likely to be exacerbated by climate change (Jha and Shrestha 2013). The anyway high vulnerability is amplified by population growth and migration to the urban area, often precarious housing standards and an unplanned urban layout, social exclusion of different societal groups and the unstable political situation after ten years of civil war between Maoists and government that had only come to an end in 2006 (Bhattarai and Conway 2010; Gautam 2015; GFDRR 2012; Jones et al. 2014; Titz 2012).

**Fig. 5.1** Kathmandu Valley

Due to the central location at the important trade route between Tibet and India the settlements within the Kathmandu valley became very important trade places in early centuries. Architectural and town planning traditions are more than 500 years old, the appearance of villages and cities hardly altered over the centuries, design and town planning concepts as well as the building materials have remained almost unchanged (Korn 1977). For a long time the word ‘Kathmandu’ was even used synonymously for ‘Nepal‘ as a whole (Amatya 2008), underlining the importance of Kathmandu as centre of Nepalese culture and civilization.

Traditional architecture of the Kathmandu Valley is brickwork built with mud mortar and timber-framed structures. The traditional urban layout and fabric of Kathmandu Valley can be called “*a shining example of energy and space efficient building techniques with a distinct community harmonization component*” (UN-HABITAT 2010a: 99).

The settlements are very dense, with narrow streets and small courtyards behind the houses, surrounded by the agricultural land (Hollé 1998). In urban areas like Kathmandu houses are joined together to form a ‘*Tol*’ (blocks of buildings), which at the same time defines the local community. The main roads of most settlements are former trade routes which people settled around. Different casts settled at different places, according to the Hindu cast system which was adapted to the Nepalese multiethnic and multireligious society. The lower casts settled on the periphery of the town. A little closer to the city centre but still near the farmlands one could find the farmers. Artisans and craftsmen who played leading roles in the Nepalese society settled in more central areas. Traders and government officials lived near the palaces in the centre. Typical elements of the temple architecture are pagodas, the multitiered roofed temples, which are locally regarded of Nepalese origin, having influenced the development of similar structures in China and India (Parajuli 1986; Dangol 2007).

During the reign of Rana dynasty the urban appearance changed because of the Neoclassical architectural influences from Europe since the end of the eighteenth century (cf. Fig. 5.2 with neoclassical palace part on the right). The Royal family started constructing huge palaces on large compounds which formerly were agricultural land (Amatya 2008). The higher casts followed, this way the traditional very dense form of settling was lost more and more. After the country has opened its borders in the middle of the twentieth century these ancient town planning and building patterns are even more endangered (Sengupta and Bhattarai Upadhyaya 2016).

Today’s rapid and mostly unplanned urbanization process, fostered by weak institutional arrangements and an administration system on the difficult journey towards democracy is shaping land use and settlement patterns in Kathmandu Valley. The population of 1.18 million (United Nations—Department of Economic and Social Affairs 2014) is still growing, among others with annually approximately

**Fig. 5.2** Kathmandu Durbar Square, with Neoclassical part of Hanuman Dhoka Palace on the *right side*, Swayambunath Stupa on the hilltop in the background

54,000 migrants moving to the urban area (UN-HABITAT 2010a; Gutschow and Kreutzmann 2012).

The valley itself had nourished its inhabitants for centuries. Until mid of the twentieth century more than half of the agglomeration's inhabitants used to practice farming on the fields surrounding the densely built residential areas (Gutschow and Kreutzmann 2012), leaving a maximum of fertile land for agricultural production (Dixit et al. 2014). Forests served as source for firewood and burning material for cooking, heating and production of construction materials. The soil itself was and is used for brick production, the rivers crossing the valley as source for drinking and irrigation water. Water resources from stone spouts were managed by local communities ('*guthis*', see chapter on Kathmandu heritage policies). The local population was rooted deeply in this surrounding, as can be seen, e.g. in the large number of temples and shrines for religious functions on hilltops or riverbanks, or recreational sites near springs (Slusser 1982).

Immigration and natural growth have led to an enormous urban expansion within the past decades, leading to severe impacts in the whole Kathmandu Valley. Garbage disposal, water supply and sanitation are working partly at the best. A gridlock is bringing the traffic almost to a daily standstill, contributing to the heavy pollution of the valley. Institutional arrangements are weak and have encouraged the rapid and uncontrolled urban sprawls which have contributed to dramatic changes in the urbanized landscape (Bhattarai and Conway 2010; Muzzini and Aparicio 2013; British Red Cross et al. 2014; Shrestha 2011; Jigyasu 2014; Nehren et al. 2013). Urban sprawl and middle-class preferences for 'bungalow' style housing with compounds have led to increasing pressure on agricultural land and skyrocketing land prices. The inability to plan and manage the rapid urban growth has resulted in uncontrollable real estate markets (Brown and Dodman 2014). Considerable doubts concerning the limits of urbanization processes regarding availability of resources like space, water or building materials are warranted (Gutschow and Kreutzmann 2012).

Urban planning and policy guidelines to manage these challenges werc enacted but at city level they are not working effectively. Unplanned urban development and poor enforcement of regulations is leading to an even more increasing density and uncontrolled urban sprawl with a growing number of squatter settlements, that has already augmented from 17 in 1985 to 40 in 2010 (UN-HABITAT 2010a). Agricultural land and open spaces are built up and divided to plots up to the size of 15–45 $m^2$ only. Urban development is hardly considering cultural, environmental or risk aspects, even if the formal planning does, informal developments are likely to have changed urban appearance before a plan is decreed (Muzzini and Aparicio 2013; British Red Cross et al. 2014; Bhattarai and Conway 2010). Housing standards are not followed, the existing legislation is not adequate (Bhattarai and Conway 2010) and is followed neither as the consequences like fines are rather weak.

The unlimited construction activities contribute to the changing sociocultural habits of the local people. City growth at the outskirts, the squatter settlements in former rural areas around the city centres and the replacement of traditional buildings by new apartment buildings within the historic city centres have led to a loss of the traditional land-use systems and of the sociocultural traditions (Sandholz 2006). Until mid of the twentieth century more than half of Kathmandu's inhabitants used to practice farming on the fields surrounding the densely built residential areas (Gutschow and Kreutzmann 2012). Today, tourism is one of the main sources of income for Nepal and Kathmandu (Muzzini and Aparicio 2013; Nyaupane et al. 2015). While the country itself is mainly known for the Himalayan range and trekking tourism, the main attractions in the valley itself are the cultural ones, mainly the historic city cores.

Traditionally, residential areas in the valley were three-storied rows of houses grouped around semiprivate courtyards, accessed by a network of lanes and few wider roads (Gutschow and Kreutzmann 2012; Korn 1977). Today, this is hardly visible in the urban areas that have been transformed dramatically. The latest development is the construction of multistoried apartment houses (Thapa and Murayama 2012), and illegal additional storeys added on top of non-suitable houses all over the place, harming the stability and changing the appearance of whole quarters. Merely the seven spots and squares declared UNESCO World Heritage Sites seem to be mostly excluded from this tendency, due to their legal status. The historic core of Kathmandu, the Durbar Square, is one of them.

### *5.1.3 The 'Historic Urban Landscape' of Kathmandu*

As mentioned above, the urban heritage of the Kathmandu Valley was predominantly created between the thirteenth and the eighteenth century, when the Kathmandu Valley was divided into three kingdoms. Each of them tried to outperform the others by building more artistic constructions and temples, resulting in the valley's three historic urban centres, the Durbar Squares. This heritage can be ascribed to the ethnic group of the Newars, the original inhabitants of the Valley (Gutschow 1982); however the valley was inhabited already some centuries before Christ, being founded in a strategically important location along the trade routes between China and Tibet. The earliest inscription found in the valley stems from the year 464AD, the Licchavi period, during which many religious sites were determined. Buildings and structures were later reconstructed and embellished during the Malla reign, famous for its fine wooden and stone carvings as well as its bronze metalwork (UNESCO Kathmandu 2004; Weise 2012; Korn 1977). The architecture of Kathmandu Valley then remained almost unaltered until the Rana Prime Ministers introduced a new neoclassical style in the second half of nineteenth century, copying decorative elements from the West (UNESCO Kathmandu 2004), as, e.g. shown in some parts of the Hanuman Dhoka Royal Palace at Kathmandu Durbar Square or the Singha Durbar Palace complex which is now housing almost all major ministries.

Looking from the outside of any old building, anyone can identify what type of house it is. The facade identifies the type of house, either as private, public or as a palace (Ranjitkar 2007). Rows of houses enclose semiprivate closed courtyards (*bahal*), open courtyards (*chowk*) and public squares are part of the often very narrow street network. They play important roles during festivals, and are used as gathering or market places throughout the year. The main construction material for both, road surface and construction of residential houses and temples, are burnt bricks, giving the city a unique layout. At most street corners, aside temples and on public squares platforms (*dabali*) and resthouses (*pati*) can be found, to rest, play, chat or simply meet. The squares are also the starting and end points for chariot processions during different festivals. Significant places are protected by a demon (*chhwasa*), adding cultural meaning to streets and squares (Shrestha 2011).

Temples and shrines can be found all around the city, which itself is following a mandala layout, intermingling Hindu and Buddhist deities and rituals. Until today the cornerpoints—marked by Buddhist Stupas—of this mandala are visible, foremost in Patan and Bhaktapur. The richly ornamented traditional architecture of the valley is of high artistic value. Temples and houses are revealing the impressive richness of craftwork in Kathmandu Valley, namely woodcarving, metalwork and stone carving (Dangol 2007; Ranjitkar 2007; Weise 2012; Ellingsen 2010). Up to the present time it has been religious dogma that dwellings should not be higher than the houses of the gods (Parajuli 1986), only recently this tradition has been given up with the construction of high-rise buildings. Urban waterways, in particular Bagmati River, are of spiritual and emotional significance to the Nepali people (GoN/NTNC 2009). Even infrastructure like the stone spout water supply is of high artistic value (Historical Stone Spouts and Source Conservation Association (HSSCA) 2007).

The Nepali interpretation of the Hindu caste system (including even non-Hindu groups, e.g. some Buddhist communities) defines status and even jobs, also implies spatial segregation, with the lowest casts settling in the most remote places (Müller-Böker 1988; Gellner and Quigley 1999).

Traditional houses were following a certain scheme, based on the same ideas as the urban layout itself, although number of storeys and richness of decoration varied depending on the owner's budget and status. In front of the house there is a spot worshipped every morning for the blessing of the house and its inhabitants. Inside the houses the floor plan is following a clear distinction, with a more private back and a more public front part. Kitchen and family chapel are located on the top floor (Korn 1977; Ellingsen 2010). Such kind of houses can be found all over Kathmandu, with a higher concentration in the urban centres. Houses are grouped to '*tole*', small areas comprised of houses, temples, etc. around a square or yard denoting a compact neighbourhood or community unit (Parajuli 1986). These communities are the backbones of the social network, related through place of residence, cast and profession, as in the Nepalese cast system the three of them are very often inseparably combined. The cast system also impacted on the emergence of a variety of cultural practices, rituals and celebration of various festivals, enhancing the feeling of ownership and sense of belonging to a certain community. This social division resulted in the spatial division of different housing types, sizes and styles within town, creating "*a homogenous community within a heterogeneous city*" (Shrestha 2013b: 127) or a "*heterogenous cluster of relatively homogenous zones*" (Liechty 2010: 271).

After the 1950s then tall reinforced concrete structures became popular and almost replaced the traditional wood and brick architecture. The traditional architecture still can be found, however, often it is in bad shape, either affected by poor maintenance, lack of financial means for proper renovation, (illegal) adding of floors, or vertical division among brothers (Ranjitkar 2007) due to inheritance legislation that determines the equal distribution of house and real estate among all sons of a family. At the same time Kathmandu has become more multicultural, as the different ethnicities that formerly settled in distinct regions of the country are now present in the city which is also the home of many Tibetan refugees. In addition, international tourism has shaped the urban appearance in particular in the centre (Ellingsen 2010).

Throughout the centuries the Newars, the original inhabitants of the Kathmandu Valley, developed a strong 'cultural entity', under pressure from the mostly foreign ruling classes. Initially the Newars have been Buddhists, but after the introduction of the cast system most of them adopted the Hindu religion. However, the two religions do coexist and many deities are worshipped by both of them. This high level of tolerance and acceptance was and usually still is very typical, as can be seen in the fact that Hindu and Buddhist Newars take part in the festivals and celebrations of the other religion, despite having own distinct rituals (Weise 2012; Müller-Böker 1988; Hutt et al. 1994).

Kathmandu's natural environment and urban structures are closely associated with legends, rituals and festivals (Government of Nepal; Department of Archaeology 2007; Slusser 1982; Hutt et al. 1994). May myths exist, e.g. around the origin of Kathmandu, which is told to have been a big lake until Manjushree, Buddhist bodhisattva of wisdom and learning, cut the Chobar ridge in half with his mighty sword after he saw a lotus flower blossoming in its middle. That way he allowed the lake to drain, the valley thus became habitable and at the lotus flower place Swayambhunath Stupa was built. Mythology also is linked to wells and waterways, where nine *nagas* (sacred serpents) are believed to reside. Legends in Kathmandu are not mere recordings of fiction; many of them are still practised (Ellingsen 2010; Maharatta 2012), one example are the *naga* decorations that can be found in all major ponds or spouts. In some areas, forests patches remain untouched because they are considered to be sacred. These sacred places are protected and conserved because of faith in or fear of deities and often serve as recreational sites for picnics or hiking (ICIMOD et al. 2007). In this context Ellingsen (2010) mentions that even a leisure activity like kite flying has a certain time and meaning, as it marks the end of monsoon season, when the air gods shall blow away the rain clouds that have covered the sky for months.

Religion and rituals are incorporated in daily life and the time, e.g. by worshipping practices (cf. Figs. 5.3, and 5.4), whether in front of the house, at the local communities' shrine or temple (traditionally each *tole* was protected by a statue or temple devoted to Ganesh, a Hindu Deity usually represented with the head of an elephant) or in one of the cities' major temples. Various objects or spaces in the city are delineated by religious conceptions (Ellingsen 2010).

**Fig. 5.3** Women queuing along the ancient trade route to worship at Dattatreya Temple, Bhaktapur (*Source* W. Lange)

**Fig. 5.4** Gai Jatra Festival in Bhaktapur, commemorating the death of people during the past year, in the background: Nyatapola Temple, the tallest temple in Nepal

Ellingsen (2010: 7) describes thatThe Newar citizens have long associations with at least the old part of the city and whose images are soaked in memories and meanings, the city with its sacred objects represents a meaningful order. Arguably, their perceptions of the city contain a relatively stable core despite the idiosyncratic differentiation and new developments, which is re-affirmed in certain fields of social activity as for instance in religious rituals.

The Newar society also boasts of one of the most highly developed craftsmanship of brick, timber and bronze in the world. Weise (2012) practised at and intertwined with arts and architectural heritage in the valley (Hutt et al. 1994; Slusser 1982).

The actual motto of Kathmandu Metropolitan City (2015) is 'My legacy, My pride, My Kathmandu', indicating a strong or at least desirable bond to the city. Worshipping and other religious rites are still practices throughout the city, including annual festivals at certain times, e.g. at the end of monsoon time or to worship a certain deity. During monarchy the King himself had to follow certain ceremonies as the God's representative, at least some of these rites were then taken over by the now democratic leaders of the state.

Even today, despite all changes, the original urban structures are still present, endowed with religious meanings and the basis of various rites and rituals of the urban population. The basic principle of the Mandala form is still structuring the now city centres (Ellingsen 2010). Amatya (2008: 282) sums it up when stating

> What makes today's Kathmandu Valley unique is the combination of the magnificent natural environment with its man-made environment - its towns, settlements and cultural sites blending together, the people dwelling harmoniously together as a part of what is a living culture, not just the remains of a bygone civilization. [...] It is not just that the Newa people feel a deep, sentimental attachment to their cultural heritage: they live and breathe it, as part of their daily lives.

During the expert interviews most of the interviewees emphasized on unique features in Kathmandu's and Nepalese planning. Interviewees K09 and K03 elaborated on particularities of urban layout, with the towns constructed as culture centric (K03), and a priest giving the template for the urban layout to the people. Also for K11 reflecting the culture of the place is among the most important things to consider in urban planning in Kathmandu. This cultural focus was true until 1965 (K09) while interviewee K3 finds today's cities business centred.

The traditional layout of the Nepalese cities was based on a network with three roads from North to South and another three from East to West. At the central crossing the central square will be located, in other corners there are market and community squares. The city would be located by the side of a river, with the river on the North side of town, near the place for the dead and the semi-dead (deities, spirits, ghosts), while the south area is for the living beings. This layout is, e.g. still visible in Patan with its four stupas at the four city corners (K09). Another particularity in Kathmandu urban planning is the hierarchy of open spaces, from public to private; usually a number of courtyards are surrounded by houses (K08).

To interviewee K04 "*Kathmandu itself is the best practice and role model for the other Nepalese cities*", there is no need for foreign ones. K09 likewise states

> City design cannot only be in international style, the planner has to look at what the ancestors did. It has to be done according to the culture of the place to arrive at an identifiable city that also people from the outside enjoy.

To do so one needs to study the culture of the ethnic groups living in the respective area and include their culture into the planning, e.g. in terms of colours and materials (K09).

Other characteristic aspects of conservation in Nepalese context are its interconnectedness with society and religion, as interviewee K03 experienced during his work life: Once he was asked to give recommendations for the replacement of a fine lattice window in one temple in Swayambunath temple are. That temple burnt down and all the woodwork was gone completely, there was a fine lattice window to hide the god's image behind: But there was no drawing or picture, nothing, so we asked the priest of that temple what the window had looked like. But even the priest did not know as he was not supposed to look at the god's image, so he did not even look at the window in front:

> As a solution we asked different carpenters if they could replace the window, asked them about their ideas how to design it and what would be the special assets of that god. Depending on their answers we could pick an 'educated' carpenter who had some ideas

how to design the window because he had learned it from his ancestors, he knew which motives would be appropriate. The lesson learned is that craftsmen are very important, they give the knowledge to the next generation and so the family preserves it.

In this context interviewee K03 also mentioned the Nara document that allows more building enhancements, thus coming closer to the Nepalese reality where the 'change aspect' is very important: "*Ideas are coming from the Western side, such as permanency in material; they are not Nepalese style and can cause problems.*"

Interviewee K03 sees a fundamental difference between the Nepalese context of conservation and the Western one, which to him means:

Freezing of the monument in time. By contrast the Nepalese conservation context allows adding or changing things during the conservation process itself, as it is lengthening the life of the object and thus preserving it for future generations. There even is a certain Sanskrit word for this process, which translated to English would mean a combination of enhancement and conservation.

According to him conservation in a Nepalese context is practiced since the late sixth century, proven by one of the earliest inscriptions in *Changu Narayan* temple.

### 5.1.4 Inner-City Patterns and Dynamics

Like in the other two case studies a survey was done to assess urban patterns and processes in inner city, beyond the margins of the historic core area itself. Kathmandu is somehow a special case, as this city has three historic core areas, Kathmandu, Patan and Bhaktapur, as explained in the chapter on urban history. As Kathmandu is the main hub in the centre of the urban agglomeration and the administrative centre of the valley this area was selected as case study area. The core area basically consists of the Kathmandu Durbar Square and its vicinity.

To the east one of the main roads, Kantipath, forms the research area boundary. As the map (Fig. 5.5) indicates the borders towards South, West and North are less apparent, due to the very winding road network. In the south the research area ends with Dharahara tower, one of the urban landmarks. To the west Gangala Marg and its extensions towards north and south delimits the research area. To its east are much less central services or attractions, less touristic activities and more housing functions. Towards north the area until Tridevi Marg was analyzed. This street forms the northern boundary of the most frequented central area, and is also the northern end of the main tourist quarter of Thamel.

The Kathmandu Durbar Square (cf. Fig. 5.6/1–2) developed on the ancient trade route and is composed of temples, palaces and open spaces, which are also the location for various ceremonies and festivities. Kathmandu itself is said to be named after one temple found at the Durbar Square, the *Kasthamandap*, dated back to twelfth century (UNESCO Kathmandu 2004). Beside this temple there are many other temples and shrines. These constructions belong to the finest in the country and mainly stem from sixteenth and seventeenth century, devoted to various Hindu

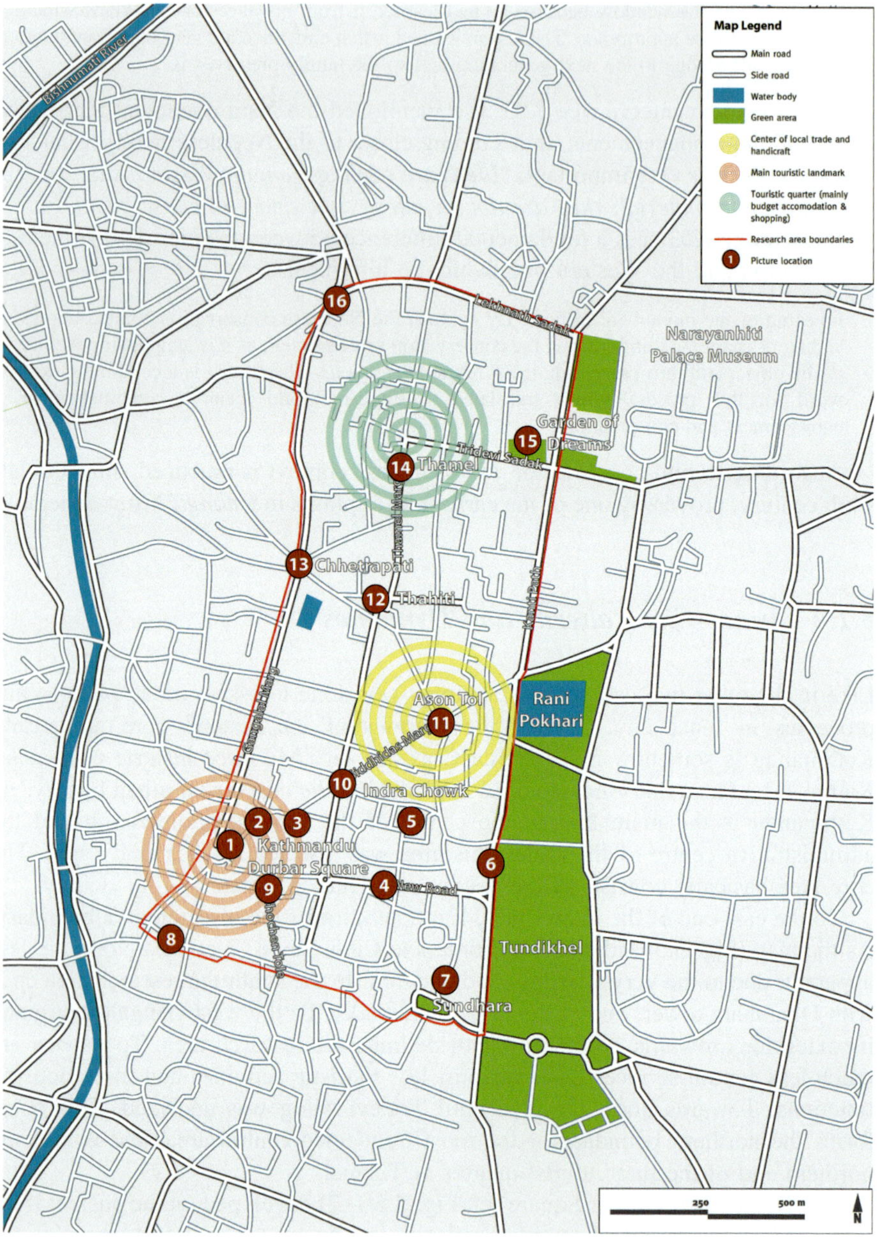


Fig. 5.5 Map of Kathmandu centre area with research area boundaries

**Fig. 5.6** Streetscapes in Kathmandu centre area, 1–8

Gods and Goddesses. Also the former royal palace, the Hanuman Dhoka, is found here, the reason why another name for Kathmandu Durbar Square is Hanuman Dhoka Durbar. The Kathmandu Durbar Square also houses the Taleju Temple, one of the most sacred spots in the city and only accessible for a limited time throughout the year. Erected in 1549, it was the first building with three roofs in town (Parajuli 1986).

As explained earlier the main architectural and urban heritage stems from the Malla Period, to be more precise the late Malla period dated 1382–1768. During the subsequent early Shah period 1768–1846 mainly modifications were made, e.g. some extensions to the Hanuman Dhoka. Further modifications were then done during the Rana rule (1846–1951). These modifications are usually very easy to distinguish, as they adopted European stylistic elements (UNESCO Kathmandu 2004), as can be seen, e.g. in the white stucco part of Hanuman Dhoka (cf. Fig. 5.2). Also at the Durbar Square the Kumari Ghar or Kumari House can be found, another masterpiece of seventeenth century Newar architecture. It is the place where the 'Living Goddess', the Kumari, stays. The Kumari is a young girl from the Newar community which is selected in a rigorous ritual and then considered as reincarnation of Goddess Durga. She is worshipped by parts of the Hindu and also Buddhist community and worshipped during distinct festivals.

Despite the comparably high level of protection Kathmandu Durbar Square areas are suffering some serious problems (Maharjan 2012; Muzzini and Aparicio 2013): traffic is not restricted and causing congestion as well as parking problems, infrastructure facilities are weak (Surendra et al. 2011), traditional houses in the buffer zone are increasingly modified and particularly the residential area south of Durbar Square is economically very weak, leading to a precarious and highly vulnerable housing situation.

The same issue was mentioned several times during the expert interviews, e.g. when interviewee K01 emphasized on building bye-laws that are the most restrictive in the monument area and its buffer zones, but that are however often not obeyed. This fact is easily visible when taking a look at the surrounding buildings that almost all are recently built ones with rooftop restaurants for visitors.

The Durbar Square is also one of the major tourism destinations in Kathmandu, therefore a huge variety of uses can be perceived, ranging from worshipping and performing of certain ceremonies in distinct occasions by locals and Hindus, leisure activities like sitting on the stairs of a temple while chatting and snacking, to tourists' souvenir shopping, sightseeing and museum visits. It is most likely the area with the largest number of functions and uses in town. To interviewee K11 it is important to keep life in the area, and not to 'pickle it' and make it an open-air museum: "*The purpose of Durbar Square is to maintain our own culture and not being attractive for the tourists*" (K11).

K05, K06 and K07 worry an increasing deterioration and vulnerability of the buffer and outside area: where traditional houses are deteriorating mainly because of vertical division, a lack of financial means and awareness, the preference for multistoreyed concrete buildings. As a consequence project proposals for the reconstruction of whole quarters are developed, like the NSET project in Sundhara and Jhhoche area mentioned before. K07 is worrying about this change of urban appearance:

> Right now people in Kathmandu core are adding one storey to their houses every year, before it used to be max. 3 storeys, now you find five or six, even in the core areas you find new houses in some parts, the brick facade has changed to plaster ones, and the core is losing its character.

To interviewee K07 the core area needs some renovation; he names examples from other countries he knows, e.g. Germany and Mexico, where urban cores were renovated and preserved and made attractive for people.

New Road (Fig. 5.6/4) is one of the busiest streets in town, and linking the Durbar Square with Kantipath, one of the arterial roads. As the name indicates the road is of relatively young age. In fact it was built after the 1934 earthquake that destroyed the majority of houses in the area. Today the road is flanked with buildings of 6–8 storeys and the urban hub for electronics. The area north of New Road (cf. Fig. 5.6/5, Guchcha Marg, one of the commercial streets there) is as well a shopping area, mostly for clothing. Figure 5.5 gives an impression of the many narrow lanes lined with small shops, opening out into open spaces and small squares, used for (motorbike) parking and trade as well. Recently, the area is under growing economic pressure as real estate prices along New Road are raising, as a consequence construction activities are going on and in the vicinity to New Road some larger shopping complexes are constructed.

To the east Kantipath is marking the border of this trade and commerce area. To cross this main road different flyovers have been constructed. This way Tundikhel, the main green and open square of town, and the quarters to its east side can be accessed. The grass-covered ground was used for various purposes like military parades, horse races or religious festivals. Today it is mainly used for leisure activities like football.

In the area southeast of the crossing Kantipath/New Road a mixture of housing and commerce can be found. While in New Road mainly electronic devices are sold, the area southwards is more for the supply and repair works. Houses are a mixture of traditional and modern ones, with a gradient from the corners (modern) towards the block centres (old). The southern endpoint of the area's main street, Sundhara Marg, is Dharahara Tower (Fig. 5.6/7). The 60-m-high tower is also known as Dharahara or Bhimsen Tower, and was built in 1832 by then Prime Minister Bhimsen Thapa. The area has been upgraded in the past years, park at its bottom has been renovated and is not free of access, in the vicinity a row of souvenir shops can be found. In contrast to Durbar Square the majority of visitors seem to be local ones or Asian tourists.

The area south of Durbar Square is mainly a housing one, where still traditional communities are settling (cf. Fig. 5.6/8). However, increasingly the original owners are renting out their houses as they prefer to build new houses outside the centre. Usually they belong to the Newar community and would therefore keep their family plot and house. The area foreseen for the NSET urban renewal project mentioned by interviewees K06 and K12 is one of them. That example is indicating a major problem of these housing areas, the ongoing deterioration and on an average bad condition of the traditional houses. Trade and commerce is mainly for daily supply, small shops for the local community.

There is hardly any tourism in this area, with the exception of Jhochen Tole, better known as 'Freak Street' (Fig. 5.7/9). The name goes back to the years between the early 1960s to late 1970s. At that time Kathmandu was the endpoint of the so-called Hippie Trail and the first touristic area of town, where the visitors found accommodation and cannabis. Since then the area declined, although there are still some low-budget guesthouses, bars and shops around, the main tourist hotspot has moved to the north, to Thamel area (Liechty 2010).

When walking along Siddhidas Marg (Fig. 5.6/3), heading northwards from Durbar Square one enters a more local shopping and commerce area. Siddhidas Marg is the former trade route which was linking India and Nepal with Tibet. This can still be perceived in its actual appearance and the items traded here. They basically comprise of traditional clothing and accessories like metalwork, including the Newar water cans. Siddhidas Marg then passes Indra Chowk (Fig. 5.7/10) and Ason Tol (Fig. 5.7/12) which are two of the major squares along the road, with different goods sold around the square. The vast majority of clients are local ones; mainly products of daily need are sold. Around Ason Tol a lot of food, spice and vegetable vendors can be found, while Indra Chowk is also a hub for traditional attire. The surrounding has still a high percentage of traditional houses and narrow lanes (Fig. 5.7/10).

Moving northwards the area's appearance changes, in terms of urban pattern and traded goods. Thahiti (Fig. 5.7/12), a large square housing a large Buddhist Stupa, somehow forms the border between the more traditional parts of town. Towards north the building uses and goods traded change, from housing and daily supply to the main tourist quarter of Thamel.

The quarter of Thamel (cf. Fig. 5.7/14), located north of the Kathmandu Durbar Square, is the main hub for tourism, but also locals come to shop things they do not get in other locations, e.g. gift items. North- and southwards from Thamel there are local shopping areas, e.g. southwards the tourist area ends at Thahiti, and one can easily recognize the sharp border, as also elaborated by interviewee K10. Chhetrapati (cf. Fig. 5.7/13), nowadays mainly a traffic junction is the southwestern border of Thamel area. Shops in Thamel have to be closed only at 10:00 pm, in the main street 1 h later, while until around 5 years ago whole Thamel was opened until midnight or even longer, but that changed with a new rule and police controls. Also in these terms Thamel is quite different from other parts of town, where a 'nightlife' does hardly exist. Locals are mainly coming to Thamel during off season, and would also go to the restaurants, as there are different kinds of cuisines that one cannot find in other places (and comparably less Nepali restaurants). Prices in Thamel are sometimes higher than elsewhere, as K10 himself is experiencing. Peak seasons when many tourists visit the country are from mid-August until the end of December and from mid of February until monsoon time. He also sees a shift in the tourists' countries of origin, with increasing numbers of Chinese tourists and decreasing numbers of Japanese and partly European visitors than only few years

**Fig. 5.7** Streetscapes in Kathmandu centre area, 9–16

ago. The composition of tourists has an influence on the goods sold, as, e.g. "*many Chinese tourists only want to buy Pashmina shawls or Nepali local music, and they are not interested in getting other things that they can get copy easily in China*" (K10).

In Thamel and beyond towards south most shops are rented, not run by the owners of the houses. Most house owners moved away as the area is too loud, but with the high rents they can afford a house outside. As a consequence the area is almost not inhabited; many buildings have been converted or replaced for trade or hotel business. Until 10–15 years ago the most expensive area for renting shops and houses in Kathmandu was Thamel itself, today the prices in New Road are as high, due to the mobile phone business. The same is true now for Kanti Path area, south of Narayanhiti Palace, which is getting as expensive as Thamel after the king's resignation. Before there used to be a law that no building should be higher than the palace, now this is no longer true and recent constructions are 15 or more storeys high (K10).

The northern border of Thamel area is marked by Leknath Sadakh, another main road (cf. Fig. 5.7/16). To its north the building density gets comparably lower. Here mostly residential areas are located. In the west Thamel borders on Kanti Path Road, and the former Royal Palace Narayanhiti, This palace was built only in 1970 after the Royal Family decided to move out of Hanuman Dhoka Palace. After Nepal was declared a republic the palace became a museum, parts of the 38 ha large compound now house governmental functions. But still this large area is not accessible for the public. Another private green space is located between Narayanhiti and Thamel. Kaiser Mahal, better known as the 'Garden of Dreams' (cf. Fig. 5.7) is a neoclassical garden with pavilions, ponds and pergolas. After having been restored with funding from Austrian Government it now serves as a park. To enter the area an entrance fee has to be paid. Therefore, the main users are tourists who can afford the fee and who want to escape from the hustling and bustling Thamel area.

Overall there is a whole variety of urban patterns in Kathmandu centre, ranging from the historic buildings in and around Durbar Square to the almost pure concrete-and-steel architecture around New Road as well as in Thamel. In between there is a mix of both, with a gradient from a block's outside (comparably more recently built houses) to its inside (still more traditional ones). All areas comprise of temples and shrines, varying in size and quality of decoration. Interviewee K01 finds temples mostly in a better condition than private houses. This opinion is stated in a number of other interviews as well, and beyond that it is backed by evidence during the survey of the centre area. To K03 the conservation in the core area is very satisfactory, although to him a higher budget would always be welcome.

As explained before, the uses and functions are varying as well, depending on the area. The most 'traditional' ones can be found along the former trade route between Durbar Square and Ason Chowk. This is where daily belongings like vegetables and spices, kitchen equipment, metalwork, or traditional attires can be shopped. From there northwards to Thamel as well as on Durbar Square itself comparably more tourist souvenirs are sold, while New Road and the area south of it are known for electronics and cell phones.

### 5.1.5 Perceived Drivers of Change in the Urban Landscape

During the interviews experts were asked about processes going on in Kathmandu or impacting on the town, and how they perceive this. Furthermore they were requested to elaborate on ongoing projects in the urban area and in particular on the centre area and their personal opinion about the consequences. The aspects mentioned most often were the rapid urban growth that is difficult to deal with, and the challenging political situation that is impacting on the non-existence or noncompliance of rules and regulations for urban planning and conservation.

Throughout the expert interviews the rapid urban growth was mentioned various times as a major problem for the city that is not able to deal with it. To cope with the rapid growth interviewees K01 and K07 emphasize on a decentralization process to lower the pressure on Kathmandu. One reason for the high level of urbanization in Kathmandu is that "*In Nepal there is only one magnet, even for a simple medical test or for any higher education you have to go to Kathmandu*" (K07).

Furthermore poverty in the countryside is a major push-factor for moving to Kathmandu that interviewee K11 considers as not equipped with the right policies: "*You cannot say that no more people are allowed to come to Kathmandu, you have to make the policies in such a way, that people don't need to come any more.*" Many of these newcomers move to one of the squatter settlements that are another urgent issue "*The number of slums in Kathmandu is increasing, the number of squatters is increasing—so the urban planning in Kathmandu is not very successful!*" (K08). A large number of these squatter inhabitants then end up in informal labour, with much of that informal work in tourism (K08).

Another issue is the very limited land in the Kathmandu Valley; as a consequence there is the need to make maximum use of it, while increasing the safety standards and building up adequate infrastructure (K06). Most urban services like water or electricity supply were planned 20 years ago in 1996 plan, for the demands 20 years ago, but times have changes, so the urban planning is very challenging: "*In theory all the rules, regulations and laws are there, but in practice no one wants to take the responsibility for it*" (K08).

Also, infrastructure development does not keep pace with the urban growth, in the end even bye-laws are weakened and adapted to local project contexts, e.g. in terms of minimum distance between rivers and construction areas to protect the riverbanks that was lowered in a certain land pooling project; what to K07 is a wrong development as "*the initial intention of the law to protect the riverbank was skipped for the political pressure*".

Various interview partners named the unstable political situation and the administrative system itself as a major challenge to achieve a better urban planning and conservation. As K03 explains:

> Politics are also a problem: there are many radical changes in government and an unstable situation, if this was more stable maybe the city would look different and the conservation might be different, too.

To K03 leftist parties do have difficulties with the monuments as their ideology is different. K03 compared the ongoing road widening project—which is of utmost preference to government—to conservation, when stating "*if conservation would be as much political will as the road-widening project it would happen as well.*"

Most government officials that were interviewed stated their department is doing good job, but they however do not know about other. K08 was not surprised to hear that:

> [...] because all the authorities do their day-to-day work, but they do not take any responsibility. I do not blame the poor and illiterate people, but the rich people and the authorities for what is going wrong.

In addition people in governmental services are not always educated for the job they are doing, furthermore they are just moved around between different governmental institutions every few years so that they do not even get a deep insight into the topics: "*There is a problem with human resources in governmental places. Only in the urban department there are experts, but not in others* (K08)."

K08 claims a big need for rules and regulations but at the same time also for awareness. During Panchayat system before 1990 he found the government quite stable, also during the following years with the king ruling, but since the introduction of the democratic system things became very unstable, but even during the Panchayat system they could not do everything, mistakes were made:

> It is too easy to blame the politicians, also an inner support is needed, it is also depending on the people. Building up a civil society is very necessary; it is not only the politicians that are responsible, it is also business of the people themselves.

He also claims the lack of governance, as there is no elected government at the moment:

> Until today the planning in Nepal is very much centralized, still the planning is hardly considering what the people want. The will of the people has to be considered somehow, they also have to be made responsible (K08).

Conserving traditional urban layout and individual buildings is a major problem to different interviewees while interviewee K04 rather see this as part of a modernization process which is welcome. K03 notices a failure to preserve the traditional buildings:

> There is a failure to control the loss of traditional residential buildings, the Department of Archaeology tried to do by putting regulations, but they only apply for reconstructions, not for the traditional buildings themselves. Many things are gone meanwhile.

To him it would be good to keep this vernacular architecture, but what is needed is a control of land use and building height as "*there is failure of urban space conservation, now it is too late to change it. Maybe there are not the right regulations to achieve it.*" Interviewee K11 shares this opinion when saying "*with a proper plan Kathmandu would not be in the present state.*"

K05 and K03 also see people forgetting the traditional knowledge. Unfortunately trust in local craftsmen was almost lost, also due to the fact that:

> The Newar/Nepalese society is a very hierarchical one, initially with the king and priests on top, and today other experts or architects are seen as 'new priests'. So even a top craftsman may not believe that an expert would follow his advice (K03).

K05 is concerned about the acceptance of conservation norms that locals may not want to follow.

> People don't want to use the traditional materials, furthermore they sometimes are difficult to get, e.g. wood for carvings is difficult to get in suitable length and quality, and a different department – forestry department - is responsible for it.

Interviewee K11 also claims a lack of commitment and determination, people think and say 'What can I alone do?', what is very common attitude among professionals also. K06 and K07 back this view when stating that the culture for housing has changed, the new culture is in favour of modern-style houses. K03 is witnessing a change of lifestyle in urban Kathmandu: "*Kathmandu has changed radically the past years, even more within the last 10 years, the change is not manageable.*"

Most loss of traditional houses is happening in the commercial areas, while the houses behind the shopping area are still often the traditional ones (K01). Interviewee K01 adds on this issue that "*planning is just not there, it only exists for individual buildings*".

In his personal opinion one problem is that people think modern buildings suit the nowadays commercial uses much better, they are easier to rent out. Interviewee K04 shares this attitude that K01 describes as to him Kathmandu and its building stock lacks modernization.

Another issue is the perception and preferences of the people that according to the interview with K01 often favour concrete buildings that are said to be more stable and that are felt to be more 'modern'. To them old houses are more difficult to maintain, and who can afford will have a new house. "*Money is destroying more than it helps*" (K01).

Even with the Patan palace and temple area people think that rebuilding old and sometimes damaged wooden structures with concrete would be more stable, as he experienced in the case of a certain temple: "*People's intention is good, having a temple is nice, but what they do is completely wrong*" (K01). K01 as well as K06 identified a change of lifestyles, one of the reasons for changed construction preferences and badly maintained traditional buildings.

**Fig. 5.8** Examples of vertical division and addition of storeys on traditional houses in Kathmandu Centre

> Properties are divided vertically, now the inhabitants have only very limited space inside, they even sometimes remove some load-bearing walls, what is very dangerous. Also some families started raising the floor heights or add additional storeys, due to family growth. What you find is old buildings being demolished and replaced by new, concrete ones (K06).

In fact the bad state of various traditional houses is due to vertical division (cf. Fig. 5.8). Family sizes are growing and house as well as land property gets more fragmented. To K02 there is a big need to address this problem in particular. The heritage buffer zones and outside areas are deteriorating, due to the division of properties, but according to K03 DoA cannot do anything about it as it is private properties. The only thing that could be done with support from DoA is to preserve the facades. K08 links the social aspects to construction and tourism when stating that:

> The tourists like to go to the cities, but the maintenance of the heritage is also a social problem, because the inhabitants often are poor, and due to inheritance the houses are divided into small parts, they even take out walls or reduce the diameter so that they become unstable, therefore they have to be demolished as the structure gets very poor. The problem then is that they replace the houses with concrete structures which are poor as well, with sometimes only four inch walls.

The weak economic situation and dependency on tourism is another topic mentioned by different interviewees, though the interpretation differed, in particularly concerning the importance of tourism. K08 states that 70 % of local employment is in the informal sector, with the consequence that there are only few people with high income and too little money to finance infrastructure.

Tourism in Nepal is flourishing, but 50 % of the employment in tourism is informal according to interviewee K08 (cf. Amatya 2008; Muzzini and Aparicio 2013) all the people working in restaurants or elsewhere as service are informal; no taxes are paid for this. He assumes that only 5–10 % of all people in Nepal are paying taxes, mainly people in governmental jobs, the rest is at least partly doing informal work without being taxed for it. The country relies on only few sources of income which are not very stable, as tourism can decrease within short time as it already did during insurgency. Interviewee K08 mentioned that living costs in the city are comparatively high, also due to lots of remittance money which nowadays remittance is one of the main income factors for urban areas, after tourism, this could be used wisely for the urban planning.

K07 does not see any development of the private sector and an untapped potential of cultural tourism which could be much more. Tourism is an important economic and development factor for the core areas, which is seen differently by the interview partners. K05 is positive about this as it is bringing economic benefits for the heritage zones, in case of problems "*it is different department which is responsible*" (K05).

K11, however, does not understand the hype about tourism: "*I am not keen on tourists*". He claims that it is not more than 5 % of the GDP, but its income goes to the middle and upper-classes, therefore they have an interest in keeping and extending it, and also in promoting it. The main share of the population gets nothing out of the tourism, what is needed is a diversification of tourism and the touristic areas:

> There is a big need for changing the tourism so that more people can benefit from it, this was done in the past in areas such as Bali or Chiang Mai, Nepal should do something similar: tourists should not only go to the monuments but to the people (K11).

Tourism and international donors also impact on the prices, e.g. "*once I went for shopping vegetables at the market, and the vendor told me 'this is not for you, sir', and only to prove that I can afford I bought it, but was then angry with myself*" (K11).

All four interviewees that are university staff in addition claim a lack of cooperation between politics and academia in the realization of urban planning. Usually government officers "*prefer to work with the private sector because then the governmental officers get some money back as provision when they give a job, at least 10 % of the overall budget*" (K06). This is also why many university staff members are having their own consultancy aside competing and under-pricing each other. Such way there is no development of the private sector according to him.

Another point made is that higher education is not focusing on traditional Nepalese constructions, during their studies the students learn on concrete and technology but they do not learn about traditional techniques (K01). Most faculty members in urban planning graduated from abroad, and are educated in '*Western concepts*' (K07). K03 shares this opinion when stating:

> Professors in university are educated in the Western system and thus follow Western paradigms; therefore they think that the community and their approaches are not right. For our buildings there is no knowledge from the West.

In K03's opinion the local view should be incorporated more in teaching on conservation, but right now teaching is dominated by the 'global knowledge system', also to equip the students for the global market, else they would have no chance to get any position. Architectural history covers mainly the Roman and Greek one instead of Nepalese. Another concern is the lack of research on conservation and traditional architecture in university itself, leading to a lack of knowledge.

Unfortunately, this results in negative impacts on heritage, as many buildings, temples, patterns and other items have gone meanwhile. K01 gives an example from his work:

> Take the case of a traditional house in Patan, where the owner wanted to have more storeys, what is understandable. KVPT suggested adding more storeys to the old house as the structure looked very stable. The owner agreed to do so, the plans were made and went to municipality, where all files then also need to be checked by the earthquake department, in this case the one of Patan Municipality. They then said that this is seismically not stable as it is an old house! Even the head of department said so who at that time was an architect trained in conservation and coming from Pulchowk [note: Pulchowk Campus is part of Tribhuvan University and the place where architects and engineers are educated].

The economic pressure on the monument zones is high as well, people there and around are asking for taller buildings and do not understand the area's regulations that prohibit such constructions (K02). Interviewee K07 gives an example to underline such processes that to him are wrong: "*Boudha Stupa in Patan is no longer visible from the outside, only from the airport or from the plane you can still see it, but before that was different*" (cf. Fig. 5.9). To him "*all this is the fault of the government offices.*"

Different interviewees again emphasized on the need to involve Nepalese planners and experts as they are more aware of local needs. In the opinion of interviewee K08 "*the urban planning in Kathmandu should be made by Nepali planners, not by borrowing Western things.*"

He believes that the Western way is not the best one, but that local assets and customs have to be considered to "*Find the Nepali way*". K09 elaborates on the same topic when stating that "*any urban planning should be rooted in the local culture. But now the planning is following the modern style and this often is getting very repetitive.*"

**Fig. 5.9** Buildings surrounding Boudha stupa and view from the airport

**Fig. 5.10** 55-Window Palace during and after renovation: (*left*) Craftsman carving lattice window, 2006; (*middle*) senior craftsman presenting carvings in upper storey during renovation phase, 2007; (*right*) after the renovation, 2011

K03 reports on a specific case, the renovation of 55-Window Palace in Bhaktapur (cf. Fig. 5.10) where he was involved in the renovation measures as an advisor:

> The problem was how to design the lean-out on the top floor; three different suggestions were elaborated by experts. I myself suggested to ask the craftsmen, as they should know best, so I involved the senior craftsman in the decision-making process. He was shown all three designs and picked the one he found most suitable. This was the way the building was re-constructed then. The idea behind was: let the people who made the building take the decision. Later another meeting was done with around 200 senior craftsmen, experts in brick, wood and carpentry, to include their advice and knowledge in the processes. But at first they didn't believe that the experts really wanted their support. They didn't think their opinion was accepted. Even the top craftsman didn't believe that the experts would follow his advice.

What is needed in K03's opinion is a better recognition of the craftsmen; to him any (Western) theory is less important and valuable than their knowledge, as "*for such buildings there is no knowledge from the West!*"

Most interviewees claimed a lack of knowledge of local people concerning the real costs of construction measures, as new buildings are considered to be more cost efficient. In this context interviewee K01 mentions that when the KVPT renovated their office building, located aside the Patan Durbar Square, costs were around 16,000US$ at that time. This is to verify that the renovation of such a traditional house is affordable and not more costly than the construction of a new one.

Different statements are suggesting ways to enhance ongoing conservation efforts, e.g. Interviewee K03 who sees some potential in cultural tourism that can contribute to the preservation of traditional quarters. He mentioned the Mahabouddha area, southeast of Patan Durbar Square, as best practice, because here a traditional neighbourhood was preserved, by the community itself:

> No architectural control is applied here, but the inhabitants preserved their brick houses due to their business and cultural sense. This shows that any conservation rule would succeed if it makes economic sense, if the people had an economic benefit (K03).

K01 and K03 emphasize on the need to take people's views more into consideration so that they accept planning, e.g. by highlighting economic aspects:

> Maybe the locals might start redoing things when the culture and perception changes – this was done in the West when people perceived they could have a profit out of it – but it will take time until that point is reached (K03).

Concerning the monument zones all interviewees agreed that they are in a comparably good shape, as the protection level is highest. Being asked about the impact of urbanization on the conservation of heritage K05 states that "*of course urbanisation is a problem, but overall it is fine, the core areas are in good shape and well preserved, problems are manageable.*"

However, things change already when taking a look at the buffer zones and in particular the rest of the town, which is under completely different legislation, as illustrated in the next chapter.

### *5.1.6 Assessment of Urban and Heritage Policies*

Preservation of urban and cultural heritage in Nepal has to be seen in the context of the respective political system. During the reign of Rana Prime Ministers (1846–1951) very little attention was paid to the preservation of traditional architectural heritage. This brought about a gradual process of deterioration, which was further accelerated by the disastrous 1934 earthquake. The beginning of democratic movements under the Shah Kings from 1951 onwards gave new life to heritage conservation. At the same time the new political system brought new social values and contact with the outside world, "modern" or Western style attracted Nepali. Traditional land use and socioeconomic system was undergoing changes due to new laws and regulations.

Before 1982, maintenance of natural and cultural sites was done by the '2, local community organizations or associations based on caste and locality. Their duty was to maintain their surroundings, including open or public spaces like courtyards or temple areas which play an important role in public life. The *Guthis* are at the same time a system of community land ownership, responsible for endowing land for religious purposes and charity. Besides, the ruling royal family, influential ministers or rich people occasionally granted restoration and maintenance measures of religious or other important sites (Chapagain 2008). Such community-based maintenance and ownership decayed for different reasons, with most of the change occurring after 1950 (Tiwari 2015). The most apparent consequence is probably the deteriorating traditional water supply of the valley. More than one-fourth of the initially 400 stone spouts, most of them beautifully carved artworks, have already gone—with more to be lost in near future due to deterioration, construction and infrastructure development (UN-HABITAT 2008).

Simultaneously much of the community land, formerly managed as a commons, got lost. Local maintenance of lands and waterways, linked to religious values is no longer in place or was meant to be replaced by other policies (Ellingsen 2010). "*Ultimately the Land Reform after 1962 gave a death blow to this system.*" (Parajuli 1986: 10), although the *Guthi Sansthan,* the umbrella organization, still holds ownership rights to certain buildings (UNESCO Kathmandu 2004). As a consequence *Guthis* declined, restoration was no longer a priority, many monuments were no longer taken care of, and government did not have the means to take over this duty (Parajuli 1986). One remnant of the guthi system on the local level are different area development trusts or development committees, with certain rights and authorities for specific areas, particularly monument areas like the Pashupati Area Development Trust for Pashupatinath Temple area (UNESCO Kathmandu 2004).

Formal preservation mechanisms of heritage started only after the establishment of the Department of Archaeology (DoA) in 1952, which achieved its legal status in 1956 with the issuing of the 'Ancient Monument Preservation Act' (His Majesty's Government of Nepal). This act is still the main legal document on heritage conservation in Nepal, not surprisingly putting emphasis on 'ancient' and 'archaeological' sites (Chapagain 2008). The DoA's responsibilities comprise the formulation of policies, programmes and guidelines pertaining to conservation, preservation and restoration of cultural heritage, as well as the later monitoring of the implementation (Parajuli 1986).

The definition of an 'Ancient Monument' is a:

> Temple, monument, house, abbey, cupola, monastery, stupa, bihar etc., which have their importance above one century, from the point of view of history, arts, science, architectonics or art of masonry, and this word shall also mean the site of the monument as well as the human settlement or place, and remnant of ancient human settlement, relies of ancient monument, cave etc. having specific value from the national or international point of view irrespective of the fact that such settlements or places are adjoining with each other or are separate in the same area. [...] These monuments belong to different periods from the Ancient period to the Shaha period and are located in different part of the country (Department of Archaeology 2011a).

According to interviewees K01 and K05 in Kathmandu itself the DoA is responsible for the seven monument zones, which are also heritage sites according to Nepalese legislation:

> Overall there are 7 monument zones within the Kathmandu Valley, corresponding to the sites inscribed as World Heritage. Within the monument zones permission is needed to rebuild or repair any house, to do so there is a certain format in the municipality. Outside these monument zones there is no interference and influence from the Department of Archaeology, only within they are responsible and will interact (K01).

The responsibility comprises the core area and buffer zones, outside the municipalities are responsible for any planning. Inside the area buildings must follow certain bye-laws, e.g. on maximum height, cantilever, materials, etc. The law also states which materials are allowed to be used, e.g. that wooden carvings should not be replaced by cement what is a popular trend nowadays. They plan to launch a public awareness program to teach about monuments, 'people public partnerships' are carried out. In addition the DoA published notes now to build according to the renovation rules. In total the DoA has a staff of 95 experts and receives an annual budget for conservation from the central government which the DoA can spend according to its priorities. Sometimes there is support from outside, e.g. from the Japanese Government, which is supporting certain projects. K05 sees the biggest problem in the lack of manpower and budget: "*People need to be skilled on the different building types, which is difficult; there is a need to have different experts*" (K05).

In practice any project is structured the following way: a DoA technician or a group of experts makes a recommendations for any measure, subsequently the municipality has to approve these measures formally as they are legally responsible for this process. Although its area of intervention is mainly in the heritage core areas, the DoA is offering consultancy and can do intervention outside of the monument zones, e.g. "*if people want to destroy an old, traditional window, then the DoA tries to convince them to preserve it, and in worst case would buy it and would take it to the museum*" (K05).

The DoA is also responsible for any intangible heritage, e.g. to preserve the traditional festivals, like the Machendranath festival. To K05 this is important as this is very much linked to the tangible heritage and "*unfortunately people are forgetting the traditional knowledge*" (K05). Therefore DoA also set up an educational program to preserve the traditional craftsmanship, as also emphasized by interviewee K03.

This exclusive responsibility of DoA in the monument zones is mentioned as problematic by other interviewees, e.g. K01 and K06, who find that there are monuments of value outside the heritage zones as well, which are not protected by law. K03 identifies a failure to control the loss of traditional residential buildings.

> There is a failure to control the loss of traditional residential buildings, the Department of Archaeology tried to do by putting regulations, but they only apply for reconstructions, not for the traditional buildings themselves (K03).

When talking about the urban heritage of Kathmandu it is indispensable mentioning international and in particular UNESCO activities. Nepal joined UNESCO in 1953. It then took some years until the first restoration and conservation project in Nepal (Amatya 2008: 285) was carried out in 1972. The 'Hanuman Dhoka Renovation Project' (at Hanuman Dhoka or Kathmandu Durbar Square) was financially and technically supported by UNESCO and UNDP, in later phases the German Government became main donor. The successful project led to the establishment of a more integrated conservation programme with national and international support, realized in the preparation of a UNESCO Master Plan for the conservation of the cultural heritage of the Kathmandu Valley in 1977 (Amatya 2008).

In 1978 Nepal ratified the World Heritage Convention, 6 years after it was issued. Only 1 year later the Kathmandu Valley became a cultural World Heritage site, inscribed because of its exceptional urban fabric, tangibly associated with the unique coexistence and amalgamation of Hinduism and Buddhism, representing an exceptional testimony to the traditional civilization of the valley, as expressed in UNESCO's justification of inscription. Overall seven groups of monuments and buildings were inscribed under this name, including the Durbar Squares of Hanuman Dhoka, Patan and Bhaktapur, the Hindu temples of Pashupati and Changu Narayan as well as the Buddhist stupas of Swayambhu and Boudha (UNESCO 2015b). The inscription was preceded by a protective inventory in 1975, that listed temples, stupas, palaces, houses, different shrines and other religious buildings, wells and water fountains, and other examples of Nepalese architecture, up to the impressive number of 888 monuments and 13 monument zones in the urban areas of primary importance within and in the vicinity of the Kathmandu Valley (Amatya 2008). This UNESCO-supported undertaken was the first concerted and large-scale protective measure in the valley.

Some years later in 2003 the World Heritage site was put on the List of World Heritage in danger ('red list'), due to the loss of traditional elements of heritage as well as uncontrolled development (UNESCO 2012). It was removed then only in 2007 after some modifications of site boundaries and the adoption of the Integrated Management Plan for the Kathmandu World Heritage Property (Amatya 2008; Weise 2012; UNESCO 2015). UNESCO was and still is a major player in conservation issues in the country and Kathmandu in particular, e.g. by providing information on the value of urban heritage in general and how to preserve it. One example is the 2007 'Heritage Homeowner's Preservation Handbook', the first guidelines for the repair of traditional old buildings to homeowners in Kathmandu. UNESCO is also involved in campaigns against numerous thefts of Buddhist and Hindu sculptures. They are stolen from temples, courtyards, fountains and fields and are usually sold abroad. For the local communities their loss is going beyond a mere material one, as they actively were worshipped as living deities (UNESCO Office in Kathmandu 2015).

But even the interaction with UNESCO, in particular in the World Heritage sites, is not free of conflicts as "*even the international organisations don't play fair, they change the rules*" (K07). To him giving more time to achieve something to developing countries is fine, but it is not correct to simply lower the benchmark, and this has happened in his opinion.

Beside the authorities there is a number of NGOs actively involved in urban development in general and the preservation of Kathmandu's heritage in particular. The Kathmandu Development Preservation Trust (KVPT) is among the most important ones. KVPT is an NGO funded in 1991 with the mission to safeguard the architectural heritage of the Kathmandu Valley. It is active in the whole city area, their main concern is the value of a certain building, no matter whether it is within a monument zone or not. As K01, a senior of KVPT elaborates, most of the KVPT projects in fact are taking place in narrow squares and places where there rarely are tourists. The KVPT principle is not to rebuild but to save old things, rebuilding is only acceptable if absolutely needed. Despite cooperating with the affected municipalities he admitted KVPT often has difficulties with the authorities. Additionally also communication and cooperation with the communities themselves is time-consuming. In his words "*I spend more time talking to the communities, explaining things, trying to convince them than with the building itself.*"

Beyond the heritage scope there is a large number of regulations, policies and related authorities on urban planning and construction activities that is implemented. The first one that was explicitly dealing with urban areas was the 1963 Town Development Committee Act, followed by the 1973 Town Planning Implementation Committee Act (TPIC) with the goal to implement physical plans of regional development centres. It was later replaced by the 1988 Town Development Act that empowers local governments to control, regulate or prohibit use of land and construction works. However, many issues were not answered (Joshi 1997).

Of utmost importance for the urban development in Kathmandu Valley are the 1976 'Physical Development Plan of the Kathmandu Valley' and the 1988 Town Development Act, enabling the Kathmandu Valley Master Plan which is renewed every 5 years, and the Nepal National Building Code. The building code was completed in 1994, meant to regulate construction and planning activities on building and urban scale, e.g. minimum amount of open spaces or minimum width of roads. However, it is rarely enforced or fined, as can be seen in today's Nepali townscape which is dominated by continuous building with few open spaces and narrow and encroached roads (British Red Cross et al. 2014; ICIMOD et al. 2007; UN-HABITAT 2010a; Sangachhe 2008) although some simplifications in the process of acquiring construction permits (electronic building permit) were recently implemented (World Bank 2014b).

According to K11 even the first formal Master Plan for Kathmandu Valley from the 1970s was hardly implemented, "*foremost because of rich powerful people who still did what they wanted to and who did not follow the rules.*" This somewhat was a continuation of the situation before. At that time the Ring Road was built, in the initial plan trees and green open areas around the road were foreseen, but *the government found it easier to use the land* and hardly any plot is left: "*The plan was perfect, but in the end the belt is a dump yard.—And this is only one example of what is going wrong.*"

The next Kathmandu Valley Development Plan which was developed with support of UNDP and Japanese planner Kenzo Tange in 1996, is a good plan to K02 and K07, however not detailed enough and still lacking full implementation. Furthermore the housing types proposed cannot always be provided in all locations, e.g. garden city housing types. To K08 the biggest need is for execution of the 1996 plan and development of subsequent zonation as well as land use plans: "*In theory all the rules, regulations and laws are there, but in practice no one wants to take the responsibility for it. All the authorities do their day-to-day work, but they do not take any responsibility*" (K08).

There are different bye-laws and regulations for monument zones and buffer or outside zones, e.g. in terms of the floor height which is limited to 35 ft in the core areas while outside 45 ft are permitted, but even in the core areas often the height is more (K01). In reality this height difference would mean 2 storeys and thus less income from renting out, so there are lots of conflicts and questions not answered (K02). Another issue is the implementation and interpretation of existing rules, which according to the government does not permit compromises to preserve old buildings, e.g.:

> Plans for a renovation of old houses with floor height of 35ft. are approved, no matter if the old structure is preserved or not, the only thing that is done is to check if the drawings match the bye-laws, but no control during or after the construction process, so that in the end all the buildings are different in reality, with floors of more than 35ft (K01).

Furthermore some regulations may not be adequate for historic buildings and even contribute to their destruction, when authorities insist on their compliance. In theory there should be supervision during construction processes, in practice there is none, so that there is no chance to prevent any false action. In the interview K01 argued that the municipalities even consider new buildings to be more stable than the traditional ones, "*and they are the ones who advise the people and who supervise the building stock*".

Interviewee K03 proposes that planning approaches could also work to do conservation and preservation, what is not done very often. Same with others, to interviewee K07 conservation is as well a political thing, there has to be a political will to conserve, then it would happen (in a different way). Additionally he emphasizes strongly that "*to do so knowledge is also very important, you need to know what to preserve and why, and what not*".

In reality urban development projects are carried out by various agencies, the interaction among them and with responsible administrative authorities sometimes is at least questionable—although answers on this question were ambivalent. The Department of Urban Development and Building Construction is responsible for implementing the national housing and urban policy, and working as consultant to the municipalities, with offices in the different districts. Interviewee K02 admits some untapped potential for interaction with the municipalities, also in Kathmandu Valley, however, to him the key limitation is the lack of interaction with and between the municipalities which is not predetermined, "*additionally there is a lack of enforcement, as the municipalities always neglect the plans coming from my Department*".

Site supervision is another concern, as there are different responsibilities: K01 describes that if something goes wrong the municipality will only report on it and will not take action. Practical action in this case would be duty of the police under the chief district officer, but there is hardly any communication between them. If some building or building process is illegal this will be under responsibility of the legal department in the municipality, the owner then has to pay a fine, what will be the only consequence. To him this signifies the lack of communication between all actors, involving the Department of Archaeology, municipality, police and community and more: "*In case a building or part of it is not complying with the regulations it is reported to the legal department of municipality and in worst case the owner has to pay a fine, there is no other consequence*" (K01).

Till today the urban land use planning is not clearly regulated and the institutional responsibilities are unclear between different authorities and levels. The establishment of new organizations such as the Kathmandu Valley Town Development Committee (recently converted into the Kathmandu Valley Development Authority) and the Ministry of Housing and Physical Planning in 1988 (now the Ministry of Physical Planning and Works) are promising but not yet very effective (Shrestha 2013a, b). The existing building by-laws are the only legal tool for urban growth management—alarmingly they are still based on the land use map of 1976. One major constraint is their applicability for new constructions only. Therefore, it fails to address ongoing and alarming activities like the vertical division of houses (harming stability and in the end destroying historic fabric) or the often wrong or even lacking maintenance and renovation measures in the historic core areas, conversion of residential houses to other functions, etc.:

> In many cases, they are conflicting with the existing other legislations. For instance, the 'Ancient Monument Preservation Act 1956' empowers the Chief District Officer after getting request from Department of Archaeology, to give order for the destruction of the houses or part of it that are constructed against the prevailing law whereas the 'Local Self-Governance Act-1999' gives power to the Mayor to punish defaulter either by imposing a fine or by demolishing the building or part of it. The city cannot take any action unless it is informed by the affected party (Shrestha 2013a: 29).

Kathmandu Valley houses five municipalities that liaised only in 2014 under the umbrella of 'Kathmandu Valley Development Authority' to jointly tackle problems, e.g. proper land use planning and a lack of open spaces. It is supposed to take care of the valley urban development, but to interviewee K02 "*human resources and budget allocation is not enough*". So the valley is still a huge space with several governing authorities and no formal and overarching plan. K04 does not even see any urban planning executed since 1976 at all, interviewee K08 gave a similar comment. Recently a new 'national land use policy' had been published:

> But so far it has not been detailed due to ongoing discussions and negotiations in the previous parliament. There is still the need for a formal approval, but as there is no parliament right now the policy will hopefully be approved by the next elected parliament (K06).

One promising activity for a more sustainable land use planning is a system of voluntary 'land pooling' in the valley, including the creation or preservation of public open spaces, implemented by the Ministry of Physical Planning and Works (International Federation of Red Cross and Red Crescent Societies 2011), following up on a land pooling program in the late 1980s (Shrestha 2011). In Kathmandu itself, the most recent and holistic undertaking is the Kathmandu Metropolitan City 'Risk Sensitive Land Use Plan', that is based on a cooperation of urban, national and international actors (Earthquakes and Megacities Initiative 2010; German Federal Foreign Affairs Office and EMI 2010).

The main problems in the current policy system can be summed up as

- Non-matching—sometimes even contradictory—rules and regulations,
- Weaknesses in the administrative and legal levels cooperation,
- Resource constraints, and
- A challenging coordination among the various governmental and semi-government agencies and local bodies as well as
- A lack of awareness and participation.

In addition, any urban planning project carried out in Nepal is deeply mingled with interventions of development assistance coming from international donors. The country is one of the major global recipients of official development assistance (ODA), in 2012 foreign aid represented 26 % of the national budget, with more than 40 donors (Government of Nepal et al. 2013; Jones et al. 2014). Being both, a blessing and a curse, international donor agencies surely support national and urban development, however, the curtain was drawn over negative side-aspects for a long time (cf. Lama and Job 2014, for the case of a protected area in Western Nepal). Aspects like lack of long-term planning reliability due to limited project lifespans or approaches adopted from donor countries without adapting to local culture and habits can be even more crucial for dynamic urban areas.

K07, K08 and K11 emphasize on the difficulties of foreign agencies and NGOs and their projects which follow different approaches than the Nepalese ones:

> The problem is that the international organizations are responsible to their own tax payers back home. Also they are coming from a different school of thinking, they have different backgrounds than the Nepali, but they all want their own vision to be implemented (K11).

The responsible national and urban authorities and further international actors are yet to come up with improved policies, programmes and projects to realize a more sustainable urban development (Shrestha 2013a). Nevertheless, some action is taken, also considering urban heritage. The 'Kathmandu Valley Development Plan 2020' has clearly set the objective of promoting traditional cities of the valley as cultural hubs. In 2012 a Ministry of Urban Development was newly established. Paragraph 17.3 of the Nepalese Interim Constitution states that "*Every community residing in Nepal shall have the right to preserve and promote its language, script, culture, cultural civilization and heritage*" (Government of Nepal 2007).

## 5.1.7 Phases of Urban Renewal in Kathmandu

Urban renewal as such only was carried out after the 1970s in Kathmandu. Figure 5.11 is giving an overview of the main phases. By far most urban renewal projects were concentrated on the central temple areas and Durbar Squares (main squares) of the three districts, almost all of them in cooperation with UNESCO

**Fig. 5.11** Main phases of urban renewal in Kathmandu Centre

and/or other international actors. Usually such projects are considering touristic purposes as tourism is a major source of income.

The 'Hanuman Dhoka Renovation Project', which started in 1972, was the first large-scale and more holistic attempt, not exclusively aiming at the historic fabric (named after the former Royal Palace at Kathmandu Durbar Square, which is at the same time the focal area). It was followed by the Bhaktapur Development Project (carried out in three steps between 1974 and 1986), which is seen as a best practice for renewal and tourism in developing countries. The Bhaktapur Durbar Square forms part of Kathmandu World Heritage Site and has been restored over time with main funding from GTZ (today GIZ) and German government. The project was carried out with local craftsmen, bringing local employment opportunities and new job for the traditional craftsmen, in particular in wood and stone carving.

During the first phase of the project there was a clear focus on the restoration of temples and buildings, in a more top-down approach than in the second phase after 1980. In fact the participation of the local communities and craftsmen was only implemented in a later stage of the project, also because of some protests of the locals against the somehow imposed planning. Today, fees to enter the area are graded depending on the tourist's country of origin and are used for maintenance and local projects. Overall more than 200,000 tourists visit this place annually (Luger 2008; Parajuli 1986). This project was a role model for any further renewal or renovation project, of which most were carried out with international support.

In the subsequent years, between the end of the 1980s and 2007 not much of urban regeneration was done, except safeguarding, for reasons like lack of budget, lack of cooperation among different actors and the difficult political situation. Additionally, renewal of private-owned houses is far more difficult than the preservation of state-owned temples and other buildings, due to their legal status. This is easily visible in the ongoing transformation of the traditional residential neighbourhood of the historic core area (Shrestha 2013b). Major losses of historic patterns and fabric in Kathmandu occurred during these decades, e.g. in the core areas what led to the inscription on the 'red list' of World Heritage in Danger from 2003 to 2007 as mentioned above.

Only since the end of the Maoist conflict and the 2007 removal of Kathmandu from the 'red list' urban renewal returned on the agenda. The elaboration and adoption of the 'Integrated Management Plan for the Kathmandu World Heritage Property' is at least supporting the protection of the sites itself, while the buffer areas are already of a lower protection level (Weise 2012). This unfortunately results in the loss of the sites' surroundings where many traditional buildings are replaced by new concrete buildings with more storeys and often with rooftop restaurants for the tourists.

Main action is taken in or around the seven World Heritage Site spots, where monitoring and maintenance is taken more seriously than in the rest of the city. Some upgrading was done, e.g. in the 'Dharahara—Sundhara Public Plaza', surrounding Dharahara Tower, where shops were introduced, what has led to a loss of public space (cf. Fig. 5.12). Access to the park and restaurant area is now restricted to those who can pay entry fee. However, "*People's sentimental attachment with the public space was not considered, while converting such public spaces into commercial uses*" (Shrestha 2013a: 30).

Shrestha (2013a) also claims the preference of short-term economic benefit over cultural concerns, as can be seen in local government projects like the conversion of '*Te-Bahal*' (one of the biggest Buddhist Monasteries of Kathmandu), 'Bhugol Park' (Earthquake Memorial Park) and parts of '*Tundikhel*' (the biggest urban open space located in the city centre) into parking spaces. Increasingly, the private sector is making use of public spaces for commercial activities. Therefore renewal is often

**Fig. 5.12** Park at the foot of Dharahara Tower, only accessible when paying entry fee

seen in context with economic aspects, in particular in the centre areas that are the touristic hotspots. The biggest ongoing initiative is on Kathmandu Durbar Square area, including a waste management plan and traffic calming, which is badly needed as the square is highly frequented and a traffic junction.

On the other hand small-scale local projects are increasingly carried out, usually combining the upgrading of communities and buildings in form of bottom–up initiatives, often supported by NGOs. The involvement of NGOs in renewal can also be found on a larger scale, e.g. in a planning of NSET (National Society for Earthquake Technology) that proposes a renewal project in the quarters of Chikanmugal and Jhhoche area south of Durbar Square, linking risk reduction and renewal—in their understanding.

NSET has set up a project to rebuild a traditional housing area in the core of Kathmandu, comprising 6 ha with 1,600 houses and 116 courtyards, as reported by interviewee K06: This project is claimed to be one of the major ones on 'urban upgrading' depicting very well one common understanding of upgrading. The plan is to preserve the temples and other heritage sites (statues, places of worshipping, overall the NSET assessment found 94 heritage buildings among the 1,600 ones) but to replace the old houses by new ones with facades following traditional style (K06). This idea had been discussed already with the community and different authorities from ministry to municipality level. In any of the three planning alternatives the traditional temples and *bahals* are preserved, while the rest of the houses is dismantled and replaced. Furthermore houses there are subdivided so much that families often only have left 2 m of facade and live in very poor conditions. K06 adds on that when stating that to him one reason for the already mentioned difficulties in maintenance and preservation is the changing society, as "*owners of houses move away, they rent out, the community is less intact than before.*" The same project was mentioned by K12 in a positive way, as both of them mentioned the high level of risk in the area, due to poor constructions and a lack of maintenance.

To sum it up, the most recent trend in renewal is heading towards a more integrated approach, explicitly combining different aspects like risk reduction, renewal and housing standards in the case of the NSET project. However the protection of traditional neighbourhoods or public spaces (e.g. in Sundhara) is not necessarily a part of it, what is indicating the comparably lower relevance or different understanding of 'conservation' in 'renewal'. Urban regeneration projects in the historic core are either in the World Heritage sites, aiming at the conservation of historic fabric, or outside that area, aiming at the improvement of building structures and infrastructure, potentially by replacing the historic building stock.

### *5.1.8 Perception of Inner-City Processes*

As explained in the introduction to this chapter, students enrolled in Master courses at Tribhuvan University were asked to express their impression of different spots in Kathmandu, in particular the centre area. The questionnaire investigated the values

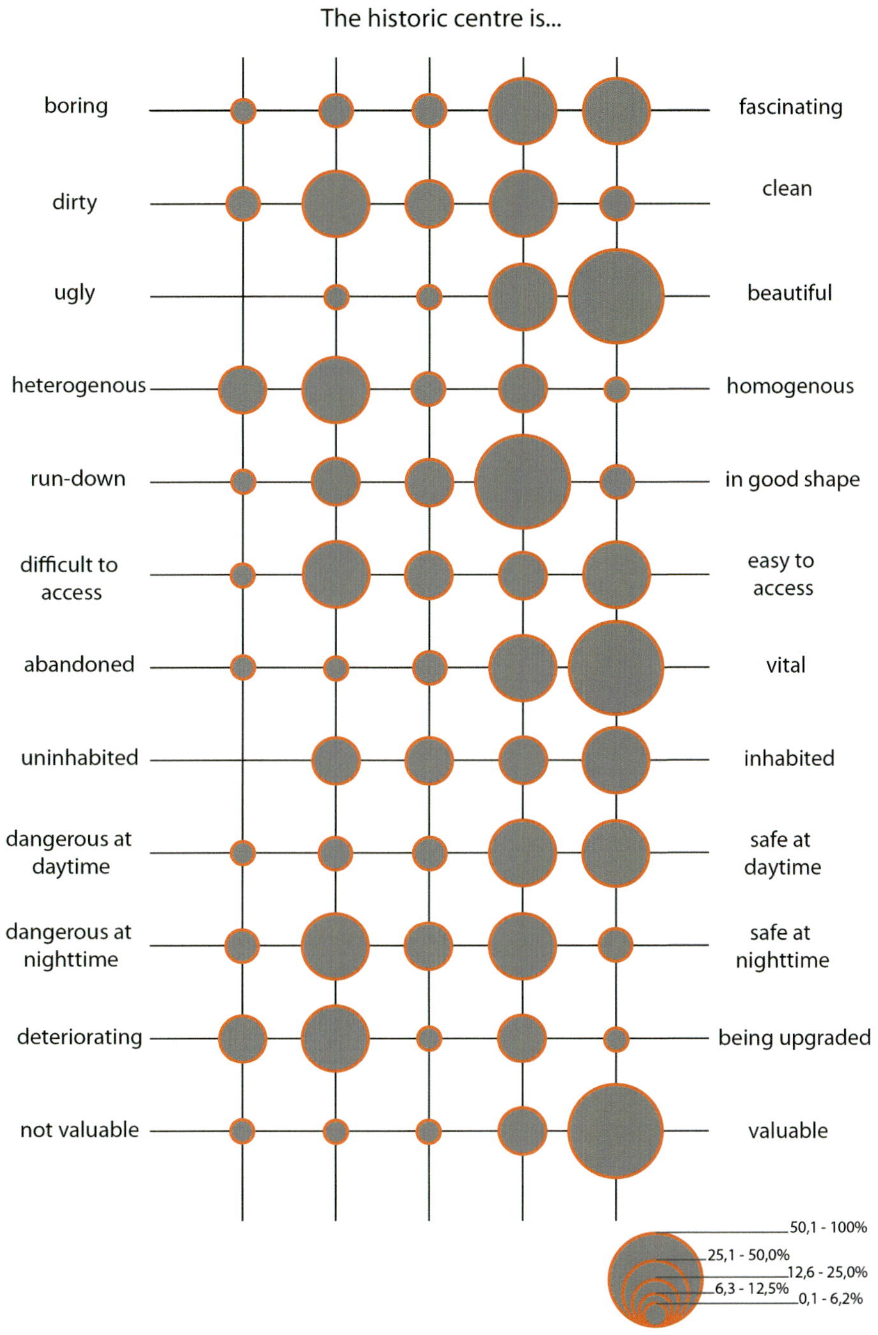


**Fig. 5.13** Responses to the question: What is your feeling towards the historic centre of Kathmandu? ($n$ = 80)

respondents to the historic city centre of Kathmandu and perception of processes perceived there. Basically the majority of respondents attach (very) positive values to the area (see Fig. 5.13: Responses to the question: What is your feeling towards the historic centre?). The area is found fascinating (overall 74.3 % answers tended to) and vital (84.5 % have a positive or very positive impression). Overall the impressive number of 93.3 % of all respondents describes it as beautiful or very beautiful. For 76.0 % the area is very valuable.

However, looking at the answers on the current condition results are little less positive. Although the absolute majority of 53.4 % find the area in good shape, it is striking that almost one fourth (24.7 %) find the area run-down or even very run-down. Even more remarkably, 50 % assess the area deteriorating, 19.4 % even very deteriorating while still 19.4 % see an upgrading process. A large majority find the centre safe during daytime (overall 82.2 %). However, this positive impression changes during night time when only half of the former positive answers (42.1 %) are reached.

Concerning accessibility answers are somewhat disperse. While 53.3 % find the area easy or very easy to access, still overall 28 % have a more negative impression. A similar kind of biassed result is true for the question if the area is found more dirty (36.8 %) or clean (47.4 %). A majority of respondents find the area heterogeneous (overall 64.3 % responses), what is depicting the variety of building styles, uses and ongoing changes in the centre. Almost the same number of respondents (66.2 %) agrees that the area is inhabited, what somehow surprisingly little considering the high building density of the area. One possible conclusion is that at least parts of the area are not inhabited (Durbar Square) or serve as commercial (New Road) or tourist (Thamel) area, with comparably lesser residents. In addition, the deterioration of the area has already led to some outflow of inhabitants to peri-urban areas, as stated in interview K12.

One can conclude, that there is a distinct consent about the overall value, beauty, fascination and vitality of the area, as while the shape and current state of upgrading is seen more critical and at variance. Subsequently the respondents were asked how far they agreed or disagreed to a predefined list of processes they possibly perceived to happen in the historic centre (see Fig. 5.14). 83.5 % disagree or even definitely disagree with the statement that the centre is only frequented by tourists but not by locals. To turn the argument on its head, it means that more than four fifth find the centre used by locals. This matches the number of 55.8 % not agreeing (74.0 % not and definitely not agreeing) to the assumption that respondents find the area outmoded.

Processes the respondents perceive to happen in the centre area are the replacement of old buildings by new ones, leading to the loss of historic ensembles (only 6.5 % disagreed or definitely not agreed to this), people preferring to live in more modern buildings instead of the traditional ones (only 3.8 % of disagreement but more than one-third definitely agreeing), and the preservation of old buildings but the loss of their traditional uses with only 6.3 % disagreeing but 72.2 % (definitely) agreeing. Lastly more than 90 % agree at least partly to the assumption

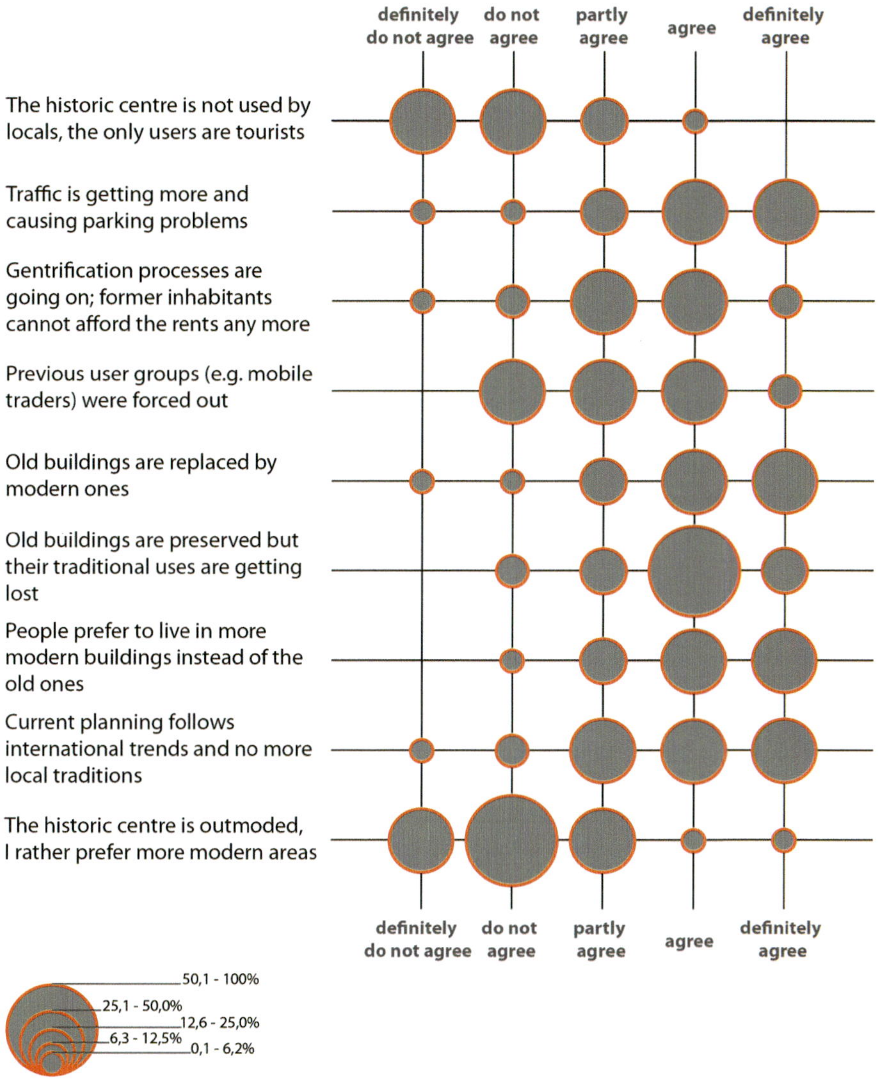


**Fig. 5.14** Responses to the question: Which processes do you perceive in the historic centre of Kathmandu? ($n = 80$)

that current planning in Kathmandu is following international trends and no more the local traditions that have shaped the urban layout in the past.

In addition, traffic and a lack of parking areas is causing more problems in the core area for overall 77.2 % (out of them 41.8 % definitely agree this is an ongoing process).

Concerning the question if there are user groups that are forced out of the area, e.g. mobile vendors or others, answers are very diffuse, as many agree as disagree with most people agreeing partly. More clear responses are given to the question on gentrification. 44.4 % of respondents agree or definitely agree that such processes can be perceived and that former inhabitants cannot afford paying rent and staying in the area any more, and another 53.1 % agree partly.

Overall, the results obtained from the questionnaire back the results of expert interviews and site visits. Traffic is perceived as a big problem, the same is true for ongoing changes in terms of building structure and uses. The peer group is witnessing an ongoing replacement of old buildings and even more of the traditional functions and agrees that most people prefer to live in modern areas. The latter answer is somehow surprising as the vast majority does not find the centre outmoded and prefers it over more modern areas. A potential explanation could be that they like the centre for visiting or leisure but would rather prefer to live somewhere else, as the housing conditions often are precarious according to different experts.

In addition to the closed questions explained before, an open question asked for personal suggestions on project to upgrade the historic areas of Kathmandu (not focussed on the centre exclusively, to allow for comparison, multiple answers allowed). It can be said that all the suggestions made for practical upgrading projects match the urban problems stated in literature as well as stated during expert interviews and in the closed questionnaire questions itself.

Overall, eight suggestions are made for the Pashupatinath Temple area, ten for Patan Durbar Square and another nine for Bhaktapur core area, all three also part of Kathmandu World Heritage. Further on projects were proposed to protect old (Newari) settlements or historic places in general, including open spaces, and to protect local culture and traditions like festivals or clothing. Overall twelve students proposed projects in this context. One claimed "*the proper implementation of land use plan, bye-laws and guidelines in Kathmandu*".

The most frequently mentioned answers circle around Kathmandu Durbar Square and its surrounding (31 people), proposing mainly the adequate protection or restoration of buildings and a proper traffic management, varying from improvement of the parking situation to a complete traffic calming of Durbar Square area, claiming, e.g. a "*pedestrianization and prohibition of vehicular flow*".

More practical suggestions concern the improvement of sanitation and the waste management system, the instalment of public toilets and an improved drinking water supply. Economic upgrading was requested inform of "*community based economic programs for poor inhabitants*" and removal of the slum-like condition near Durbar Square. Related to urban planning on a larger scale students propose to, e.g. install community-based heritage walks "*so that the cultural values can be portrayed to the visitors who come to feel the importance of heritage structure*". Also, intangible values and awareness rising are found important by several respondents, one of them, e.g. promoting to "*upgrade and maintain the values and keep the belief in such a historical foundation and raise awareness of all people towards the conservation of heritage sites and historic places.*"

## 5.1.9 Importance of Tangible and Intangible Assets

In addition to the questionnaire section dealing with tangible places and the centre itself another section dealt with values attached to these areas. As demonstrated throughout the previous chapters' intangible and tangible assets are closely

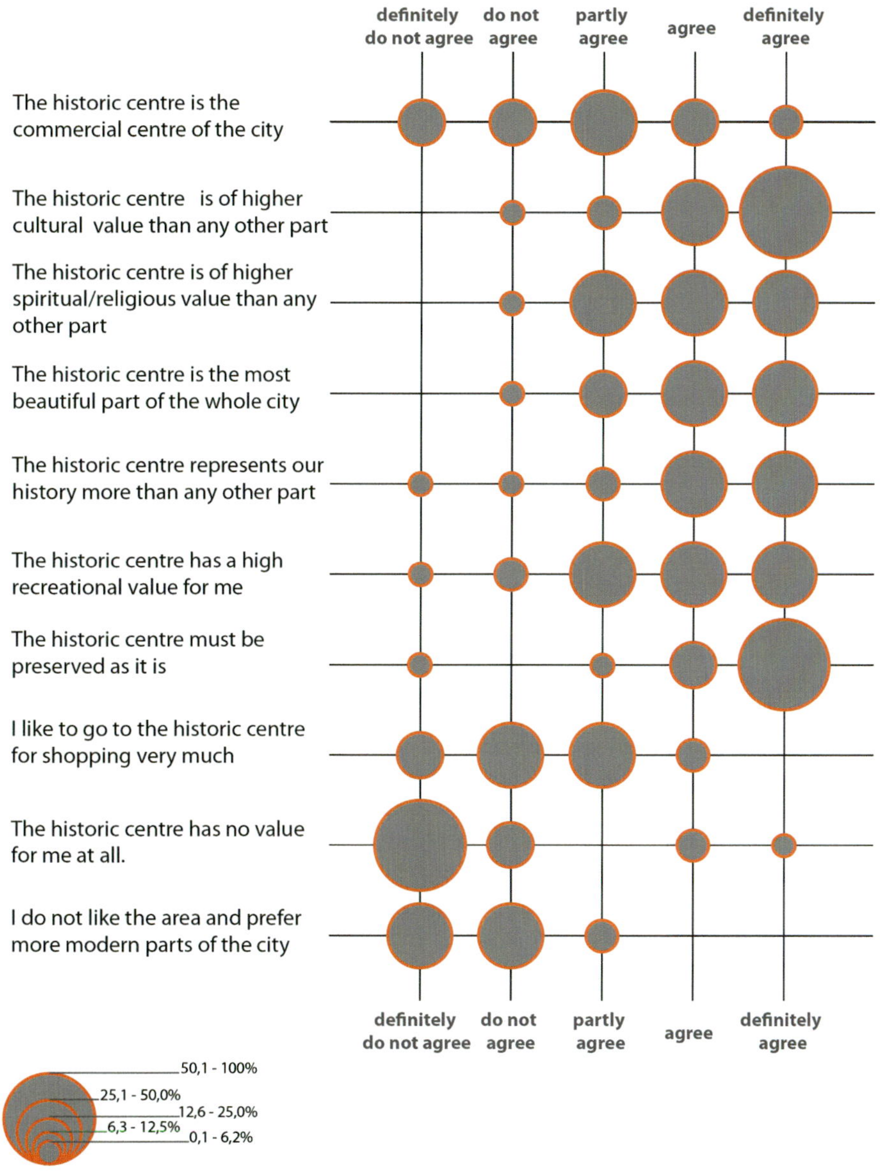


**Fig. 5.15** Responses to the question: Which values do you attribute to the historic centre of Kathmandu? ($n = 80$)

interrelated and becoming manifest in the historic centre in particular because of its accumulation of uses, meanings and places.

After giving their opinion on ongoing processes the respondents were requested to assess values attributed to Kathmandu centre, giving even more distinct answers (see Fig. 5.15). No one agrees to prefer more modern parts of the city instead of the traditional city centre, while only 2.6 % do not attribute more cultural value to the area compared to other parts of town (with even 56.4 % definitely agreeing to this statement). Also, high spiritual value and beauty are associated with the area (with more than 72 % attributing higher spiritual value, perspective 80 % agreeing or definitely identifying the area as more beautiful than other parts of town). However, agreement to the higher spiritual value of that area is comparably less strong than to the other attributes, with more than one fourth agreeing only partially. The recreational value of the area is a bit lower, although still 57.7 % agree or definitely find the centre has a recreational value for them. The least important asset of the area seems to be shopping, as less than 10 % would go to the area for this purpose while 56.0 % denied or even definitely neglected such thing.

85.5 % of all answers agree or definitely agree to the statement that the historic centre of Kathmandu represents the history of the respondents' more than any other part. The clearest position is observed when asking on the need of preservation, with 72.7 % definitely agreeing that the area has to be preserved as it is.

Summing it up, the impressive number of 91.0 % (definitely) disagrees with the statement that the area does not have any value for the respondent. Taking a look at all answers it can be assessed clearly that the historic centre does have a very high cultural value, and respondent's feel it represents their history more than any other part of town. Also, respondents stated they like the area and that it has to be preserved as it is. These answers again are matching the statements of most experts during the interviews. To K11 the protection of cultural aspects is as important for development as economic development and political stability.

Even presumed 'profane' activities like maintenance of a building or temple is endowed with a meaning, as it is part of worshipping. Therefore, local communities play a major role in keeping such traditions alive. In the opinion of K03 "*preserving the intangible association to the built environment is the best way to preserve the buildings: let the Newari do it, let them preserve it, and get their advice for any measures.*" Therefore he also elaborates different times about the importance to protect and appreciate local craftsmanship, in particular wooden carving, but also metalwork and stone carvings, as these are the traditional techniques and "*a living culture*". K09 claims that any renovation measure has to reflect the culture of the place. K07 emphasizes on the same aspect when stating "*you foreigners don't know about Nepalese culture*" (K07).

Surprisingly, intangible assets and their protection was rarely emphasized on during the interviews, to most interview partners they still are alive and need less attention than the tangible ones. Intangible heritage is not mentioned in any legislation. In practise, the DoA is responsible for the preservation of traditional festivals. To K05 from DoA this is important as it is directly related to the tangible heritage, but "*people are forgetting the traditional knowledge.*" K06 even states that "*when people move out also the local community goes down*" (K06). Beyond that K06 states that the preservation of the cultural heritage is for two reasons, for the people themselves but also for the touristic and thus economic value.

Also he finds lifestyles changing, e.g. women working and having less time for daily worshipping of the local deity, or formerly common caste devotion to certain areas, including worshipping and regular maintenance. As consequence maintenance is less voluntary work anymore but is increasingly seen as employment: "*There is no more community devotion as in my grandparents' generation*" (K06).

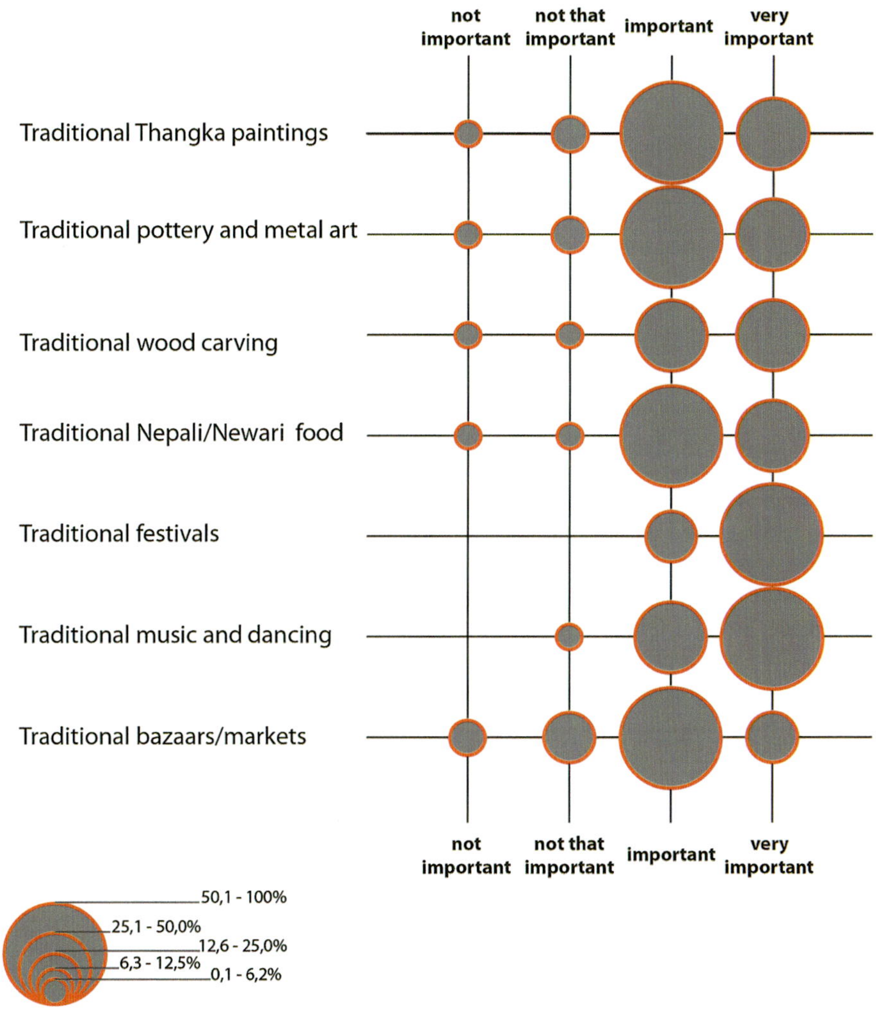


**Fig. 5.16** Answers to the question: Which intangible things or activities do you consider as typical for Kathmandu area, how important is it to preserve them? ($n$ = 80)

Similar kinds of statement were given in different interviews. The interviewees claimed a loss of cultural traditions and even appreciation of the past. In this context the results of the questionnaire are of particular interest as they somehow give a different result. When the students were asked if they are interested in the history of Kathmandu the impressive number of 49.4 % said they are interested, and another 34.2 % even very interested (cf. Fig. A.2: History interest stated by the peer groups of the three case study cities in Annex IV).

Following the questions on values and processes attached to the tangible heritage a last query was posed on the intangible heritage of Kathmandu Valley in general with the aim to assess the importance respondents attached to certain intangible values (cf. Fig. 5.16).

The traditional handicrafts that Kathmandu Valley is famous for are of high importance for the students that filled the questionnaire, with traditional wood carvings being valued little higher than pottery and metal art (93.5 % of respondents find it important or even very important to preserve this skill that is displayed at each single traditional building). Also local Nepali and/or Newari food is of high value, with even 95.0 % considering it as typical for the area and worth being sustained. This goes together with the desire to also preserve traditional music and dancing (94.9 % find this important, out of them 56.4 % or very important) and most importantly, traditional festivals. The impressive number of 85.0 % finds this very important, another 15.0 % important, what means that there was no respondent negating the value of this intangible heritage. The least priority was given to traditional markets and bazaars (still an absolute majority of 51.9 % found it important but 27.3 % said it was not that or not important).

In total, the assessment of intangible values compared to tangible ones is even more impressing, in particular the values the interviewees attach to traditional festivals, music and dancing. This data leads to the conclusion that overall there still is a high appreciation of such intangible assets. Much of the intangible assets listed in Fig. 5.16 are linked to the case study area, as many of Kathmandu's big festivals take place here (at least partly), involving also music. Furthermore here the masterpieces of wood carving and metal art are displayed, although many of the craft workshops are not located inside the area. A large portion of traditional marketplaces are as well located within the area, Thangkas (the traditional Buddhist paintings on cotton or silk, usually showing Buddhist deities) are surely sold in the tourist shops and the Buddhist area around Thahiti, but can also be found in other Buddhist quarters. The same is true for dancing which surely is not restricted to the centre area, but still linked to certain festivals.

Overall the attachment to both, tangible and intangible assets is quite high. Despite the fact that many experts were concerned about the ongoing urban change and the loss of societal and community bonds to certain places and rituals people still attach importance to heritage. A similar picture emerges when taking a look at Fig. 5.17, which depicts the peer group's appraisal of different predefined places or buildings within Kathmandu. Here, interviewees were asked how far the interviewee considers these spots as important place of remembrance of urban history. The answers are unequivocally revealing a very strong linkage to certain spaces. Out of the eleven areas four are located in the city centre, being Kathmandu Durbar Square

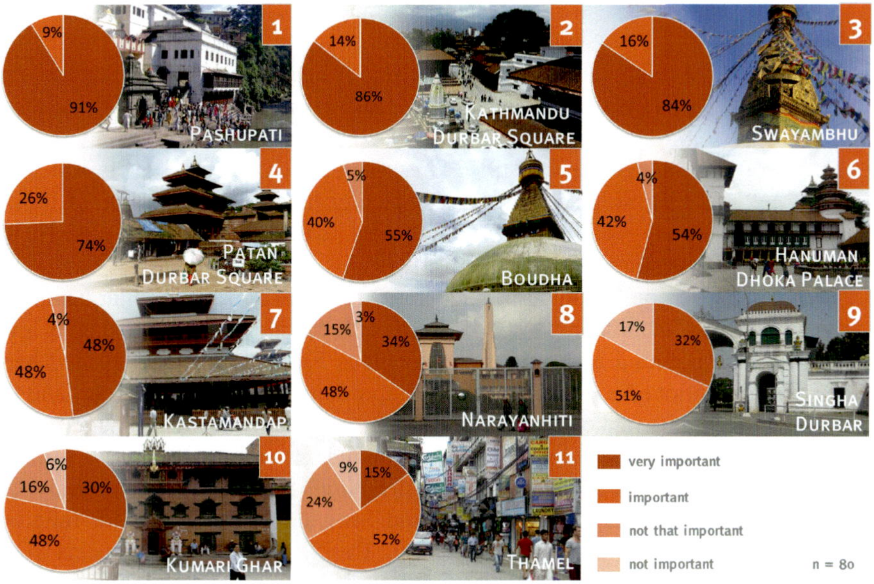


**Fig. 5.17** Answers to the question: Which places/buildings do you consider important as place of remembrance of Kathmandu history?

and three distinct temple or palace buildings. Another three are other World Heritage Sites within Kathmandu Valley. Furthermore Singha Durbar, a Rana palace that houses most of governmental functions today, and three more sites from more recent times are put on the list. These four sites are ranked among the last five, while the historic sites are ranked on top. In particular the sites of Pashupati, Swayambhu and the Durbar Squares of Kathmandu and Patan are all seen as either very important or important, with no voice at all finding them of no or little importance. The three distinct buildings or temples that are ranked at Kathmandu Durbar were found of comparably less importance than the Square itself, in particular the Kumari House is found much less relevant.

In this context it is interesting to take a look at the reasons the respondents gave for their assessment. Besides the ranking itself they were asked to put a justification. In fact only few comments are given for the Kumari House, 18 out of 80, mostly just stating 'Living Goddess'. Few comments were rather negative, saying that there are many chowk (complexes) like this or calling this one a later development. Comments on the Hanuman Dhoka Palace and Kastamandap are comparably more positive. 22 respondents commented on Hanuman Dhoka, most of them emphasized on its historic value as it depicts Nepalese history. The 24 comments on Kastamandap Temple are quite similar; most comments deal with its historic value or mention that the name of Kathmandu emerged after this temple which is said to be constructed out of the wood of a single tree. Kathmandu Durbar Square itself is commented 29 times, most answers deal with its historic and heritage value again, the Square is showing the "*heritage of our Nation*", the "*history and identity*".

Another few answers deal with its beauty and in particular its monuments and fine architecture.

In the case of both, Hanuman Dhoka and Kathmandu Durbar Square, few answers deal with its importance to depict the history of the past kings. The same is true for Narayanhiti Palace, which is found comparably less important (ranking eight out of eleven). However, still 24 respondents commented on it and most answers call it a historic place or find it depicting history as the palace of former kings. Some few answers also deal with its recent history, as the palace is where the Royal Family massacre took place in 2001.

Comments given on Patan Durbar Square (overall 21) are quite similar compared to the answers on Kathmandu Durbar Square, with the additional value that its level of preservation is found high and that the unique Krishna temple is located here. Pashupati is mentioned even 28 times. This spot is ranked most important and the comments reveal it is ranked that high due to its religious value as Pashupati is one of the holiest places for Hindus. Therefore respondents call it a "*religious icon*" or "*one of the most scared temples of the entire Hindu world*". Simultaneously, the respondents find Boudha (25 responses) and Swayambhu important for religious reasons, in this case for Buddhist religion. Beside the religious aspects some respondents found Swayambhu important for its mythological background (see Sect. 5.1.2 on Kathmandu's history) and due to its location on the hilltop with nice scenery.

Thamel is ranked last. This is understandable when going through the 23 comments that mostly find it a "tourist area" for shopping and foreigners. The shopping function is partly mentioned with a positive, partly with a more negative connotation, while the nightlife is mentioned twice as positive.

Overall, the question reveals a very high degree of agreement among respondents concerning the importance of Kathmandu Durbar Square and other sites. The ones ranked highest are those sites with either a religious value (no matter if Hindu or Buddhist, although most probably there are many more Hindus among respondents if the denomination of students is comparable to the one in Kathmandu in general) or a high historic value and quality of stay (Durbar Squares). Summing it up, the importance that the respondents ascribe to these sites is quite similar to the statements of experts during interviews or compared to the justification of WHC when including the Kathmandu Valley on the World Heritage Site. It can therefore be concluded that historic sites are still of a distinct value.

The answers to this question match the result of another—open—question on places the respondents would visit when taking a friend out in Kathmandu. Although overall 54 different answers are given, (Fig. 5.18 depicts the 15 most frequent replies, all others were given only once or twice and are uncared for here), most of the spots the respondents would visit are the urban heritage sites. Kathmandu Durbar Square is ranking no. 3 on that list, but it has to be added that no. 7 on the list was 'all Durbar Squares' that people wanted to visit, among these 3 sites is again Kathmandu Durbar. That shows the importance of the area. Here again space was left to give a reason. The 38 answers are as well quite similar to the answers given when asking about the place remembrance. Most responses deal with

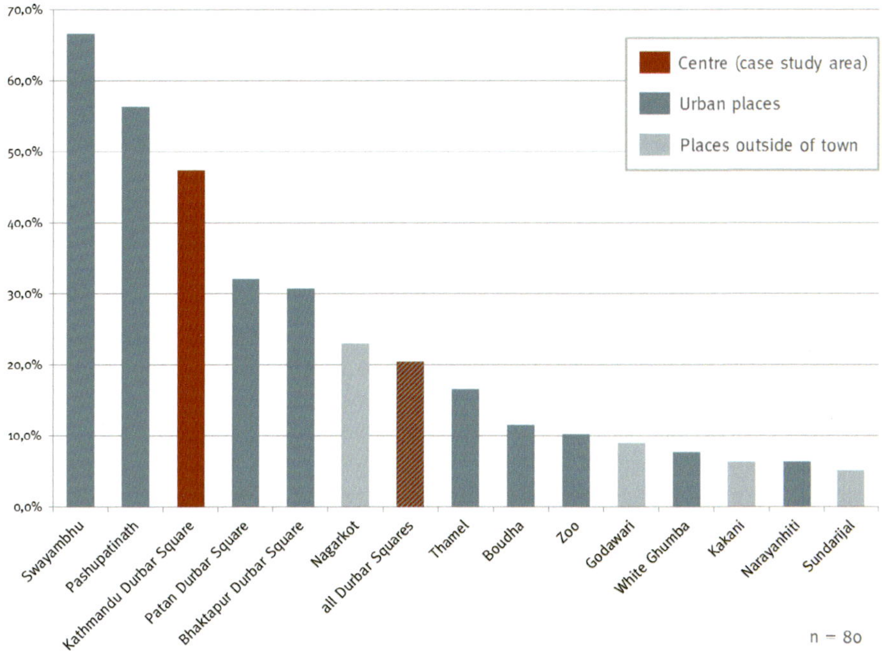


**Fig. 5.18** Answers to the question: Imagine some friends from outside visit you in Kathmandu. Where in and around the town would you take them for sightseeing and leisure?

the area's historic value that is worth a visit to take a look at the urban and national history, e.g. when saying that "*it is our history*" or "*it does not only reflect the history of Kathmandu but also the social life of the local residents here*". Other answers deal with the local culture that can be perceived in and around Durbar Square or emphasized on the comparably higher level of conservation in this spot.

Questionnaire respondents reveal a high level of attachment to the cultural heritage of Kathmandu. The heritage of the Kathmandu Valley is appreciated and people are as well aware of the need to protect it as was revealed during interviews and comments given in the questionnaire.

### *5.1.10 Conclusions on the Role of Heritage in Shaping the Urban Identity of Kathmandu*

Overall, the evaluation of the questionnaire results reveals a high value ascribed to the historic urban centre as well as to other historic fabric in the urban agglomeration. At the same time questionnaire respondents perceive a degradation of the centre area, with traditional houses getting lost and also traditional uses getting lost even if the building itself is sustained. This is matching the perception that current planning is following international trends and no longer the local traditions. This loss of traditional urban layout is visible all around the townscape, where almost only reinforced concrete structures with brick noggins are erected that hardly match the historic layout. This makes the seven World Heritage Sites including Kathmandu centre unique features in a growing ocean of houses. However, while respondents perceive that people prefer to live in more modern houses instead of traditional ones 91.0 % of them claim to like the area and favour it over more modern parts. It seems likely that the historic areas and traditional buildings are something to look at and to visit for certain purposes but that for residential use modern houses are preferred.

Literature poses that local communities such as the Guthis have always been in charge of maintenance of the tangible heritage (Weise 2012; Chapagain 2008), many of the regular maintenance measures themselves became part of the annual festival calendar and thus part of the intangible heritage. The decline of these community-based measures and the social cohesion thus impacts negatively on both, tangible and intangible values. The questionnaire results allow drawing the conclusion that intangible assets like traditional festivals are nevertheless still of very high value for the inhabitants. The vast majority finds the preservation of traditional festivals and traditional music or dancing of high importance. Concerning the tangible heritage, almost three fourth of the respondents claim that the centre area has to be preserved as it is.

Despite the fact that most of the questionnaire respondents did not grow up in Kathmandu but moved there from outside (only 35 % are of Kathmandu origin), they attach great importance to the historic area(s) and emphasize on the preservation of its tangible and intangible values. Obviously neither the number not the sample of respondents is representative for Kathmandu; however, it sheds a light on a potential not yet fully tapped.

The questionnaire results indicate clearly that respondents are very much aware of the risk of irrecoverable losses. The appreciation of intangible values is as high as or even higher than the one of the tangible urban heritage. At the same time the questionnaire results credit the historic centre with a higher cultural value than any other one while respondents express their concerns that these values and uses get lost. Conversely, this attachment to the city centre and its values could be taken advantage of for developing more holistic conservation.

Two questions of the expert interviews dealt with the vision the experts are having for Kathmandu, how they imagine the city to be in 20 years from now. In addition they were asked to give recommendations to improve the problems they described in their own assessment of today's situation. Answers differ a lot between optimistic and pessimistic scenarios. While K01 is lacking awareness for a proper urban planning other experts (K02, K04) find the planning adequate and already existing. They however have different opinions regarding the need for heritage conservation. While K02 emphasizes on the maintenance and extension of monument zones, K04 hopes for a modern city.

Interviewee K01 mentions the wish to preserve the old buildings also in future. To do so he finds it important to accept little changes, e.g. adding one storey. He sees a deep need to take people's view into consideration, as any planning has to be done in such a way, that they accept it. However, "*the government doesn't do this way; they set up rules without further explanation or discussion so that people don't follow*". K03 mentions the wish to keep vernacular architecture, but is rather sceptic about it, as it is already probably too late. "It will only happen if people become aware of the importance and also the economic potential" K03 also states a need for communication among all actors to find a compromise, also more confidence in own capabilities and less trust in only outsiders.

K01, K06, K07 and K09 share a more sceptic view and emphasize on the high level of vulnerability. K06 and K07 therefore propose a maximum height for houses, e.g. 4 storeys (K07). To implement this and for many other long-term strategy political stability is of utmost importance (K07, K11). K05 claim the need for a risk sensitive land use planning for the Kathmandu Valley, which is there in theory but hardly implemented. To him, houses in core area should be built in traditional architecture. He sees it quite possible to make such shift, but what is needed to do so is a vision and commitment. "*It is not a social problem to achieve it.*" In addition, "*There is a big need to manage the urbanisation; the question is how to do it, the high level of urbanisation is crucial to conservation*" (K03).

To K08 conservation has two aspects: the preservation of traditional architecture, this needs incentives and employment generation, and preservation of the rivers and streams: "*City design cannot only be in international style, the planner has to look at what the ancestors did. It has to be done according to the culture of the place to arrive at an identifiable city that also people from the outside enjoy*" (K08).

In conclusion, the results are rather ambivalent. Although most of the experts and the interviewees agree on the importance and values of tangible and intangible heritage, current urban planning and outlooks for the future development look rather negative. The lack of proper planning instruments as well as an even bigger lack of planning implementation is likely to impact negatively on the preservation of urban tangible heritage. Intangible heritage by contrast is not even protected by any laws or policies, although it is appreciated even more than the tangible one. To sum it up, tangible and intangible heritage does still play a role in the life of Kathmandu's inhabitants and even in the life of young people as proven by the questionnaire. Place attachment is there, so is the wish to preserve the historic centre and other heritage areas. However, hardly any of these aspects is considered in urban development.

Kathmandu Summary

- The urban outline of Kathmandu—influenced by Buddhist and Hindu beliefs—is changing rapidly. Additional storeys, vertical division, poor maintenance and changing preferences are resulting in ongoing losses of historic buildings. Triggered by an unstable political situation since the end of twentieth century, unplanned urban growth and a growing number of informal dwellers have led to densification, losses of open spaces and the decay of historic buildings.

- Conservation legislation is focussing almost exclusively on the World Heritage Sites in Nepal as well as on state-owned properties. Conservation in reality is challenging, due to the large number of formal and informal actors involved on the different administrative levels, particularly in case of private buildings outside the monument zones, where building bye-laws are hardly enforced.
- The maintenance of constructions like temples or water spouts is part of traditional worshipping rituals and was carried out on regular intervals by local communities. Disappearing community ties impact negatively on the maintenance, which is additionally hampered by Western understandings of conservation which is not permitting change.
- Intangible values like rituals, beliefs, arts and craft are still practiced and have a very high significance for the local population. However, they are comparably less considered in anyway upgradable conservation schemes.

## 5.2 Case Study II: Yogyakarta, Indonesia

During the field research in Yogyakarta, which took place January–March, 2013, overall 10 in-depth interviews were conducted, with experts from urban and regional planning, conservation authorities, practice and academia. Interview length ranged from between 30 min to 2 h in case of an interview that was done in two meetings. The interview language in most cases was English, two interviews were fully or partly done in Bahasa Indonesia with the help of a translator. In the following chapters the interview partners do not appear with their full names but encoded with Y1 to Y10. Annex II is providing detailed information on names, institutions and position. One interview (Y08) was done with two representatives of the Indonesian Public Works Ministry (among others responsible for regional development and infrastructure) at the same time, according to the interviewees' wish. In this case an 'a' or 'b' is added to the interview code to indicate the respondent.

Two interviews were carried out with representatives of urban or provincial authorities; Y01 is division manager for the Disaster Management Agency of Yogyakarta Special Region (*Pusat Pengendalian Operasional Badan Penanggulangan Bencana Daerah Provinsi, DIY—BPBD*) and used to work for the Regional Development Planning Agency (BAPEDA DIY) before. Y02 is staff of Yogyakarta's conservation authority (*Kantor Balai Pelestarian Cagar Budaya (BPCB)*). Overall six interview partners are employed fully or part-time at different sections of Gadjah Mada University (UGM), indicating the university's importance in the urban area. All of them are working or researching on topics in Yogyakarta, Y02 in the Department of Public Management and Policy Studies and Y04 as researcher in the Center for Disaster Studies which is a major actor in studying the vulnerability of Yogyakarta urban area to different natural hazards which are also regularly harming the city core. Y05 is head of the Center for Tourism Studies, which is working on cultural and heritage tourism in Yogyakarta itself. Y06, Y07 and Y09 are working on different aspects of urban development, in research and practice. Y09 is human geographer and working on aspects of urban vulnerability and urban growth. Y06 and Y07 are both staff of Architecture Department. Y06 is head of the Master of City and Regional Planning and working on kampung development. Before, he had worked in practical urban development projects. Besides being coordinator of the Center for Heritage Conservation at UGM Y07 is member of ICOMOS and the UNESCO-ICCROM's Asian Academy for Heritage Management and chairman of Jogja Heritage Society. Furthermore her family owns a hotel in a historic building within the city centre. Finally, Y10 is project manager for an NGO which is active in different spots in Indonesia, including socioeconomic upgrading projects in Yogyakarta city itself.

Overall 120 questionnaires were filled by UGM students. Respondents are mostly from Master Level, plus few PhD students. The study programmes comprised different programmes within geography, furthermore ecology, population studies, architecture, history and a recently introduced programme on 'Religion and cross-cultural Studies'.

All respondents were residing in Yogyakarta agglomeration when being asked to fill the questionnaire, with the majority (50.8 %) staying there for 5 years or being Yogyakartanese. Another quarter (26.7 %) is staying in town between half a year and two years (cf. Table 5.3), indicating that they came to study their Master Programme. This is reflecting the Indonesian educational system which is providing ambitious students from other islands the chance to study at UGM. On average the respondents are 28.6 years old (standard deviation 6.4 years), ranging from 22 to 52 years.

Like in the other case studies the questionnaire asks about a distinct 'centre area'. In Yogyakarta this area is defined as the Kraton (palace) compound and the adjacent quarters to the south, east and west, mainly the ones within the Kraton area outer walls. Towards North, the case study area includes the Malioboro Road (or Jalan Malioboro) and its neighbouring kampungs to the east and west until the railway line. Figure 5.21 gives a more detailed overview on the area.

**Table 5.3** Period the questionnaire respondents are already residing in Yogyakarta ($n$ = 120)

| Grown up in Yogyakarta | For more than 5 years | Between 2 and 5 years | Between 6 months and 2 years | Less than 6 months | No answer |
|---|---|---|---|---|---|
| 24.2 % | 26.7 % | 5.8 % | 26.7 % | 14.2 % | 2.5 % |

### *5.2.1 A Brief Country and Planning History of Indonesia*

The Republic of Indonesia is a multiethnic and multi-language country (Kötter et al. 1979). The Indonesian national motto is *Bhinneka Tunggal Ika*—Unity in Diversity. The state philosophy is following 'five pillars' or *Pancasila*, namely the belief in the One and Only God, a just and civilized humanity, the unity of Indonesia, democracy through unanimous deliberations, and social justice for the Indonesian people. With overall 252.8 million inhabitants Indonesia is the fourth most populated country (World Bank 2014a). It has the largest Muslim population on a global scale; however due to the country's very diverse history some islands or regions have large Christian and Hindu communities. The island of Java is the most densely populated part of the country, with more than 60 % of the population living in urban areas (Zahnd 2005). Indonesia is considered a lower middle income country, forms part of the G20 and ranks number 108 out of 187 in the HDI index (UNDP 2014). The country comprises of a tropical climate with a dry and a rainy season and is prone to different natural hazards, in particular earthquakes, volcanic eruptions and tsunamis, as well as landslide and flooding (Hadi 2008).

Due to its multi-island geography the country is very diverse in cultures, beliefs and traditions (Röll 1979). This is also true for arts and architecture. The first humans arrived over a land bridge more than a million years ago and colonized different islands until first small kingdoms with permanent settlement structures emerged around the first century AD. During the next centuries first Hinduism, then Buddhism, and lastly Islam was brought to the archipelago (Vickers 2013). Due to its strategic sea-lane location and the richness of natural resources and agricultural products, particularly spices, the Indonesian archipelago was of high interest for colonial powers. In the seventeenth century the Dutch arrived and became the leading power for the next centuries, by means of the newly established Dutch East India Company (*Vereenigde Oost-Indische Compagnie*—VOC). Dutch rule ended only after the end of World War II in 1945 when the later President Sukarno declared the country's independence. After 4 years of struggles, during which Yogyakarta became the capital city, finally the Republic of Indonesia was formally recognized. Jakarta became capital city, a manifestation of the postcolonial unitary nation at the location of the former colonial capital of Batavia (Lademacher 2001). General Suharto, who followed Sukarno as president from 1965 to 1998 then implemented his New Order policies, focussing on economic development and a strong role of the military by means of an authoritarian regime. During his presidency the *transmigrasi* programme that sought to transfer families from densely populated regions like Java to other less populated islands peaked. It declined at the latest with the fall of Suharto's regime in 1998. Since then democratic processes, including decentralization and regional autonomy, were pushed forward (Vickers 2013; Salazar 2010). In 2014 Joko Widodo, better known as Jokowi, became the first president of Indonesia that had never had a military affiliation. He became popular all over Indonesia due to his advances in urban development and his fight against corruption as mayor of Surakarta, Central Java, and governor of Jakarta.

Indonesia's urbanization rate is 53.7 % (United Nations—Department of Economic and Social Affairs 2014) and is predicted to grow. The first phase of urbanization occurred only after Indonesian independence in 1945, until then the country was predominantly rural (Rutz 1985). In this context Evers (2007) talks about a cultural construction of Indonesian urbanism, as it almost did not exist before the arrival of the Dutch. Until then the Kraton areas served as local centres around which people settled. This so-called "focal urbanism" of Asian cities is different from European cities and their "local urbanism" as they mostly evolved from older settlements (Böhme et al. 2003; Siauw 2003) Evers claims that during the very early stages of Indonesian globalization, the new urbanism "*was glossed over by a symbolism of the pre-colonial past, as if the elite were afraid to face the challenges of globalization whilst losing the connection to the Javanese past that had hitherto been the source of their inspiration*" (Evers 2007: 59). This symbolic return to local traditions resulted in a mix of modern architecture and traditional forms of art, e.g. modern architecture that is still following the traditional planning of Yogyakarta. Here, for example the museum dedicated to the national revolution, Monumen Yogya Kembali, is aligned with Merapi and Kraton, as if it was part of the initial urban outline (Evers 2007).

Much of the country's early urbanization of the 1960s and 70s took place in an informal way (Wolfram-Seifert 1986) "*without leading to the modernization of built structures, modes of transport, industries and occupations*" (Evers 2007: 52). Since then urbanization trends continued resulting in urban growth beyond the administrative borders and along infrastructure lines, denoted by the Bahasa Indonesian word '*desakota*', joining the words for village (*desa*) and town (*kota*) (Kraas and Nitschke 2008). Most city governments are still poorly prepared for such rapid urbanization, resulting in inadequate urban infrastructure and services, the need for slum upgrading and affordable housing for the urban poor, inadequate land-use planning and development control, aggravated environmental problems as well as a lack of security of tenure (UN-HABITAT 2010b; Brillantes and Flores 2012; Puppim de Oliveira et al. 2013). Tenure and land titles are a main concern in Java, where the original land use system did not foresee individual land properties that could be sold. That system was only introduced with the Dutch resulting in a still often unclear situation with different individuals holding legal rights of the same plot of land (Zahnd 2005; Sosiawan 2009).

The traditional housing patterns are so-called kampungs. Ford (1993: 392) defines a kampung as "*mostly unplanned, primarily low-income residential area that has gradually been built and serviced*". Kampungs are the traditional settlement form, for most Indonesians it therefore is a synonym for 'home community' and the lowest level of (informal) community organization (Sullivan 1986). After having been considered as outdated form of housing during early urbanization (Larasati 2007) in the 1970s massive need of housing space for the growing population led to the so-called "Kampung Improvement Program—KIP". The KIP can be seen as a success story in Indonesian urban planning (Böhme et al. 2003). Despite its initial focus on provision of low-income housing and upgrading of living quality the KIP became an important component of government-initiated upgrading strategies in dense urban areas, bringing road construction, social services and sanitation to the traditional (urban) settlements (Larasati 2007; Tunas and Peresthu 2010; Minnery et al. 2013).

The provision of housing for Indonesia's growing population is a crucial issue. As the National Housing Corporation (*Perumnas*) is not able to fulfil all demands construction still done quite often in a more informal approach, following the principle of *gotong royong,* an Indonesian phrase referring to mainly informal community self-help activities. Although being comparably more common in rural areas, the approach is still practiced in urban areas as well, particularly in case of non-engineered structures or communities settling on ground not belonging to them (Bowen 1986; Larasati 2007). At least since independence in 1945, *gotong royong* is an essential part of the ideological basis and the construction of a national Indonesian tradition (Bowen 1986; Evers and Korff 2003).

Since 2001 Indonesia is undergoing a rapid and extensive process of decentralization, usually referred to as 'Big Bang' decentralization. Much of regulatory authority was given to local governments, to the districts and municipalities, including a wider control over natural resources. Another aim was to give more room for public participation and equality principles (Brillantes and Flores 2012; Phelps et al. 2014). One result of the rapid decentralization was a fragmentation of local governments, as the large number of newly founded district and provincial government units caused problems in financing, monitoring and evaluation, particularly for urban agglomerations. Local governments were not always well prepared to take over the newly gained autonomy, resulting in tendencies to exploit local resources more excessively to maximize incomes, facilitated by the partly weak institutional capacities (Firman 2010; Brillantes and Flores 2012).

## *5.2.2 Introduction to Yogyakarta*

Yogyakarta is located in the Central Java region and was founded in 1755, following a planned layout stretching from Mount Merapi in the north (cf. Fig. 5.19) to the Indian Ocean in the south. It comprises of a tropical climate with a dry and a rainy season. Fertile soils are used for doing agriculture with rice and different vegetables being the main crop. A large share of the province's rural population is still doing agriculture, although ongoing urbanization is building up fertile land, converting it to housing areas (Hizbaron et al. 2013).

Until and throughout the nineteenth century the city was one of the political and economic centres of Indonesia, however its importance declined latest when Jakarta became the capital city after gaining independence. Today it is known as an educational hub on national level (Subanu 2008; Rana and Marwasta 2015) with more than 90 universities and higher building institutions. One of them is Gadjah Mada University, the oldest and one of the most renowned universities within Indonesia (Firman 2010; Prabawa 2010; Harsono 2011). Furthermore, Yogyakarta is a centre of Javanese culture (Sugiana 2008; Salazar 2012). There is no major industry

**Fig. 5.19** Aerial picture of Yogyakarta's north with the slopes of Mount Merapi in the background

located in or around Yogyakarta, the leading economic sectors are tourism, higher education, and small and medium manufacturing (Sugiana 2008). Services, trade, restaurants and hotels, transportation and communication account for the vast majority of the urban gross domestic product, while in the surrounding districts agriculture is still predominant (The Consultative Group on Indonesia 2006).

The city is the capital of Yogyakarta Special Province (*Daerah Istimewa Yogyakarta*—DIY) of around 3,200 km$^2$ (Siauw 2003) and 3.68 million inhabitants, 70 % of them urban (Badan Pusat Statistik 2015b, c), making it the second most densely populated province in Indonesia (The Consultative Group on Indonesia 2006). The urban population is mainly of Javanese origin (85–90 %). The rest is partly of Chinese origin, often engaged in business and trade, beside that there are inhabitants with Arab, Dutch, Indian roots or from other Indonesian islands (Siauw 2003). DIY is one of only two provinces in Indonesia holding this special title, the second one is the capital city of Jakarta itself, indicating the importance of both cities throughout Indonesia. The title was granted in 1950 in return to Yogyakarta's role during Indonesia's struggle for independence, also granting the Sultan the title of Governor. Despite being a democratic country its special status permits the Province to still keep the Sultanate. The Sultan as hereditary monarch is in parallel holding the title of DIY Governor granting him the same authority and responsibilities as any other governor in the country, with the difference that he is not bound to legislative periods (Höflich de Duque 2006; Salazar 2010).

The Sultan was set as governor by law, creating a certain dualism in his figure, as a symbol of the traditional Yogyakartanese culture and at the same time as legitimated ruler under the Republic's law. As a consequence his sphere of influence is going beyond formal political power, as many traditions and customs are still related to him and the traditional societal system (Harsono 2011). In the Javanese traditions land rights and inheritance almost did not exist, with the exception of the Sultan court. Until today a large share of Yogyakarta's land is belonging to the sultan, like riverbanks or most of the UGM compound. With the arrival of the Dutch the land title system changed as Dutch implemented their system. As a consequence until today different systems of land certificates exist in parallel. In reality more than one land owner may hold a certificate for the same plot of land (Siauw 2003; Sosiawan 2009).

The urban area is densely populated, with an average of 2,464 units/ $km^2$ it is ranking third in whole Indonesia. More than 20 % of the houses in Yogyakarta are found relatively vulnerable to different hazards (Hadi 2008). Yogyakarta itself is prone to different natural hazards. Particularly the riverine areas are highly affected by annual flooding during rainy season, and the eruptions of nearby Mount Merapi, one of the most active volcanos in Indonesia, which can result in lahar floods impacting severely on the urban area (World Bank 2011). The last eruption in 2010 blanketed many houses and heritage sites like Borobudur Temple Compound with volcanic ash which had to be removed to avoid damages (UNESCO 2012). In addition, the Yogyakarta has to prepare for the high earthquake risk. During the last major quake on May 27, 2006, which reached 5.9 on Richter scale, at least 6,000 people died, and many more lost their homes (Adishakti 2008; Hadi 2008). Only in Yogyakarta City more than 4,800 buildings were completely and another 3,500 partly destroyed (The Consultative Group on Indonesia 2006).

In this event Yogyakarta has lost important parts of its cultural heritage, tangible and intangible one. The earthquake affected in particular the Kotagede Heritage District, remains of the old Mataram Kingdom (today an urban quarter located southeast of Yogyakarta city centre). Additionally it also seriously disrupted core activities of local industries such as sterling silver, batik craft, pottery, wood craft, *ikad* weaving, and other traditional crafts practiced in the southern part of Yogyakarta agglomeration (Adishakti 2008). Particularly the volcanic hazard is part of daily life and culture. The Sultan himself plays a fundamental role in traditional risk management of Mount Merapi, and volcanos are essential part of *Wayang Kulit* (puppet theatre) performances (Mercer et al. 2012).

The relatively high risk level calls for an institutional framework for disaster preparedness as well as post-disaster rehabilitation and reconstruction which is involving national, regional and urban authorities. In DIY the Disaster Management Agency of Yogyakarta Special Region—BPBD—is responsible for elaboration and implementation of such policies, including the heritage zones (Y01). The foundation of this agency on provincial level is a quite recent undertaken, while before most initiatives were on national level. This can be seen as one major outcome of the country's decentralization strategy, which has led, among others, to the creation of metropolitan regions. This way planning and administration also tries to cope with the ongoing growth of urban agglomerations. In 2001 three local governments within the Special Province of Yogyakarta (DIY Yogyakarta), namely the city (*kota*) of Yogyakarta, and the two districts (*kabupaten*) of Sleman and Bantul, joined under the umbrella of Kartamantul metropolitan area. The agglomeration covers as much as 234 $km^2$. As Yogyakarta itself is almost built up, the urban growth is taking place mostly in Sleman and Bantul areas that were formerly used for agricultural activities. In addition the centre areas are densified and increasingly verticalized, impacting on the urban appearance. Therefore the Kartamantul administration is working on urban infrastructure, transport, risk management and land use for the whole agglomeration, with the (initial) support of international NGOs (Höflich de Duque 2006; UNDP Indonesia 2006; Firman 2010; Yap 2012).

Yogyakarta is the second most important Indonesian tourist destination after Bali (Evers and Korff 2003; Sugiana 2008; Salazar 2012). Tourism plays a leading role in urban economy and urban development since President Suharto's New Order regime (Salazar 2011a). Dahles (2001: 54) calls the city "*a pampered child of the New order government*" that supported and promoted in particular the Javanese court tradition. During the 1970s and 80s Yogyakarta became one of the spearheads of the national government that promoted the city excessively. The city was promoted as showcase of typical 'Javaneseness', until today part of Yogyakartanese self-image and the base of regional cultural development planning (Suryanti et al. 2014).

> The image favoured by government and tourism industry was the image of a city with a multifaceted cultural heritage: as a historic city, Yogyakarta represents the diverse religious and cultural traditions that have characterized the area through the ages; as a heroic city, Yogyakarta represents the struggle for independence and national unity; and as a cultural city, Yogyakarta represents the uniqueness of a 'traditional Javanese community' (Dahles 2001: 54).

As a consequence domestic as well as international tourism became major sources of income while a huge number of hotels were constructed within the city's boundaries. Tourism then declined in the mid-1990s, due to the economic crisis and an unbalanced choice of hotels with a preponderance of five-star ones, while particularly international tourism meanwhile had changed to a more budget and backpacker one (Dahles 2001; Hampton 2003; Sörensson 2008).

Recently, with an again rising number of visitors (Badan Pusat Statistik 2015a), the construction of luxury hotels has continued, with a large share of them nearby the historic urban centre. The most important tourist attractions in the region are Borobudur and Prambanan, both of them are UNESCO World Heritage sites (Salazar 2010) and one of the major destinations since the early days of Indonesian tourism (Helbig 1949a, b). Yogyakarta serves as the transportation and accommodation hub to visit these temple complexes. In the city itself tourists are mostly interested in getting to know the Kraton and the surrounding area, as well as Malioboro Road with its shopping faclilities. Besides that local handicrafts and living cultural traditions are major attractions (Timothy 1999). The hub of budget tourism as well is located within the centre area, in the *kampung* of Sosrowijayan, west of Malioboro and south of the train station. The relatively small area is full of budget hostels, housing, *warung* (local restaurants), bars, small shops and workshops (Dahles 2001; Hampton 2003; Sörensson 2008).

As Jakarta is increasingly suffering from regular flooding and congestion Yogyakarta became popular as business, leisure and conference site over the past few years (Y10). As a consequence hotel construction activities are increasing, while the traffic problems are continuously growing. To escape the traffic gridlock and to cope with the problems of rapid urbanization and potential impacts of climate change, Yogyakarta municipal government introduced a public bus system, the 'Transjogja', in 2008 (Bulkeley et al. 2011; Hook 2005).

Urban growth has led to a strong suburbanization process and subsequent losses of agricultural land. It accelerated with the construction of the outer ring road in the 1980s and subsequent developer-led housing projects (Rana and Marwasta 2015). Today, the urban area is a mix of traditional layout and modern, globalized, urban appearance. While particularly the central quarters around the Kraton have preserved a traditional appearance, shopping malls, hotels or restaurants belonging to international chains are shaping the outer areas (Salazar 2008). Within the past decade the number of malls in the urban and peri-urban area has grown. Partly, they are co-owned by the Sultan himself and his entourage. At the same time the Sultan in his function as provincial governor encouraged the mall construction, officially for the sake of economic development:

> Traditionalists, however, argue the malls obstruct the spiritual line between the sea and the volcano. They claim the sultan is destroying Yogyakarta's heritage while his divine mandate is to protect the cradle of Javanese culture. Both the northern and southern powers are showing their disagreement, and the sultan does not have enough spiritual authority to negotiate a truce (Salazar 2012: 36).

Urban transformation is hardly controlled, beyond that social changes are impacting on the traditional urban outline. In the quarter of Kotagede Adishakti (2008: 245) describes a decline of handicraft industries, decreasing interest in cultural activities, and a neglect of architectural remains, in particular of traditional houses. Nevertheless, in the urban outline still many pre-colonial and colonial elements are conserved (Siauw 2003). The different societal groups had their distinct housing preferences which are visible in the urban outline and its ethnic quarters (Evers 2007; Colombijn and Barwegen 2009), while the recent trend of modern apartment housing and hotel construction is rather following international style. Within the past twenty years many malls and high-rise hotel buildings emerged. Although individual houses are still the preferred type of housing which are now given up due to land scarcity (as stated by Y09).

### 5.2.3 The 'Historic Urban Landscape' of Yogyakarta

The urban history of Yogyakarta is multifaceted, from different époques like the ancient Hindu and Buddhist kingdoms in the area, the Mataram dynasty that actually built the city, Dutch colonial and contemporary times (Siauw 2003; Suryanti et al. 2014). The historical background of Yogyakarta dates back more than 1,000 years to Buddhist and Hindu dynasties that reigned the area and built huge the temple complexes of Borobudur and Prambanan showcasing the multifaceted history of the region. Both sites are UNESCO World Heritage since 1991. In sixteenth century the Islamic Mataram kingdom was established in the area. The Mataram kings settled in Kotagede, today a district of Yogyakarta. Modern Yogyakarta came into being in 1755, when a land dispute split the Mataram Empire into the Sultanates of Yogyakarta and Surakarta, by will of the Dutch colonial rulers. The first Sultan, Hamengkubuwono I., then spent 37 years building the new capital of Yogyakarta, in the Code River area, between Mount Merapi and the sea (Subanu 2008).

Yogyakarta is considered an outstanding example of Hindu-Javanese inland residential cities, where the city was seen as reflection of the four world areas and their divine centre, characterized by the close linkage between political power and religious importance, which are reproduced in social and spatial urban hierarchies (Siauw 2003). Yogyakarta is an outstanding example of Javanese court culture, with the Sultan and his Kraton as symbols of Javanese culture (Salazar 2008, 2012; Prabawa 2010). The ancient planning is still perceived in present-day urban pattern. Yogyakarta is built around the Sultan palace, the Kraton (Zahnd 2005). The Kraton residence is where political and spiritual power was and is located, situated in the centre and surrounded by a wall of 6 m height. The compound of 9 ha comprises of seven courtyards surrounded by numerous buildings, also following a distinct scheme. Until today it is home to the Sultan and his family, although a part of the area has been converted to a museum.

The urban layout is following the regular ground plan of Javanese cities with four elements in the centre: the Kraton palace area, a ceremonial square with two Banyan trees, called alun-alun, a mosque and a pasar (market). All these elements follow the planning scheme of *catur tunggal*, 'four in one' (Siauw 2003). The only major exemption from this scheme is a second palace complex southwest of the Kraton, the Taman Sari or 'Water Palace' which today is a tourist attraction. Outside of this centre were residential areas for different classes. It can be said that the status of the settlers defined their place of residence, the closer to the Sultan the higher in hierarchy. Basically the urban layout followed the hierarchical structure of Javanese society, with the inner Kraton for the Sultan and his family as inner circle, surrounded by nobility within the walled larger Kraton compound, urban dwellers in the third circle, surrounded by agricultural land with rural population (Sullivan 1986; Zahnd 2005).

Near the Kraton the major mosque was built, representing the religious power. Around the mosque usually a Muslim quarter called Kauman emerged. Chinese traders settled and founded their Chinatowns near the market while the noble people settled around the palace in a quarter called Dalem. After the arrival of the Dutch most of the Javanese towns changed their appearance through adopting western elements, so did Yogyakarta. The most apparent change was a fortification, Fort Vredeburg, beside it facilities like banks, post offices and churches became part of the urban landscape near the Kraton. The Dutch urban elements were added to the initial urban planning, e.g. the residence of the Dutch resident was built north of the *alun-alun* along the main axis, displaying the colonial power. School, church, hospital and the residences of important Dutch officials were built nearby, somehow a mirrored image of the Javanese centre layout but with Dutch-style buildings (Sullivan 1986; Ellisa 2010). The last modifications under Dutch rule stem from a plan elaborated 1936 by Dutch engineer Herman Thomas Karsten, including among others the upgrading of major roads, introduction of office and shopping facilities, and the construction of Kota Baru, the Dutch quarter northeast of the centre (Yunus 1991).

In Yogyakarta the different ethnic groups are still visible in the urban outline of the different quarters (Ellisa 2010). Due to ethnic segregation Dutch or Europeans settled apart from Chinese, who themselves had their own quarter, not living together with Javanese. Chinese migrants brought their traditional shop houses. This housing type was introduced by Chinese immigrants that came to Indonesia during colonial times in sixteenth century. Usually they are built in a row, with the front for shop, while back and top floors are housing or storage area. In Yogyakarta shop houses are mainly found in the centre area around Malioboro Road, north of Beringharjo Market. The majority was built under Dutch colonial regime and concentrated in this area due to ethnic segregation and political regulation prominent at that time. Over time the outer appearance has partly changed, however, the interior room division mainly remained the same (Siauw 2003; Zahnd 2005; Anggraini 2012). While the Dutch preferred villa-style single houses on spacious green plots, Chinatowns consist of almost uniform blocks. Javanese housing types on the other hand are characterized by their outward orientation towards front and back spaces. Traditional Javanese houses consist of a front part for welcoming visitors, the decoration and size depends on the owner's social status—representing the underlying philosophical concept of Javanese society. The inner part is used for sleeping, eating, gathering and religious activities while the back part is for cooking and bathing (Marcillia and Ohno 2012). The densely built blocks have wide streets to the outside and small alleys in the inside, connecting the small building lots (Ellisa 2010).

There are even differences within Javanese building styles, as e.g. the quarter of Kotagede southeast of the centre, is known for its traditional houses (*Joglos*). The home town of the late Mataram kings is famous for its buildings and handicraft, in particular silver jewellery. Sterling silver craftsmen settled in their own kampungs, which exist until today. However, this district was among the most affected by the 2006 earthquake. Many of the traditional houses, which are an embodiment of folk heritage, have been destroyed, craftsmen have lost their resources. In total 88 of the approximately 150 traditional *Joglo* houses were harmed or destroyed (Adishakti 2008). As a consequence, major action was taken to restore Kotagede in an adequate way, paying attention to local traditions and heritage. Also the communities' economic situation was considered by analyzing the potential for local economy and tourism. The project was executed by different institutions and organizations, including local communities, NGOs, university and governmental as well as international organizations. It is remarkable that one local initiative called *Pusaka Jogja Bangkit!* (Jogja Heritage Revival), meanwhile also active in Kraton area, was set up only 2 days after the earthquake. The project's main aim can be summarized as "*building the local economy through strengthening both tangible and intangible heritage for economic and sustainable development*" (Adishakti 2008: 242), indicating the importance of the area and its heritage.

Currently Yogyakarta has one of the best-preserved city centres in Indonesia. Since the 1970s the KIP was implemented in Yogyakarta. Although the preservation of traditional buildings in inner-city areas and thus the place attachment of local people were not the main approach (it was rather road construction, social services and sanitation), the KIP contributed substantially to preserving the kampungs as historic elements of Yogyakarta (Böhme et al. 2003). However, conservation projects are mainly taking place in the upper-class or temple and palace areas and not in the historic middle- or lower class residential areas, as they are of less value in Javanese planning scheme. As a consequence, mainly the palaces and representative buildings are taken care of, while much of the middle- or low-class constructions are either in a less good shape or even replaced. This is true as well in the urban centre; where despite existing heritage protection frameworks (see Sect. 5.2.6) construction activities are taking place, including the replacement of buildings, e.g. in the busy area around Malioboro Road (cf. Fig. 5.20).

The diverse urban history is reflected in buildings, monuments, places, street names and other artefacts, much of them reminding of past events and époques: "*Only some have meaning for the urban population or its rulers. These artefacts often remain even during urban renewal when the process of urbanisation continues. Looking back, we have, therefore, to engage in a sort of 'archaeology of meaning'*" (Evers 2007: 56).

Bonds between culture and nature become apparent in traditions, rituals and beliefs, but also increasingly in more recent activities like the 'Jogja International Heritage Walk' which is carried out for the seventh time in 2015 (Jogja Walking Association). As stated previously Yogyakarta is a centre of Javanese culture, visible in various cultural expressions, such as traditional music (*gamelan*), puppet

**Fig. 5.20** Malioboro Road

theatre (*wayang kulit*, intangible UNESCO Cultural Heritage since 2008, see Sect. 3.5), batik (intangible UNESCO Cultural Heritage since 2009, see Sect. 3.5), silverworks, local food and festivals (Hampton 2003). In this context Sunaryo et al. (2011: 296) state:

> Some of the physical artifacts and rituals in the city's public spaces have survived to this day and have found new meaning in the context of the present era. It can be said that the public space in the city of Yogyakarta is still trying to preserve the core values that shaped its identity.

Over the last years activities to promote local culture and values have increased, what can be seen in line with the recollection of the national 'Unity in Diversity' motto (cf. Sect. 5.2.1) as a counter-reaction to ongoing independence movements of certain regions (as stated by Y09).

Many activities on urban scale tend to particularly keep local customs and traditions alive as a part of Yogyakartanese life. Also markets are essential parts of Yogyakartanese life, from local neighbourhood markets which are explicitly promoted by the Urban Spatial Plan to the major traditional markets under the Traditional Market Municipal Agency (*Dinas Pasar*), including Beringharjo Market in Malioboro Road, where many goods of local supply but also handicraft and batik are sold. Particularly batik is still practised and sold all over the city, in different qualities and styles; at least a part of the fabrication is still done around the Kraton, where the businesses were located originally. Tradition is very important and the Kraton claims to be the true "*inheritor of the great tradition of Java*" (Shiraishi 1990).

## 5.2.4 *Inner City Patterns and Dynamics*

Yogyakarta's centrepoint is the Kraton, the palace of the Sultan and his family. Today large parts of the Kraton serve as a museum, while the inner parts are still inhabited by the Sultan family and not accessible. It is mostly composed of one- to two-storey buildings with large roofed pavilions (cf. Fig. 5.22/4). The Kraton area opens onto two large squares towards north and south, the alun-aluns. While the southern one is located within the Kraton walls, the northern one "*represents the interface between the court and the city*" (Sunaryo et al. 2011: 287). In the middle of both squares there are two fenced banyan trees as landscape element and as symbols of the places' sanctity. Both squares served as important meeting places as well as a venue for ceremonial occasions, military demonstrations, festivities and markets (Siauw 2003; Sunaryo et al. 2011) (Fig. 5.21).

Today particularly the southern alun-alun is still a picnic and gathering place for young Yogyakartanese in the evening time while it is less used during daytime (cf. Fig. 5.22/1). One favoured leisure activity is to find the way through the Banyan trees with their eyes blindfolded, which is said to bring luck to the one who succeeds: "*The square's sacred atmosphere of the past has been greatly reduced*

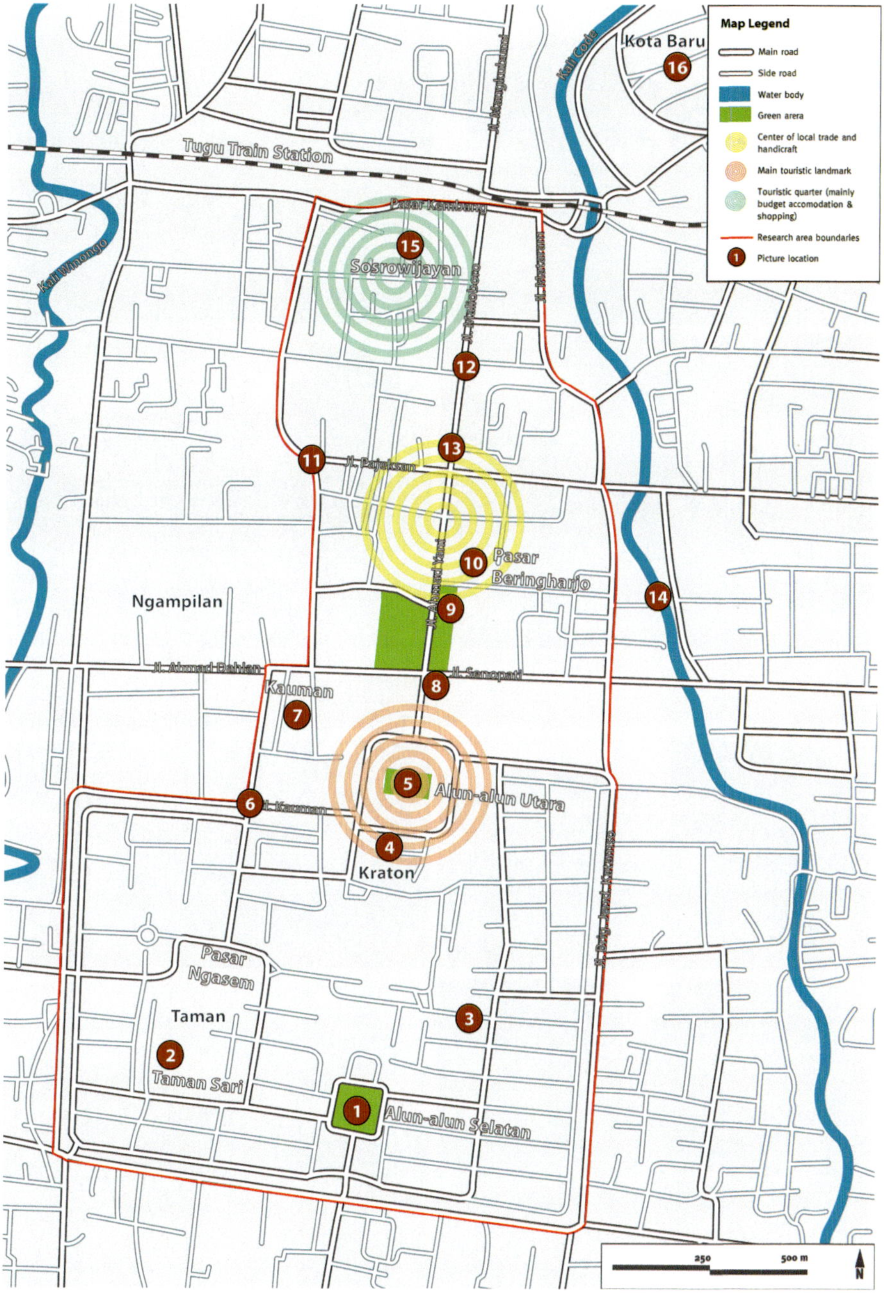


Fig. 5.21 Map of Yogyakarta centre area with research area boundaries

**Fig. 5.22** Streetscapes in Yogyakarta centre area, 1–8

*but today the square serves as a true public space*" (Sunaryo et al. 2011: 294). The northern alun-alun is the entrance area for the museum part of the Kraton. During daytime it is filled with buses and cars, souvenir and food vendors (cf. Fig. 5.22/5). Informal trading is popular in both places and "*able to attract people coming around to the town square during the evening particularly*" (Nitisudarmo 2009: 524).

The outer walls of the Kraton (cf. Fig. 5.22/6) enclose a quarter which initially was the living place of the Sultan's inner circle and high-ranking servants. It is composed of a grid of narrow roads and alleys (cf. Fig. 5.22/3; here the lane is used as parking for the illuminated rickshaws which are another popular leisure activity at southern alun-alun in the evening). Until today the houses are still often following Javanese models, surrounded by small gardens. The ground itself belongs to the Sultan, until today more restrictive building codes are applied here to impede uncontrolled alteration and verticalisation of the area (Y03, Y07):

> As new development is unavoidable there are rigid rules for any measure, like houses surrounding the Kraton area should only have one floor. They can have two ones but only if the overall building height is not exceeding the height of Kraton roofs, it is not allowed to have buildings higher than the Kraton, people are also obliged to follow the traditional architectural design codes (Y03).

One particularity of Yogyakarta already laid down in the initial planning is the Taman Sari palace area south of the Kraton (see Fig. 5.22/2). Although being a ruin today it still shows off its former beauty, particularly the richly ornamented bathing pools that once were used by members of the royal family. Today it mainly serves as a tourist attraction. In both areas, Taman Sari and Kraton, Y03 is witnessing similar problems with new constructions not following the rules, but different in size, shape and appearance.

The Kauman area west of the Kraton is the 'Islamic' quarter of Yogyakarta, where also the central Mosque is located (cf. Fig. 5.22/7 with the Mosque in the background). It emerged as part of the initial Yogyakarta planning scheme in the eighteenth century. Until today the Sultan himself is the formal head of the religious community. The quarter is famous for Batik, although production itself has declined while showrooms emerged. Most of the area is not accessible by car; the narrow and contorted lanes allow access only for pedestrians and motorbikes. The still homogeneous building stock consists mainly of one- to two-storied buildings and is in a comparably good condition. The predominant use is housing, commerce is much less prominent except smaller shops for daily supply (Zahnd 2005). Larger commercial areas are mainly found in the quarter's fringes and towards its north where Malioboro Road is located.

At the junction the Kraton and Malioboro areas many buildings from Dutch colonial phase are located, such as the former post office (Fig. 5.22/8), or Fort Vredeburg (Fig. 5.23/9), which today is used as a museum showcasing, among others, Indonesia's struggle for independence. The plot of land was provided to the Dutch by Yogyakarta's first Sultan in 1760. The building itself has been rebuilt in 1867 after its precursor was destroyed in an earthquake. In front of the fort the

**Fig. 5.23** Streetscapes in Yogyakarta centre area, 9–16

estate of the former Dutch resident (*Gedung Agung*) is located. The initial construction built in 1824 was destroyed in the same earthquake as the fortification and rebuilt in 1869. Later it became the residence of President Sukarno before the government was moved to Jakarta. Today the fenced complex is used as presidential guesthouse.

As typical for Javanese cities, the main market or *pasar* is located close to the Kraton, in this case right north of Fort Vredeburg. Pasar Beringharjo (cf. Fig. 5.23/10) is Yogyakarta's biggest bazaar selling clothing, particularly batik, and any other food or daily needs item. Its origins date back to the years 1755–95, the period when Yogyakarta itself was constructed. Although the initial building itself is not preserved the location is still the same as in the original urban design. Pasar Beringharjo is under a renovation scheme of the local and the provincial governments (Sugiana 2008).

Pasar Beringharjo is surrounded by a multitude of local vendors and food stands, like in the rest of Malioboro Road (cf. Fig. 5.23/13) which is linking the Kraton area with the Tugu train station in a direct South–North connection. The street is a major shopping and tourism destination, full of commercial buildings and small businesses. Most buildings comprise of a shopping area in the front and storage areas on the top floors or the rear rooms. While Malioboro road and the major side roads are accessible by car, the small and semiprivate narrow lands in the back are only accessible by foot, linking very small houses (cf. Fig. 5.23/12, the entrance to Kampung Sosrokusuman, in the northwest part of Malioboro).

Initially Malioboro was a residential area for Kraton servants; later it became the Chinese quarter with shophouses in Chinese style, which then again mixed with Javanese elements. As a result, houses to the fronts of main roads are more large and representative than the constructions inside the blocks. Shophouses still shape the multifaceted area which also comprises of some buildings of Dutch and Javanese origin. At the same time the majority of buildings are home and workplace for many people engaged in small enterprises (Siauw 2003). Since the late twentieth century the streetscape is undergoing changes, starting with the construction of Malioboro Mall in 1998 (Timothy 1999) and subsequent changes of existing houses (in terms of ornaments, window patterns and large-scale advertising panels) or even their replacement with new and more spacious constructions. Like in the vicinity of Taman Sari and Kraton, Y03 is witnessing problems with new constructions which do not follow building regulations and therefore do not match the neighbourhood in their appearance. Malioboro Road is a vibrant place from early morning until night time (cf Fig. 5.24, right side), with shops, numerous street vendors under the arcades of the shophouses, horse-drawn carriages and rickshaws. Shops are vending everything from modern items to traditional batik and handicraft:

> During the day and into the evening, Malioboro Street, the main shopping area for both tourists and locals, is lined with nearly 1,000 vendors of various goods and services ranging from brooms and radios to snack foods and souvenirs. For tourists, these vendors and their affinities for bargaining are a major attraction (Timothy 1999: 376).

**Fig. 5.24** Mobile food vendors (*left*) and band playing Javanese music at Malioboro in the evening (*right*)

During lunch and particularly during dinner time mobile restaurants (*warung*) occupy distinct areas in the northern part of Malioboro and its side roads while bands occupy the pedestrian areas (cf. Fig. 5.24). There are different forms of stationary and usually informal vendors, mostly using pushcarts, tents and small carts with umbrella. The majority is vending food, snacks and drinks in form of small-scale family enterprises, which are found in Malioboro area since the 1970s (Rukmana and Purbadi 2013).

In 1993 Yogyakarta's first mall opened at Malioboro Road, mirroring the trend across Southeast Asia (Hampton 2003; Siauw 2003). Technical and traffic infrastructure has improved within the past years, particularly with the direct connection to Transjogja bus transportation (cf. Fig. 5.23/13, bus stop at the right side). Conservation schemes prohibit replacement of traditional houses by modern constructions. This can be seen as a consequence of out-of-character buildings such as Malioboro Mall or large-scale hotel constructions to its north (cf. Fig. 5.23/14 with the back facade of Hotel Melia Purosani). Arcades in front of the buildings as well as street corners are used for mobile trade, which usually is not linked to the trade use inside, corners and sidewalks are used by mobile food stands, particularly in the early evening when the area is particularly frequented.

The northern extension of Malioboro, north of Tugu train station (Jalan Margo Utomo), is occupied by comparably different commercial uses, such as car vendors, office supplies and hotels. Here and along the East–West connection crossing Jalan Margo Utomo, Jalan Jendral Sudirman, upper-class hotels are under construction. Particularly Jalan Jendral Sudirman is already housing a number of posh hotels frequented by (foreign and domestic) tourists. On the intersection of the two streets the Tugu monument (cf. Fig. 5.32, No. 2) is located, one of Yogyakarta's landmarks. It is located along the magical line linking the southern ocean, the palace of Yogyakarta and Mount Merapi. The original monument has been built during the time when Yogyakarta kingdom emerged and thus has a symbolic value. After its destruction in the course of an earthquake it has been renovated in altered form by Dutch in 1889. Despite being located in the midst of a congested junction it is a popular meeting point of young Yogyakartanese in the evening time and the spot to drink *kopi joss*, coffee heated up by flaming hot charcoal. Besides that it is a student tradition to hug and kiss the monument to express happiness after graduation as well as the wish to return to Yogyakarta.

The area south of the train station, Sosrowijayan (cf. Fig. 5.23/15), is famous for backpacker tourism, low-budget guesthouses and restaurants. It emerged as tourist hotspot in the 1970s due to its proximity to both, the train station and Malioboro. "*To the inhabitants in Yogyakarta, Sosrowijayan is infamous as a 'black' area, as was shown in the previous chapter, and it is believed to be home to pickpockets, sex workers and various sorts of suspicious individuals*" (Sörensson 2008: 85). The area was never officially foreseen for touristic purposes, it somehow emerged by itself. Backpackers that stay in Yogyakarta have an average stay of only 2 days to see the two world heritage sites of Borobudur and Prambanan, therefore the fluctuation in Sosrowijayan is quite high, with lots of offers suited for backpackers (Sörensson 2008). Today, the area is characterized by small shops, budget and middle-class guesthouses, restaurants serving Indonesian and international food, travel agencies and other tourist facilities. In addition a smaller section of the area houses the Yogyakarta 'red light district' (Timothy 1999; Dahles 2001; Sörensson 2008).

The areas surrounding Malioboro have a completely different appearance. Middle-class hotels have been built south of the backpacker area of Sosrowijayan around Jalan Dagen, (cf. Fig. 5.23/11), then changing into a mix of major roads with commercial use aside and small side lanes into the neighbouring kampungs. Towards east the commercial uses along the streets changes into largely informal kampungs along Kali Code (cf. Fig. 5.23/14), In the northeast of Malioboro the Dutch quarter of Kota Baru (cf. Fig. 5.23/16), is following a different layout, with small Dutch-style houses surrounded by green areas, also including Protestant churches.

### 5.2.5 Perceived Drivers of Change in the Urban Landscape

*Expert interviews, literature research and surveys revealed different factors that are impacting on the historic centre*. First of all, all interviewees stressed that Yogyakarta is 'special'; it has the status of a Special Province, a special layout and a special status within Java and Indonesia. They deduced a number of consequences and responsibilities resulting from this status.

The unclear legal and owner status of many plots is hindering the implementation and execution of conservation measures. Y10 sees a major hindrance of development in the too complex land title system, where Dutch colonial and pre-colonial Javanese systems overlap. In addition, different ministries are involved in the land title issues, e.g. forests are under the Forest Ministry and agricultural land is under the Ministry of Agriculture. He therefore claims that "*the land rights are very tricky, a land right reform is needed*" as the dweller's rights depend on the land status—there are user and ownership rights, making huge differences in reality.

Yogyakarta again is a special case as on top of the complex system the Sultan himself is the owner of a large share of the urban land. As a consequence many urban dwellers built their houses informally on Sultan's land, what has been tolerated so far, however it is questionable what may happen if urban pressure and inner-city densification processes continue.

Actors in urban development and conservation are partly having conflicting interests and at the same time different social levels, beginning with the Sultan himself who is the main land owner in town. On the other hand with his symbolic power and his political role as governor, the Sultan is the determining factor not only in politics but as well in culture: "*Yogyakarta Istimewa—what does that mean? People are listening rather to the Sultan than to the president in here*" (Y05). Y07 is concerned as well:

> The problem in Yogyakarta is, the Sultan is the governor, the other lower parts of the government, and they won't say 'no', that is the worst part. And then, we have this provincial government, and city governments, and each level and government, they have their own parliament. I think this is not good. I think the mayor should be responsible, like in Jakarta. We lose a lot of money with elections and governments, we don't need these governments. The Sultan says something, and the provincial people say something and the mayor or the region will do something. But even if they say yes the parliament may still say no (Y07).

On the other hand, the Sultan is one of the few constants in urban authorities. After each election the newly elected deputies may change conservation plans, priorities, and people in charge, therefore Y07 is lacking continuity in conservation. Both, Y06 and Y07 are complaining about deficiencies in Yogyakarta's urban planning which to them is worthy of improvement. Most interviewees mentioned that the ongoing densification is influencing negatively on the built heritage and its setting. Yogyakarta is increasingly verticalised, in and around the centre many five-star hotels are under construction or have opened recently. In other cases the building location has already been allocated, partly resulting in the eviction of the current population. Increasingly, apartments buildings are constructed as high-rises as well (Y06, Y07, Y09). Prabawa (2010: 96) is concerned that:

> With rapid modernization and because of the changes in the socio-political conditions, Yogyakarta must be alert that the city may lose its main selling point for tourism as a village-like city. Among the international tourists, the salience accorded to the special characteristics of Yogyakarta as a village-like city is evident.

Different interviewees emphasized that Yogyakarta is famous for its culture; as a consequence it is attracting many cultural tourists. Cultural heritage is a major source of income. The 'traditional way of living' has become a brand for tourism and economic factor (Siauw 2003). Therefore a holistic concept to develop urban sites for touristic purposes is crucial (Y05, Y07, and Y09). To Y05 such vision does not only need budget but also a plan how to improve the living conditions of the local people: "*I want the people surrounding the area to benefit from the projects, to be involved and get employment—we have to see how to get them involved. Else they will not love it!*"

Y06 and Y09 find the city in a problematic situation, caught between conservation and economic pressure, mostly induced by tourism, as e.g. visible in actual discussions on the location of new hotel buildings. Large-scale hotel complexes developed right aside the historic centre, "*building permissions are given without considering the setting*" (Y09). Y06 sees a controversy, while people interested in conservation seek to preserve the centre area and the urban appearance, others argue that Yogyakarta as a tourist city has to develop hotels:

> Just this morning I had a discussion on the cosmological planning of Yogyakarta. Some people are arguing that today's planning should follow this scheme, that we should keep the central axis free from high-rise buildings instead of building hotels and apartments. This is showing the controversy of conservation on the one side and development pressure on the other side (Y06).

Particularly residential areas are prone to conversion and loss, as the houses of lower social classes are of less value in the *catur tunggal* scheme of Javanese residential cities. Siauw (2003) found a clear preference of key conservation actors for Sultan buildings, while the preservation of historic residential areas and buildings is found comparably less important. In addition, preferences for so-called modern living styles are harmful for the traditional housing forms.

The large variety of housing styles is challenging in terms of finding adequate conservation measures (Y03). Basically three different types of heritage buildings can be distinguished: traditional Javanese-style buildings in the Kraton area, using materials like wood or bamboo, Chinese shophouses and colonial buildings made of brick – as well as mixed forms. The conservation of this historic part is underlying more restrictive regulations than the rest of the city, e.g. "*The Kraton area requires a certain quality of materials for repair and retrofitting, e.g. wood from a certain location, of certain diameter and size, due to beliefs in unity between logic and supranational values*" (Y01).

Things get particularly challenging with constructions in mixed style which are common in Yogyakarta from the reign of the 3rd Sultan on after 1810 (Zahnd 2005; Ellisa 2010). The Javanese building style slowly changed and adapted elements from Dutch architecture (Y01). It was very common that in case a building was destroyed to rebuild it not exactly in the same style and size, but to restore mainly its function while adapting the fabric itself to present time, adopting elements from housing styles of other ethnic groups. This happened, e.g. with the Tugu monument which collapsed end of the nineteenth century and was rebuilt in a different style. Today, "*it is not allowed to change the style of historic buildings, but this is easier for buildings owned by the government, and difficult with private houses*" (Y01).

The relations between the different population groups have not always been free of conflicts. For decades Indonesia's cultural diversity was not a subject of discussion. Subsequently, neither were ethnic minorities and their buildings or quarters. As a consequence no distinct conservation schemes could be developed. Only since the end of the Suharto era things are slowly changing and decentralization policies have supported the development of more regional and local initiatives (Siauw 2003). Within the past years a change in the appreciation of non-Javanese constructions can be witnessed. While Siauw (2003) still found a clear preference of Hindu–Javanese heritage and an almost dislike of constructions from Dutch era and ignorance of Chinese ones, things have changed since then (a shift he already forecasted). Neither in questionnaire results nor in the expert interviews was such clear preference stated (see Sect. 5.2.8).

In interview Y08 the cultural diversity was stressed as a problem for planning, as each region or even quarter does have certain characteristics which have to be considered what may slow down processes. Y01 considers Yogyakarta Special Region as a special case even within Indonesia due to the artefacts from many periods, namely prehistoric, Hindu, Buddhist, Islamic, Javanese and Colonial. As a consequence, conservation as well as urban planning is comparably more challenging than in other regions.

Yogyakarta is also found special because of its status as educational city (Y05, Y09). Different from the other case study cities, education was hardly posed as a problem. University staff finds their education focusing a lot on Yogyakarta planning and conservation. There is even literature available to educate school teachers on how to teach their students on Yogyakarta's heritage (cf. Adishakti and Hadiwinoto 2010). The only improvement considered as necessary is a closer cooperation between academia and urban planning, as it used to exist previously during the 1980s and 90s (Y06). However the decline of this cooperation is at least seen partly as fault of the university staff due to lack of available time, although personal connections to the urban authorities are still there (Y05, Y06, and Y07). To Y06 the importance of the local architecture and traditions in university education is partly "*depending on the distance to Jakarta and its urban developers that want to have new buildings*". Y07 sees "*too little cooperation with the city of Yogyakarta and too little involvement in projects*", although she is doing projects on the urban core with her students, like a renewal plan for the northern alun-alun.

Interviewees Y08a/b witness that spatial planning is getting increasingly complicated with the decentralization process and growing regional aspirations. In addition, the large number of authorities and actors involved in formal decision-making for heritage and urban renewal initiatives result in lengthy processes, and are delaying implementation (Y06, Y07, Y08 and Y09). A multitude of authorities is involved in urban affairs, from national to urban level, belonging to different ministries and administrative units. In addition the lowest (and informal) three of the overall seven levels of organization are *kampung*-based and mostly informal (Zahnd 2005; Obermayr 2013). The challenging legal system with its multitude of actors and legal documents was mentioned several times as a major hindrance for improving urban planning and conservation (Y2, Y06, Y07, Y08, Y09). Y07 is lacking a heritage movement in planning, as

> There is no place or law taking care of the total system of heritage, cultural and natural heritage are under different ministries, not to talk about shared heritage or cultural landscapes – in our culture the different heritage values and types go together – but regulations don't, there are only for natural or cultural heritage, nothing for cultural landscape or living heritage (Y07).

Sunaryo (2011: 297) finds the governmental urban design approaches mostly top-down due to an "*unprepared bureaucratic system which finds it hard to deal with public participation methods*". In addition Y03 sees gaps in law compliance as many constructions are built or enlarged without paying attention to laws and guidelines:

> In former times the heritage areas were surrounded by agricultural land, this is no longer the case due to urban development and increase in population, now there are difficulties with the heritage area zonings as people want to have the freedom to develop or build their own houses (Y03).

Y07 is particularly lacking law enforcement in the heritage areas. Among others she is witnessing ongoing constructions activities in the quarters surrounding the Kraton which do not follow the regulation that no construction is permitted to be higher than the Kraton itself. In addition Y06 and Y07 see the need to facilitate conservation in general, as there are different responsibilities for spatial planning (Ministry of Public Works) and heritage conservation (Ministry of Culture and Tourism). Y08a/b claimed that the responsibilities between the different levels and between authorities on the same level are not always clear—as a consequence they witness a lack of power to coordinate, adjust and implement planning. In addition the multitude of legal documents, including different laws for natural and cultural heritage, are additionally complicating urban conservation (Y07).

The historic centre is an arena where different interests and goals are converging and potentially conflicting. The multitude of actors involved in urban development issues, as well as the lack of coordination even between governmental authorities and the different administration levels are concerns of different interviewees (Y01, Y02, Y06, Y07, Y09 and Y10). Y06, e.g. mentions that:

> So many institutions are dealing with heritage issues. As common in developing countries coordination is one of the most problematic issues. Now under the local autonomy and decentralization each mayor has its own programme, and on national level culture has become part of the Ministry of Education. The institutional framework for heritage conservation is not very clear.

Even formal processes very often rely on community-driven approaches, as Y02 describes for the 2006 earthquake rehabilitation when "self-reliant housing community groups" of 8–15 houses were set up. However, she finds communities within Yogyakarta comparably less strong than in the surrounding but yet less urbanized areas like Bantul, as: "*There people have fewer facilities than in Yogyakarta, so they depend more on each other, and have a stronger need and will to help each other.*" Nevertheless such processes are yet often semiformal or even informal and are lacking perpetuation and in some cases formal recognition (Y02, Y10).

Social networks are an important part of Javanese and Yogyakartanese society (Y02). Community-based activities are very popular and a major strategy of social as well as economic security. To Y09 public participation processes are not always functioning well, she witnessed that, e.g. not always the 'high-level' staff is sent to public hearings, but just some deputy who is not really responsible for the case. In addition, Javanese culture usually bans open conflicts and expression of disagreement. In such a case dissonances are rather solved in one-to-one discussions than in public meetings. Timothy (1999) blames Dutch colonialism at least partly for the still challenging participatory processes, as Indonesians were kept from participating in democratic decision-making during colonial times. But even after the end of colonial times planning was a purely top-down process for many years. To him (1999: 387) "*it appears that resident participation in decision making is currently non-existent in Yogyakarta. Official plans have not recognized the need for this, and it is not carried out in practice.*"

Non-governmental urban development initiatives can be divided into two main categories: community-based initiatives on local scales mostly concerned with their own quarters, and organizations founded to support heritage conservation, like the Jogja Heritage Society (Y07, Y10). The members of such organizations are mostly in high-ranking positions, often in university. Increasingly, cooperation schemes between the two categories emerge, as e.g. in the aftermaths of the Kotagede earthquake (Suryanti et al. 2014). In addition, international donor organizations are influential players in urban development, mostly concerned with infrastructure development, such as ADB, JICA or GIZ. Although they are less concerned with conservation issues per se, the historic centre forms part of infrastructure development programmes as, e.g. the traffic improvement (Siauw 2003).

Despite the promising Transjogja bus system Y04, Y06 and Y09 are witnessing a rapidly growing number of motorbikes and cars in town, as increasingly business and conference activities are shifting from Jakarta to Yogyakarta. As a result, traffic congestion and air pollution are growing problems, as, e.g. Y05 mentions: "*When I was a child in elementary school, my father showed me around in the city, there were many people with bicycles, now it is much more crowded, the city is growing.*"

Different interviewees stressed the lack of a comprehensive disaster risk management plan for the historic centre (Y02, Y07 and Y09), as climate change, environmental pollution and urbanization are likely to exacerbate urban risks. According to Y01 such plan is still under development as it only started after the 2006 earthquake which had disastrous impacts on thousands of historic buildings. In addition, Y02 expresses the need to develop a sense of locality in emergency planning, considering local cultural aspects and values. Y09 wishes for a stronger '*consideration of local wisdom*' including the right choice of building materials and types in reconstruction. Again, the 2006 earthquake and its impacts on Kotagede is given as an example where the semipermanent traditional building types (bamboo, wood) caused less fatalities than 'modern' constructions. The incident, however, revealed another major problem: house owners, deprived of necessary resources (Subanu 2008), sold valuable parts from their houses or were not able to restore them to their traditional forms (Y07, Y09). Furthermore, according to Adishakti (2008) the governmental reconstruction schemes partly resulted in the loss of form and value of traditional houses due to inappropriate design.

Y02 is lacking concerted action among levels and institutions in the historic centre, although to her this has been proven to exist in the past. Y01 gives the example of a historic building in Yogyakarta's centre used as a hospital. The planned modifications to modernize the hospital led to discussions with different authorities involved, including governmental officers, NGO's, cultural authorities, local communities and the Kraton area authorities, slowing down the process significantly. In addition, Y09 witnesses a lack of monitoring the sustainability of any programme or initiative, which should preferably be done by authorities and local communities together:

> I guess our planners have quite good knowledge in preparing the master plan, but perhaps we are still looking for good ways to prepare detailed plans, because what we need are building regulations. The issue is not only the plan-making but its implementation. Indonesia is very much lacking it (Y06).

To Y09 urban development has to respect culture and balance economy and social development. She, as well as Y06 and Y07, see the main challenge in future inner-city development in coping with the economic development, while not neglecting social aspects and the cosmological beliefs shaping the urban layout.

### *5.2.6 Assessment of Urban and Heritage Policies*

The very first national legislation dealing with cultural heritage conservation dates back to Dutch colonial times. The Monument Ordinance of 1931 defined three kinds of monument that shall be preserved, namely

- Objects fabricated by humans that are older than 50 years or that belong to a style that is at least 50 years old and "*which are considered of great interest to the praehistory, history or art*",
- Goods interesting from a paleontological view, and
- Areas containing objects of the groups mentioned before (Governor General 1931).

This legislation was replaced only in 1992 and not little of the Dutch influence and interpretation of heritage remained in subsequent legislations.

After Indonesia's independence the 1945 Constitution contained a paragraph which specifically deals with culture. Paragraph No. 32 states that "*the government will promote the Indonesian national culture*", and "*The national culture is the culture, which arises as the fruit of the entire Indonesian people*". The culture of Indonesia's ethnic groups is recognized as part of the overall Indonesian National culture and it should be protected and promoted as a means to national unity and identity (Galla 2002). In 1988 the following sentence was added:

> Tradition and historical remains which give special character to the National culture [...] should be preserved and developed in order to create the historical consciousness and devoted spirit as well as preserve the culture and continuity of the National development (Article 32, Constitution of Indonesia).

In practice this very good base for any renewal of historic city centres is very often hampered by economic arguments in favour of new constructions or attracting more tourists (Böhme et al. 2003). Subsequently the National Basic Guidelines Policy (NBGP) of 1988 elaborated further that "*Indonesian national culture reflecting cultural high value that can be used for promoting and strengthening national identity and national interest should be preserved and developed*" (Galla 2002).

Other laws and regulations that were enacted in the following years are the

- Law Number 9 of 1990 concerning Tourism, including heritage tourism,
- The Government Regulation Number 19 of 1995 concerning Preservation and Utilisation of Museum Collections, and
- The Presidential Decree Number 107 of 2000 which regulates the utilisation for underwater archaeological heritage.

However, as mentioned above it took decades since Law Number 5/1992 concerning "Items of Cultural Property" then replaced the outdated 1931 Monument Ordinance. Despite the long timespan some of the contents and definitions of 1931 survived, like defining only objects of more than 50 years as cultural heritage. According to Article 1 of Law No. 5/1992 (Ministry of Culture and Tourism) these "Items of Cultural Property" include:

- Artifacts made by man, moveable or immovable, individually or in groups, or parts thereof or remains thereof, which are at least 50 (fifty) years of age, or represent a specific stylistic period of at least 50 (fifty) years of age, and are considered to possess value of importance to history, science, and culture;
- Natural objects which are considered to possess important value for history, science, and culture.

Such Items of Cultural Property have to be declared as property of the state by Government Regulation No. 10/1993 if being of important value for Indonesia's history, science, and culture, or if having characteristics which confer particular motives and uniqueness, and/or if being limited and rare in quantity and type (Galla 2002; Ministry of Culture and Tourism 2003).

Heretofore heritage conservation in Indonesia focussed exclusively on artefacts, mostly defined by their age. With the end of the 1990s then the understanding broadened, following the global discourses. Heritage areas, heritage cities and also intangible heritage appeared on stage. However, particularly for intangible heritage it took some years until public awareness arose. The inscription of *wayang kulit* and *batik* on the List of Intangible Cultural Heritage in 2003 and 2005, respectively, helped attracting public attention (Adishakti et al. 2011; UNESCO 2014d). Throughout the 1990s and 2000s various regional Heritage Societies were founded to care about the region's natural and cultural heritage. In 1992 the Jogja Heritage Trust was established to care about the conservation of the regional culture and cultural environment (Adishakti et al. 2011). However, formal laws or policies to safeguard intangible heritage were and are still pending. But also the implementation of existing heritage policies is challenging, particularly during the Asian economic crisis, where conservation lost importance (Siauw 2003).

In 2003 then the Indonesian Charter for Heritage Conservation (*Piagam Pelestarian Pusaka Indonesia*) was issued (Indonesian Network for Heritage Conservation and International Council on Monuments and Sites). Adishakti et al. (2011) call it the first agreement in Indonesia that is following ethical and moral notions of conservation. The Charter states that:

- The heritage of Indonesia is the legacy of nature, culture, and saujana, the weaving together of the two. [...] Saujana heritage is the inextricable unity between nature and manmade heritage in space and time.
- Cultural heritage includes both tangible and intangible legacies.
- Heritage, bequeathed from the generations that precede us, is the a vital foundation and initial capital for the development of the Indonesian nation in the future, and for these reasons, must be conserved and passed along to the next generation in good condition, without loss of value, and if possible with an enhanced value, to form heritage for the future (Indonesian Network for Heritage Conservation and International Council on Monuments and Sites 2003).

In this text it becomes obvious that the Indonesian understanding of heritage links cultural and natural aspects more than the European understanding does. It considers the different forms of setting, mentioning both abiotic (natural and built up) and biotic (flora and fauna) aspects as well as sociocultural activities. The charter concludes with the wish "*that that heritage conservation in Indonesia will help to affirm the nation's identity in the world's very diverse and dynamic community [...]*". In addition, with this charter "*the notion of conservation has moved from being mainly concerned with beautification to a more holistic approach that is based on participation programs, economic analysis and attracting business and cultural activities to the area*" (Adishakti 2008: 242). It recognizes the ongoing loss of heritage, discrimination against minorities and traditional ethnic groups and lacks in preservation techniques, management, and of a clear framework (Adishakti 2008; Salazar 2011b). To Y07 particularly the *saujana* principle is of utmost importance. To her, on the one hand it is reflecting Indonesian holistic understanding of heritage very well, on the other hand *saujana* also corresponds very much to the global understanding of cultural landscapes, and therefore she uses it as a synonym.

In the course of the 2003 charter the Indonesian Heritage Trust (Badan Pelestarian Pusaka Indonesia/BPPI) was founded in 2004 to support conservation in the country. In 2008 the Indonesian Heritage Cities Network (JKPI) followed to support the protection of heritage cities and regencies (Adishakti et al. 2011). In fact cultural heritage, particularly buildings and built artefacts, are the best protected items by law. Natural heritage is under the Law on Living Environment, while there is no law yet to protect intangible cultural heritage like music, literature, dance, theatre or customs. There are even different ministries responsible for natural and cultural heritage and a third one for the coordination of Indonesia's World Heritage sites (Adishakti et al. 2011). This division is somewhat thwarting the holistic *saujana* approach. A promising approach to implement the saujana approach is the 2009 Joint Regulation of the Ministry of the Interior and the Ministry of Culture and Tourism (no. 42/2009 and No. 40/2009) on guidelines for cultural preservation, regulating community participation and the local governments' responsibilities in conservation (Adishakti and Hadiwinoto 2010). The most recent legal document that was passed is Law No. 11/2010 concerning 'Items of Cultural Property', explicitly including heritage buildings. Y07 appreciates the revised law because it allows area-based conservation, but more details are needed how to apply it in heritage cities.

It is apparent that culture, heritage and the nation's identity are closely linked in all policies and laws (Adishakti et al. 2011). In 2009 Indonesia established a Ministry of Tourism and Creative Industry, as culture, heritage, arts and economy were seen as closely linked (United Nations/UNDP/UNESCO 2013). In 2014 then newly elected president Jokowi established a standalone Ministry of Tourism, indicating the importance of tourism for national development.

Under the national level there are regional legislations and regulations governing the protection within the region itself. The government issues five-year national development plans under which provincial governments prepare their own development plans which, together with the regional development strategy formulated centrally, helped determine sectorial projects and related budgets. Such approach became possible with Indonesia's decentralization policy in the early 2000s (Brillantes and Flores 2012). Law No. 11/2010 is the basis for Regional Regulation No. 62012 about Cultural Heritage Conservation and the Governor Regulation No. 186/2010 about the determination of Cultural Heritage Districts (Suryanti et al. 2014). Regional development planning is formally done for long-term, mid-term and short-term intervals. The responsible Department of Regional Development Planning (BAPPEDA DIY) is also executing the Action Plan of the region's Heritage Districts, like the Action Plan of Integrated Management for Kotagede Cultural Heritage District while the Publics Works Department is responsible for building and environment layout plans and detailed engineering designs (Suryanti et al. 2014). Heritage concerns are also increasingly considered in disaster management and recovery, e.g. by allocating a special budget for heritage conservation in the next regional disaster management plan (Y01). The spatial plan already considers unique features like the different heritage areas from different époques: "*In the plan we did already designate unique areas, like Kota Gede, Kraton, Malioboro, the next steps will be to develop detailed plans for that areas and to do building regulations considering the special features of the areas*" (Y06). *However, being asked about the spatial plan for Yogyakarta* Y09 poses that "*the protection of cultural heritage is already somewhere in the spatial plan, but it is not specifically communicated.*"

In Indonesian traditional societies the conservation of heritage has a long history and is an intrinsic part of society's life (Adishakti et al. 2011). Various bottom–up heritage movements have emerged over the past 20 years, such as Bandung Heritage (1987), Sumatera Heritage (1998), Jogja Heritage Society (1999) and Bali Heritage Trust (2000). Mostly they were initiated by academics and professionals to promote local community awareness (Kwanda 2010). Below the umbrella of BPPI various local activities emerged, among them also youth initiatives focussing at

**Fig. 5.25** Youth Heritage Camp in Yogyakarta, February 2013, on the topic of 'creative conservation by youth' (*konservasi kreatif oleh pemuda*)

heritage conservation and sustenance of traditions, linked to raising awareness on sustainable development (cf. Fig. 5.25 of a local youth heritage camp in Yogyakarta). In this context Y09 mentions a new and promising initiative introduced by the Ministry of Social Welfare to support community-based approaches for heritage conservation, and to make local people benefit from the heritage and to foster the conservation.

Aside the national and local efforts also international organizations are actively involved in preserving the urban heritage. One example is the development of a heritage trail for the historic areas of the walled Kraton, which was planned and implemented with the support of AusHeritage, an Australian network of governmental and cultural heritage management organisations (AusHeritage 2015). Another example is the '*Pusaka Jogja Bangkit!*' (Jogja Heritage Revival!) Initiative, which was set up 2 days after 2006 earthquake. Besides taking care for the reconstruction of the affected *Joglo* houses in Kotagede it is also active in the Kraton area (Adishakti 2008).

In Yogyakarta there are tax incentives for privately owned heritage buildings, e.g. in 2014 overall 306 owners of heritage buildings in Yogyakarta region were compensated (Tavare 2014). However, Y03 and Y08a/b find private houses hardly protected, mostly it is done in case the house forms part of an ensemble, or in the palace area where the Sultan himself is the owner, what again indicates the city's special status. In Yogyakarta overall six heritage sites are protected by law: the Imogiri cemetery (royal graveyard complex), the Dutch quarter of Kota Baru, the former royal palace area of Pakualaman in the city's east, Kota Gede, the Kraton compound and Malioboro area (Y03, Y07). In theory such larger heritage sites are divided into core areas, buffer areas and development areas, following international approaches. Each site is obliged to have such a master plan, zonings are obligatory, however in practice until now such master plans are not yet elaborated for all sites. Y03 finds preference is given to the UNESCO World Heritage sites of Borobudur and Prambanan particularly.

As mentioned in Sect. 3.5 the Kraton compound with Taman Sari was listed as tentative UNESCO World Heritage since 1995. However, Salazar (2010, 2011b) claims that it has comparably little chance to become listed, as it is a living place where current political decisions are taken instead of a purely historic heritage site. To him "*such politics of heritage serve as a reminder that, ultimately, a WHS is the product of agency on the national level*" (Salazar 2010: 141). In fact, since the last revision in January 2015 the Kraton area does not appear any longer on the official tentative list of Indonesia (UNESCO 2015e), probably erased for the reasons Salazar mentioned.

Over the past few years Indonesia has made tremendous progress in overcoming antiquated conservation legislation and master plans. The country is among the signatories of the 2000 'ASEAN Declaration on Cultural Heritage' which defines cultural heritage in a regional context, with a focus on linking tangible and intangible heritage and combined human-nature creations. In the document preamble cultural traditions are declared an integral part of ASEAN's intangible heritage and its regional identity (Galla 2002; Ahmad 2006). Y07 is optimistic that the 'Indonesian heritage cities' programme will facilitate conservation, by overcoming challenges in the legal system, as there are different laws on spatial planning (Ministry of Public Works) and preservation of heritage (Ministry of Culture and Tourism), leading to diffusion in single cases, where responsibilities and communication are not clear. It was launched by heritage trusts and the Ministry of Public Works in 2013, with Yogyakarta as one of the 10 participating cities all over Indonesia. It remains to be seen how the promising approaches will be realized and how far the historic centre can benefit from it:

> For heritage conservation a revision of the old plans is important, as nowadays 'heritage' means a broader concept than previously, where only monuments and single buildings were preserved, the concept now enlarged, it is quite broader and area-based – but people are saying that still the concept is focused on archaeological sites and monuments rather than historic cities, like detailed regulations how to delineate boundaries of heritage areas – they are good for archaeological sites but not for cities. How to do in Kota Gede, how to do in Malioboro? (Y07).

### *5.2.7 Phases of Urban Renewal in Yogyakarta*

Colonial buildings as symbols of the urban colonial past were—and partly still are in some of the comparably young nation states of Southeast Asia—unpopular in the first years after gaining independence (Tokya-Seid 2003). In Indonesia rethinking and appreciation of historical quarters to the modern metropolis emerged in the 1970s in Jakarta with a series of gubernatorial regulations. They paved the way for revitalizations schemes in historic districts in other Indonesian cities (like in Bandung, cf. Agsten and Hoppe 1995) with a peak in the 1980s and 1990s. "*However, enforcement of regulations was more challenging than their promulgation, since for a long time, the colorful historical sites and relics of Jakarta still have been largely abandoned*" (Ellisa 2010: 478). Jakarta's latest major initiative is the 2006 plan to turn the Kota core area into a tourist hotspot, combining aspects of heritage conservation and economic activities. Secondary Indonesian cities like Yogyakarta then followed in such transition processes of urban transition, with comparable challenges. Mostly the original urban structure initiated by the Dutch and/or following Hindu-Javanese traditions is still visible, while much of the building stock has been altered or replaced (Ellisa 2010).

After having become capital of Indonesia in 1945 Yogyakarta witnessed a first slow economic boom which partly decelerated again after the capital city had been transferred to Jakarta in 1949. This shift in combination with the centralized state led to a focus on Jakarta and a decline of Yogyakarta's importance. As a consequence not too much of urban change occurred during the next decades and much of the historic building fabric kept untouched. With the growing importance of tourism the urban agenda changed in the 1960s:

> With the New Order government under Suharto in 1966, a new ideology of development affected the lives of the people and the development of the city. One important aspect of this ideology was the idea to beautify the city through improvements to roads and the modernization of the physical appearance of urban space (Sunaryo et al. 2011: 290).

With funding from the central government several urban regeneration and beautification projects were launched. Particularly Malioboro Road is a major topic in urban conservation and renewal since the early 1970s (cf. Fig. 5.26) when arcades in front of the shops were introduced. To do so the urban government asked the shop owners to give 3 m of their ground floor area for this purpose, a kind of conversion of private into public space. Out of the three meter two are for often informal vendors that sell their goods in front of the formal stores, the rest has a sidewalk function (Nitisudarmo 2009; Sunaryo et al. 2011). They pay a small fee to the shop owner for using his space, a way of '*formalizing the informal economy*', what to Rukmana (2013: 133) is special in Yogyakarta compared to other Indonesian cities. Interviewee Y06 is also convinced that the arcade construction

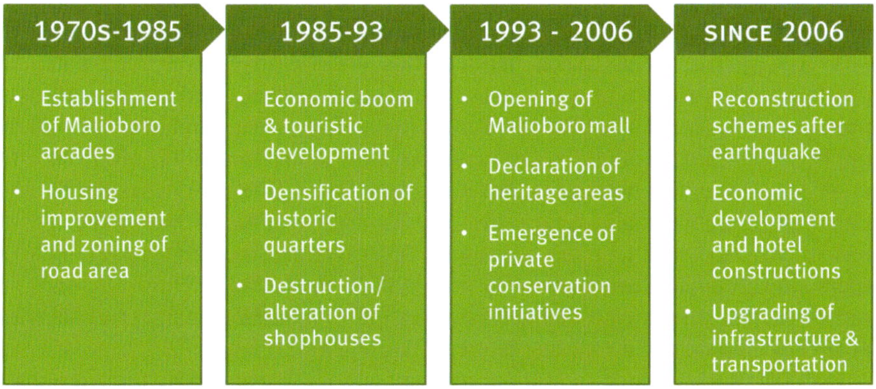


**Fig. 5.26** Main phases of urban regeneration in Yogyakarta Centre

contributed significantly to making Malioboro the bustling place it is today: "*Why is Malioboro always livable? It is because of the arcades, they protect from sun and rain!*" (Y06).

In the late 1970s and early 1980s a second project to revitalize Malioboro area was implemented, among others focusing on the road space itself. A zoning was done, allocating particular areas for distinct uses and modes of transportation, including areas for horse carriages and *becaks* (bicycle rickshaws), parking for motorbikes, pedestrian areas, and motorized transportation. Also an improvement of the houses behind the front row was foreseen to upgrade and/or create affordable habitations for low-income classes. To Y06 and Y07 this was a very important conversion:

> Interesting to me was that the revitalization can provide more public areas in the Malioboro Road and better housing facilities, showing a strengthened interaction between the kampungs and the commercial tourism area, making to me the dynamic of Malioboro (Y06).

An economic boom from the mid-1980s then resulted in transformation of the centre area where some historic shophouses were removed to construct the Malioboro Mall (inaugurated 1993) and the nearby Matahari Supermarket—despite existing legal conservation schemes. Further changes were the densification of the surrounding historic quarters inhabited by lower social stratum, road widening, an ongoing loss of historic facades and fundamental changes of historic buildings (Siauw 2003).

After 1985 commerce was no longer mostly restricted to Malioboro, areas outside the centre area were designated for this purpose (Yunus 1991). In the 1990s then malls in (yet) peripheral quarters were constructed under governmental planning. This was also meant to reduce the risk of losing more historic buildings in Malioboro, to lower development pressure inside and to provide new development outside the centre (Y06). In the same time between the mid-1980s and early 1990s many of the informal businesses located in Malioboro and also Kraton area got formal licenses and registrations. Nearby kampungs such as Sosrowijajan shifted from small-scale manufacturing to budget tourism and Malioboro area increasingly became a destination of international tourism (Timothy 1999; Hampton 2003).

Yogyakarta jumped on the bandwagon of conservation popularization trends all over Asia in the 1990s, when heritage was discovered as economic potential, resulting in publications, guidebooks or websites of aestheticized and romanticized heritage for the tourism industry (Kwanda 2010). The area around Malioboro Road has been declared heritage by the governor (the Sultan) in 1995, 10 years after having achieved the heritage status on urban level. The 1995 masterplan also identified Malioboro Road as priority area for regeneration, including the idea to pedestrianize the area to promote cultural functions and tourism with the help of international development agencies (Siauw 2003; Hook 2005). However, until the turn of the millennium no distinct conservation-focussed urban renewal schemes have been implemented in the core area (Siauw 2003).

In 2005 rampant traffic in Malioboro Road was partly calmed by transforming it into a one-way road, introducing pedestrian crossings and appropriation of public

**Fig. 5.27** Houses in Jalan Margo Utomo, the northern extension of Malioboro, picture on the *left* taken in 02/2013, the one on the *right* in 09/2014, indicating ongoing upgrading processes

space. Other ambitious plans like banning the motorcycle parking are however still lacking full implementation due to converging interests of the manifold formal and informal actor groups involved. As a consequence, still much of the public space is occupied by motorcycle parking (Hook 2005).

Conservation movements then got another push after the last big earthquake in 2006, when much of the urban historic fabric was destroyed, among them also parts of the Kraton area, which then were rebuilt from 2009 on (Kwanda 2010). Since then also increasingly local bottom–up initiatives focussing on the preservation of Yogyakarta's cultural heritage emerged. Upgrading schemes are increasingly applied to the northern extension of Malioboro Road (cf. Fig. 5.27), linking Malioboro with existing and newly constructed hotels near the Tugu.

Recent urban revitalization schemes also include the improvements of traditional markets' conditions, notably Pasar Beringharjo as Yogyakarta's market with the longest history. These plans are based on the awareness of the markets' cultural importance as well as their supply functions. This can also be understood as effort to counteract the emerging preference for newly constructed supermarkets. For carrying out such revitalization projects urban authorities are increasingly promoting PPPs, to also cover up limited public funds (Sugiana 2008; Sunaryo et al. 2011). "*Particularly at the northern part of Beringharjo Market which once was known to be material and construction supplier is now revived again with a greater variety of small scale retail goods and services*" (Anggraini 2012: 48).

Past and present DIY master plans have concerned the historic centre to different intensity. The first plan in 1971 included the phrase 'antique areas' without defining it in more detail. The 1986 plan, renewed in 1993, focussed particularly on infrastructure and commerce, in line with the 'Yogyakarta Urban Development Strategy 2019', including infrastructure development, KIP and an update of the heritage inventory (Siauw 2003).

Mostly Malioboro Road and its vicinity is targeted in urban regeneration plans, while the area around the Kraton is concerned to a much lesser extent, probably because it is a predominantly residential area and private land, owned by the Sultan himself. Here, mostly renovation measures focussing on the fabric and the urban layout are carried out, following the anyway more restrictive building and conservation regulations. Due to the different ownership status, change is more apparent in the area outside the Kraton compound, notably Malioboro, which forms part of the (extended) CBD. The urban regeneration projects therefore are an answer to the urban densification since the 1970s and the changing urban appearance of Malioboro and to a lower extent also other quarters from the 1980s onwards (Siauw 2003). The 1980s renovation scheme already included debates on local values, but to Y07 this appreciation of local assets got a bit lost in subsequent planning. An indicator is the construction of Malioboro Mall right after the completion of the renewal project: "*They did well—but now we increasingly have to think about the next step. Development pressure is increasing, now perhaps it's time to think about how to conserve Yogyakarta city for the future*" (Y06).

### 5.2.8 Perception of Inner-City Processes

The assessment of the perception towards the historic centre of Yogyakarta was done by means of a questionnaire, matching the approach in both other case study cities. A question concerning the respondents' feeling related to the centre (Fig. 5.28) revealed clear positive ratings of the centre's beauty and value. Overall 81.1 % found the centre beautiful or even very beautiful (equally distributed), 74.5 % found it either valuable (27.3 %) or very valuable (47.3 %). 68.8 % rated the centre as fascinating, out of them 43.1 % rated it as fascinating and 25.7 % even stated 'very fascinating'. The same percentage (74.5 %) was achieved for the area's vitality, out of which 30.0 % answered 'very vital' (44.5 % stated vital).

Answers regarding the centre's cleanliness are less positive, almost as many positive as negative answers were given (33.3 % negative or very negative impression, 40.5 % with a positive or very positive impression and 26.2 % 'neutral'). The centre is mostly perceived as heterogeneous (57.4 %, out of this 21.3 % find the centre even very heterogeneous), matching the expert's elaborations on the manifold building styles in Yogyakarta (cf. Sect. 5.2.3). A majority has the impression that the centre is in a good shape (overall 59.8 %, out of this 18.8 % find it in a very good shape). 50.5 % have the impression it is being upgraded 50.5 %, although 28.4 % are not sure about it and gave a neutral answer.

66.1 % answered that the centre is easy to access; to 37.6 % it is even very easy. However, in a subsequent question, 72.2 % of the respondents agree to have traffic and parking problems (cf. Fig. 5.29), out of which 27.0 % definitely agree, matching literature and experts' views. A potential reason for the positive answer on the centre's accessibility is the Transjogja bus system or simply the fact that the traffic in the centre is not worse from other parts of town and still having a comparably good road access. Overall 51.9 % find the centre is inhabited, while 18.0 % did not (29.2 % gave a neutral answer), matching results on the perceived heterogeneity, as both, uses and inhabitants differ a lot within distinct subareas of

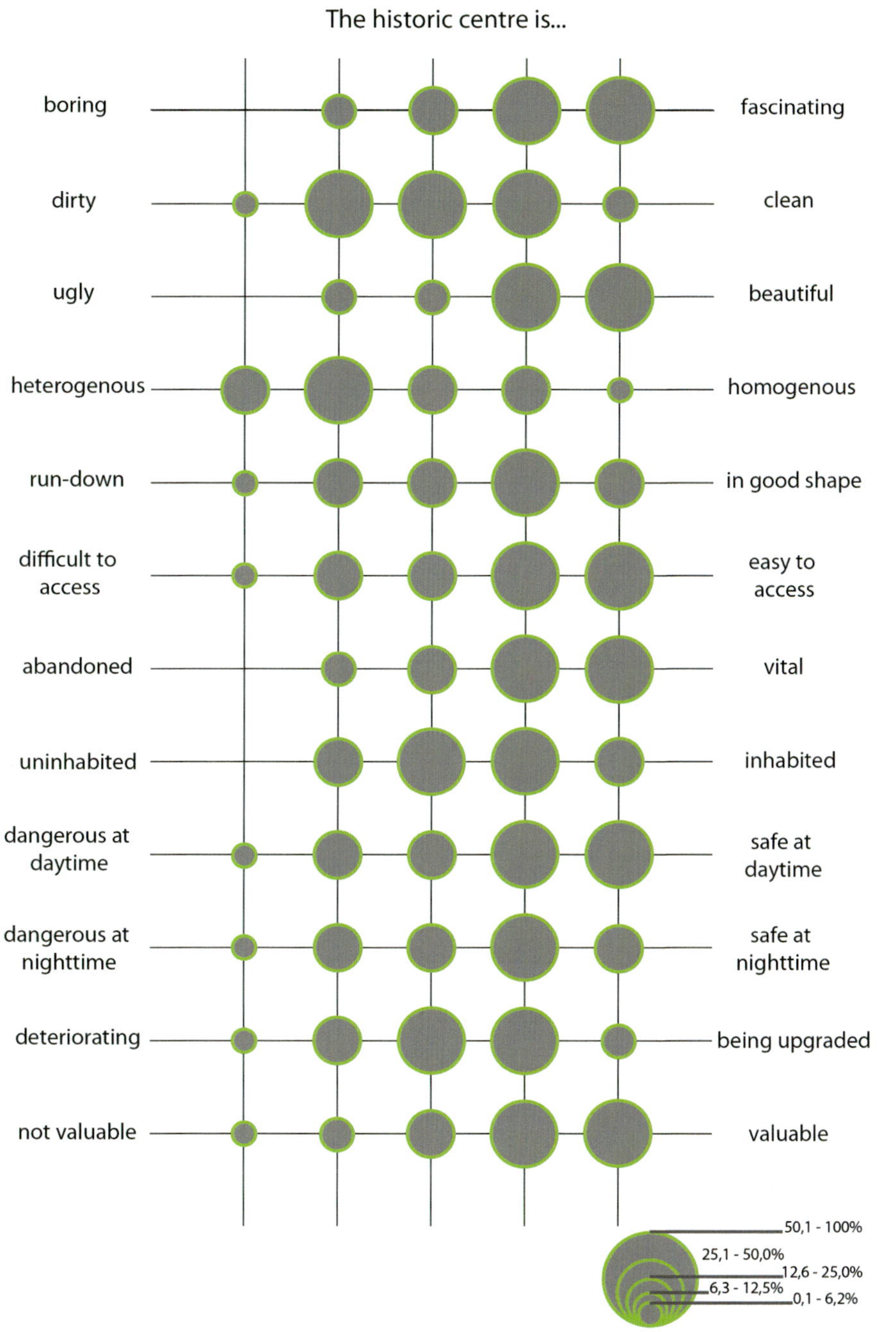


**Fig. 5.28** Responses to the question: What is your feeling towards the historic centre of Yogyakarta? ($n = 120$)

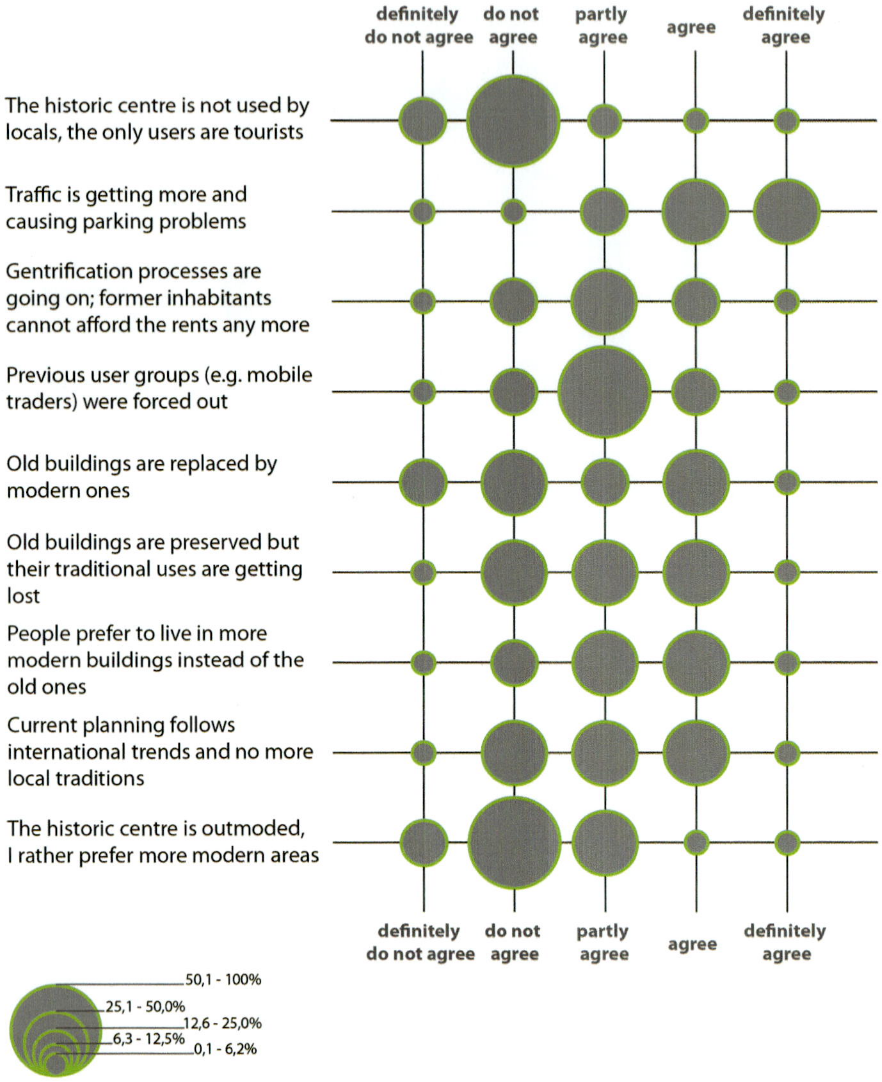


**Fig. 5.29** Responses to the question: Which processes do you perceive in the historic centre of Yogyakarta? ($n$ = 120)

the centre. Safety is not a big concern for the respondents, at least during daytime, when 70.0 % have the impression that the centre is safe during daytime (42.7 % answered 'safe' and 27.3 % 'very safe'). At night still 53.6 % regard the centre as safe or very safe (38.2 % and 15.5 % respectively), but to 25.5 % it is not (21.8 %) or not at all. As a conclusion, safety issues are not a very big concern, while the area is not perceived as very clean. Nevertheless, the overall impression is a rather good one, with a positive evaluation of the area's value, its beauty and fascination.

This result is matching the answers on a question regarding the processes the respondents witness in the centre (cf. Fig. 5.29). As many as 90.0 % disagree that mostly tourists instead of locals are frequenting the centre (66.1 % disagree, 14.8 % even disagree definitely). Answers concerning gentrification processes and the displacement of former user groups are ambiguous, with a large majority agreeing partly (47.3 % and 51.5 %) and almost the same percentages agreeing or disagreeing. In this case potentially the heterogeneity of the centre has led to a partial agreement; as such processes are much more likely to occur in the Malioboro area and much less in the southern part with the Kraton and its surroundings owned by the Sultan. Most likely for the same reason the assessment on ongoing replacements of historic buildings and their uses are as ambiguous, here 35.1 % disagree that replacements are occurring while 31.6 % agree. The loss of uses of traditional buildings is perceived a little more, 28.7 % agree partly and 26.1 % agree this is taking place, however, as many as 27.9 % of the respondents disagree to this statement. Again, it is potentially the heterogeneity of the area which is leading to such split results. A comparable picture has also been obtained when asking whether current planning in Yogyakarta is following international trends or local traditions. In this case the majority of respondents disagree (34.8 %), while 26.1 % agree and 31.3 % do partly. One interpretation is that planning itself is following two different directions, on the one hand there is urban growth and a growing number of high-rise hotel and apartment buildings on the fringes, on the other hand the '*catur tunggal*' scheme is still followed and appreciated.

The answers to the last two questions on people's housing preferences and the assessment of the historic centre in general reveal the fundamental contradistinction: while overall 35.8 % agree (30.3 %) or definitely agree (5.5 %) that people prefer to live in modern instead of historic buildings (and another 45.0 % agree partly), the vast majority of 67.8 % disagrees (out of which 13.0 % completely disagree) that the centre is outmoded. So the majority does not prefer modern areas but appreciates and uses the historic centre, while estimating that mostly Yogyakarta's inhabitants would prefer to live in modern buildings.

In another open question the questionnaire respondents had the chance to suggest some project to upgrade the historic centre or parts of it. The answers given can be grouped into six different categories, namely heritage buildings/sites, transportation infrastructure, cleanliness, culture/uses, environmental concerns, and safety issues. Overall 24 time wishes to protect or renovate specific heritage buildings or sites are expressed. Two times the respondents stress the need to elaborate an improved spatial plan for the historic centre, while the other suggestions tackle Malioboro Road, Taman Sari, Fort Vredeburg, the *alun-alun*, and the Kraton, e.g. "*upgrading the Kraton and surrounding area because it has high historical and cultural value*". One answer stressed the need to promote accommodation in historic buildings, another one the "*Renovation of old building with original design*" in general; and a third respondent wants to "*limit investors that influence on urban development without concern of the historic site*".

Overall fourteen answers dealing with the improvement of the traffic and/or parking situation in and around Malioboro Road, e.g. by introducing better public transportation facilities, car-free days or even the pedestrianization of the whole area. Eleven suggestions dealt with the need to improve the cleanliness of different areas within historic centre, namely Malioboro and both alun-alun. Five respondents wish to conserve, establish or enhance green areas within the centre and to upgrade its environmental conditions. Overall three times the wish to enhance security and reduce thievery was expressed, one time mentioning in particular the area around Fort Vredeburg. Seven proposals dealt with cultural aspects, often in combination with preserving or bringing back uses to historic sites, e.g. asking for traditional performances in selected places, or to preserve the traditional art of Yogyakarta, expressing the fear that the Javanese culture of Yogyakarta is getting lost. One respondent states that "*the Kraton as a centre of Yogyakarta culture must create some improvement related to our culture rather than building shopping centres as happening now*". It is worth being mentioned that overall six times tourism was given as a reason for carrying out some measures, e.g. improving the transportation system and parking areas in Malioboro for the tourists, to keep historic buildings for the tourists or cleaning the centre for them to make Yogyakarta's visitors feel comfortable. It seems that tourism is perceived as an important development factor for Yogyakarta, even by students that do not immediately depend on the urban tourism.

Overall the answers given above matching the results obtained in the previous questions, revealing attachment to the centre, but at the same time the awareness of problems concerning mainly traffic and the preservation or promotion of heritage buildings and cultural expressions.

### 5.2.9 Importance of Tangible and Intangible Assets

When inquiring on the importance of tangible and intangible cultural heritage in Yogyakarta, both, questionnaire and interviews revealed high values attached to both (cf. Figs. 5.30 and 5.35). 74.1 % the historic centre is of higher cultural value than any other part of the city—out of which 27.6 % definitely agree to the statement. To overall 56.4 % it is the commercial centre of the city (42.7 % agree, 13.7 % agree very much), however, 50.0 % of the respondents do not like to go there for shopping very much (39.7 % disagree/10.3 % disagree very much) while 30.2 % like it at least partly. Potential reasons for this contradiction could be the traffic problems stated above as well as a growing preference for shopping centres and the leisure facilities provided there.

47.4 % of the respondents attach a higher spiritual and/or religious value to the centre (34.2 % agree and 13.2 % agree very much)—a comparably low number given the significance of the Kraton stated in literature. Potential underlying reasons are again the heterogeneity of the centre with the profane Malioboro Road, as well as the fact that the Central Mosque is not having a higher significance or different role from any other Mosque.

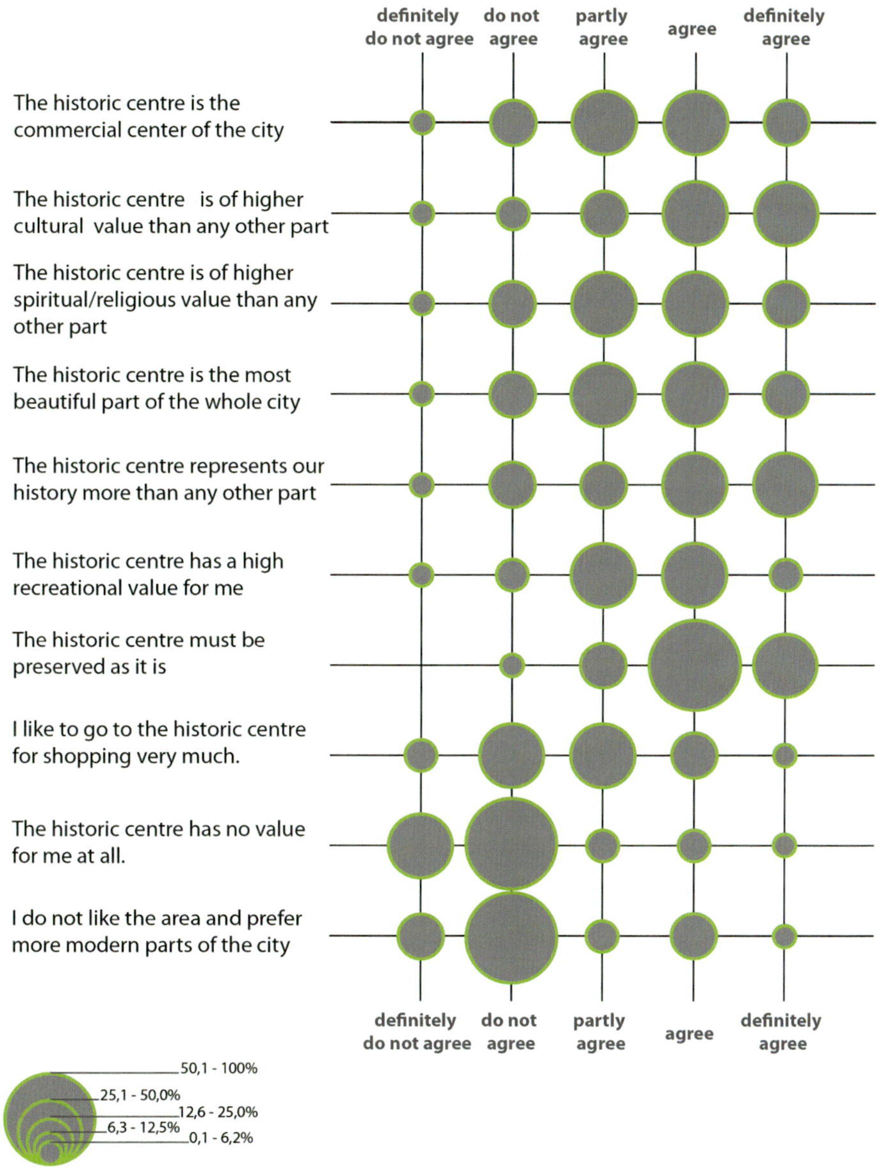


**Fig. 5.30** Responses to the question: Which values do you attribute to the historic centre of Yogyakarta? ($n$ = 120)

Overall 47.0 % state that the centre is the most beautiful part of the city, out of which 33.0 % agree, and 13 % agree definitely. However, 33.9 % agree only partly, maybe leading back to the perceived dirtiness of the area. Nevertheless, 56.1 % state a high (44.7 %) or very high (11.4 %) recreational value - and 30.7 % agree partly.

Despite the not very overwhelming impression on the centre's religious importance, its recreational value and its beauty as many as 84.1 % want it to be preserved as it is (53.1 % agree, 31.0 % definitely agree). To 77.6 % it is a representation of history, more than any other part of the city (48.3 % agree, 29.3 % definitely agree). Conversely, 81.0 % disagree to the statement that the centre has no value at all for the respondent, out of which 50.9 % disagree and 30.2 % even definitely. Overall 72.6 % disagreed (54.0 %) or definitely disagreed (18.6 %) not to like the area and to prefer more modern parts of the city. Overall, the question results reveal a high appreciation of the historic centre, particularly for its cultural values and for its representation of Yogyakartanese history.

This appreciation becomes apparent also in other occasions. During the last Yogyakarta marketing campaign in 2014 new logos and icons were selected in a competition, the winner was announced by the Sultan himself. Among the icons there are physical assets like the traditional means of transport (carriage and rickshaw) or spaces like the Tugu monument and the Kraton as well as the Banyan trees on Alun-Alun. The trees are at the same time endowed with a spiritual meaning, the same is true for Merapi volcano and the beach, the two corner points of Yogyakarta urban layout (see Sect. 5.2.2). Furthermore traditional shadow puppets are depicted, which is also declared intangible World Heritage.

The attachment to Yogyakarta is also expressed in daily life. One example is the high number of school classes carrying out surveys on Yogyakartanese cultural features witnessed throughout the field research phase. Mostly students from junior schools were asking foreign tourists why they picked Indonesia and in particular Yogyakarta as travel destination and about their impression of Yogyakarta culture and different places. As reason why they did such survey they named deepening their English knowledge and in particular arousing the tourists' interest in the local culture they were proud about. Sugiana (2008: 68) states:

> In Yogyakarta, Javanese culture and tradition is still alive and plays an important role in everyday life. The cultural inheritances are found in the expression of artifacts and spatial arrangements, customs, arts, dances, believes and local institutions.

Results from the expert interviews as well as from the questionnaire are substantiating this statement. Yogyakarta has produced very famous artists and artworks like paintings (Y07), all interviewees were proud of the city and named it as capital of education and culture. Cultural events are flourishing in the city, including a local popular culture, with popular music, local food and festivities. Also in this context various interviewees mentioned the importance of the Kraton as the entity that cares about preserving the local culture, e.g. by practising among others traditional music (*gamelan*), and *wayang kulit* inside the Kraton (cf. Sect. 5.2.3 on the historic urban landscape of Yogyakarta). As described above batik fabrication is closely interwoven with the Kraton, as there were even distinct fabric patterns only worn by Kraton officials. Yogyakarta was appointed as 'World Batik City' by the World Crafts Council in 2014, recognizing the town's importance as centre of batik fabrication in the country (Hengky 2015). Until today there are batik showrooms in the direct vicinity of the Kraton, where most of the fabrication used to happen in earlier time—while not little of the batik sold in Malioboro Road is of minor quality or even imported (Y06, Y09). Y07 is having her own batik business together with other family members and is witnessing a growing interest in the traditional batik, including colours and patterns. She even presented her batik in the occasion of fashion weeks in Jakarta and even Milan, indicating even international interest in it. Another trend she states is using traditional batik but for a global community, e.g. by using the fabric for doing clothes for colder climates outside Indonesia, worn by Indonesian expatriates but also international customers. Batik is a living part of Indonesia's and Yogyakarta's daily life, worn on regular occasions and definitely during festive occasions.

The Sultan is also involved in popular culture. Until today *becaks* (cf.) still survive in the urban area, particularly in Malioboro area much of the informal business is done with them: "*Unlike elsewhere in Indonesia where becaks have been replaced by administrative decree, in Yogyakarta they are supported by the Sultan*" (Hook 2005: 16). Other examples of local popular culture are the Yogyakartanese cuisine that local people do not get tired of promoting as being excellent, the traditional bazaars and music. The results obtained for a question on intangible assets of Yogyakarta and the importance to preserve them reveal underpin the importance of local cultural expressions. Not a single asset listed was estimated not important or 'not that' important by a majority. In each case at least a three quarters majority gave a positive or very positive answer (cf. Fig. 5.35).

Not a single respondent assessed batik fabrication as not important; on the contrary as much as 70.6 % rate this as very important and another 29.4 % as important. Impressive 98.3 % evaluate *gamelan* music as important (31.9 %) or even very important (66.4 %). For overall 95.8 % *wayang kulit* is of importance (31.1 %) or even high importance (64.7 %). Yogyakartanese food as well is rated as important (39.8 %) or very important (54.2 %) by a vast majority of 94.1 %. In each case an absolute majority was obtained, showing the high importance attached to these assets. Only in two cases the majority of questionnaire respondents voted for 'important' instead of 'very important': in the case of silver works as traditionally practised particularly in Kotagede and in the case of the traditional markets like Beringharjo. Silver works however still achieved 93.2 % positive answers, out of which 51.7 % assess it as important, and 41.5 % as very important. 83.1 % rate the local markets as important (52.5 %) or very important (30.5 %).

In total, these answers and the appreciation for intangible values is very high, comparably higher than results obtained when asking about the appreciation of built heritage. It seems that these intangible values are not perceived as 'heritage' but still as something 'lived', a part of Yogyakarta's present culture and not of its past. In addition, intangible heritage is inextricably intertwined with built as well as with natural heritage as, e.g. expressed in the *wayang kulit*, which has an own puppet for 'volcano' and which is practised and promoted in the Kraton. It is potentially because of the still active role of intangible heritage in daily life that hardly any expert elaborated on measures or specific programmes to promote these assets. In addition, the responsibility is still very often seen with the Sultan and less with urban authorities or other actors (Fig. 5.31).

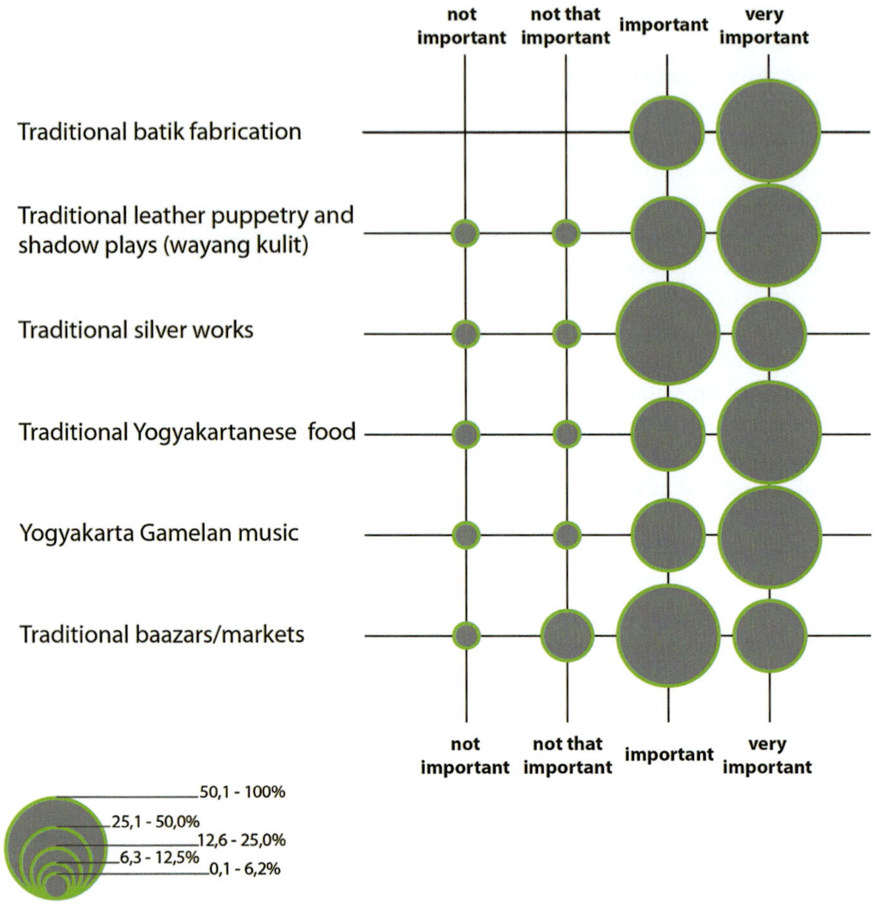


**Fig. 5.31** Answers to the question: Which intangible things or activities do you consider as typical for Yogyakarta area, how important is it to preserve them? ($n$ = 120)

Nevertheless, the questionnaire interpretation of results allows to state that urban history is of fundamental importance. Overall 58.8 % of the questionnaire respondents state they are interested in the urban history, another 21.8 % is very interested, while only 17.6 and 1.7 % are only a bit or not interested at all. The attachment to urban tangible and particular intangible heritage and the centre are high among experts and questionnaire respondents.

In addition, the questionnaire asked about (predefined) distinct historic places in the urban area and (not predefined) places the students would choose for sightseeing and leisure in case of a friend's visit. The results again emphasize the importance of the historic centre (cf. Fig. 5.32).

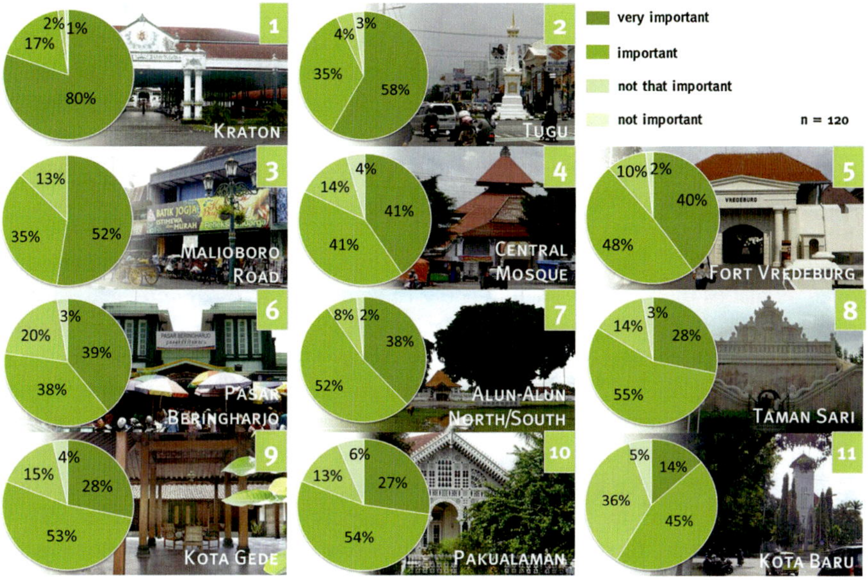


**Fig. 5.32** Answers to the question: Which places/buildings do you consider important as place of remembrance of Yogyakarta history?

Among the eleven locations provided for selection, all locations within the centre are ranked first. Particularly the Kraton (on top, with impressive 80.0 % considering it as very important place of remembrance), the Tugu and Malioboro Road are rated as very important places of remembrance, with absolute majorities. Another around 40 % assesses the Central Mosque,

Fort Vredeburg Pasar Beringharjo and both alun-alun as very important. The last positions are either sites that serve as museum exclusively (Taman Sari and Pakualaman) or quarters outside the city centre: Kotagede and the Dutch quarter of Kota Baru ranked last.

The students were also asked to give a reason for the importance. In case of the Kraton, the majority of reasons are dealing with 'culture', 'history' or 'government', indicating the Kraton's importance for the city—e.g. "*it is why Yogya is special!*"—but also the Sultan's significance until today. The Tugu monument is considered for both, its historic value, e.g. as "*Yogyakarta's icon*" and "*part of the collective memory of Yogyakarta's past*" as "*symbol of the imaginary line*", but at the same time as a place to hang out in leisure time. The same applies to Malioboro Road which is appreciated for both, its historic value as well as its function as commercial hub and leisure place. Both alun-alun are as well appreciated for their leisure function as large open spaces, but also for their historic value as a "Symbol of Sultan Power". Also in case of the Mosque the answers can be categorized in two groups, either dealing with the Mosque's historic value or with its function as a place to go for praying.

The answers given in case of Fort Vredeburg are a bit more biassed, as not only the historic or leisure value is stated, but also the fact that it is a Dutch colonial building, what for few respondents ("*It is Dutch!*") discredits the building. Comparable answers were given in case of Dutch quarter of Kota Baru, e.g. "*it is the place where the foreigners lived*". The answers given for Beringharjo are mostly dealing with its market function, either appreciating it as a traditional market and place to purchase batik or as an ordinary market just like others. Kotagede then is mentioned foremost for its artistic roots and the silver market. In case of Taman Sari and Pakualaman the places' history was given as reason exclusively, indicating the sites' restricted uses.

Taking a look at Fig. 5.33 shows almost the same locations, in this case mentioned as places the students' would show to visitors from outside. Apparently, the order is a bit different, instead of the Kraton, which is ranked no. one as place of remembrance; students would take their friends to Malioboro first. This may again indicate the dual role of Malioboro, as a historic area but at the same time a place that offers many leisure activities.

Overall 58 locations are mentioned, out of which the top fifteen (mentioned four times or more) are given in the figure. Out of the 15 seven are located in the historic centre, among them the top-three ranks, namely Malioboro, Kraton and the southern *alun-alun*—three spots ranked as important places of remembrance as well (see Fig. 5.33). Even the reasons are the same as stated above, a mix of historic value and leisure, e.g. in case of the southern *alun-alun* students wants to go there to enjoy the atmosphere and hang out at night, finding the way between the Banyan trees. The other inner-city places mentioned are Pasar Beringharjo, mainly for

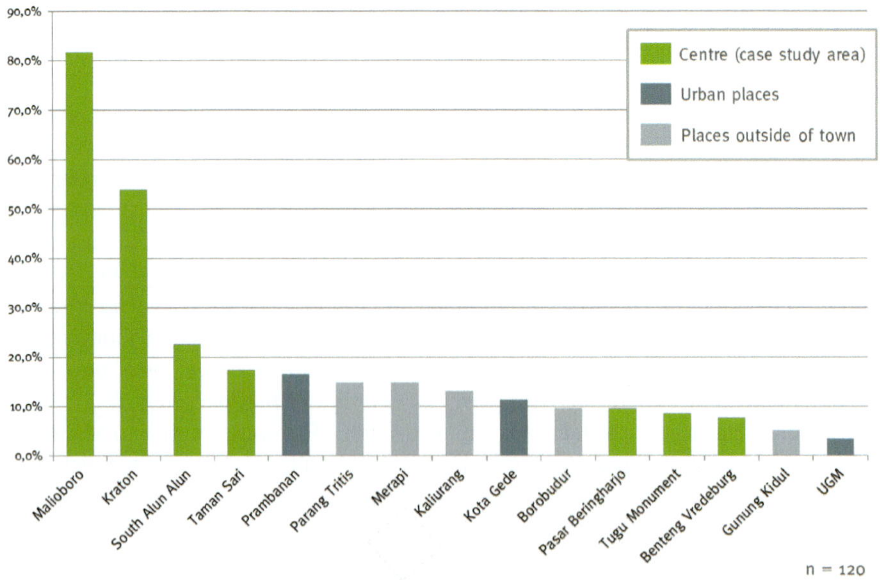


**Fig. 5.33** Answers to the question: Imagine some friends from outside visit you in Yogyakarta. Where in and around the town would you take them for sightseeing and leisure?

shopping activity, the Tugu monument and Fort Vredeburg, again for the same reasons as in the previous question, for leisure and to see iconic landmarks of Yogyakarta. Even the mentioned places outside the centre are partly related to the urban plan, as Parang Tritis beach and Mount Merapi form the two endpoints of the imaginary main axis of Yogyakarta.

In total, the appreciation for intangible assets is astonishing, including values and memories attached to sites within the historic centre. The importance of local cultural expressions and activities is particularly high in the questionnaire results. Culture in Yogyakarta is intrinsically intertwined with the Kraton and the Sultan, who is somehow regarded as keeper of the local culture. Although tourism was mentioned often as a source of income and target group of sites within the centre, the area is still a major point of attraction for the local residents.

### *5.2.10 Conclusions on the Role of Heritage in Shaping the Urban Identity of Yogyakarta*

The empirical study of Yogyakarta shows the importance of the historic centre. The initial urban layout following the 'cosmological planning' is still perceived in the urban layout, not only by planners, it is rather but a part of the urban culture. "*Yogyakarta is special!*" was stated various times in the expert interviews, and most experts expressed their wish this speciality should be given more emphasis in urban development. However, although Yogyakartanese still take pride in this planning scheme (according to the majority of interviewees) it is getting difficult to preserve buildings and endowed meanings, "*it is a challenge for the future*" (Y01). Like in other Asian countries, the value of traditional architecture is going beyond its fabric, it endows spiritual meanings and continuous repair or restoration works are part of conservation (Kwanda 2010).

Local value systems like *gotong royong* and 'Unity in Diversity' are principles that are still believed (Y02, Y08). Larasati (2007) proclaims the feasibility of *gotong royong* in urban areas to contribute to sustainable living environments. In particular the potentials to build a local sense of belonging and the focus on specific local characteristics can contribute to a stronger consideration of sociocultural aspects in urban development.

In a study published 2003 (Siauw) family roots were found a major trigger of developing emotional ties to the Malioboro area. However, the results' reveal that also other groups, like the students that filled the questionnaire, have ties to the historic centre, even if not being from Yogyakarta or the centre area itself. In fact the attachment to the centre is very high; particularly the cultural value is recognized. Furthermore the centre is found beautiful and vital. However, also some problems are stated by interviewees and questionnaire respondents, namely traffic problems, and to a lesser extent the loss of historic buildings and uses. Among the underlying reasons are urban change and densification, preferences for modern constructions and often poor policies which in addition are not always implemented properly, as stated in the interviews. Literature as well poses non-adequate renovation or constructions as main menaces for the historic core area (Siauw 2003; Zahnd 2005).

"*Yogyakarta basically is the Kraton*" (Y05). This and comparable statements were given very often, on the one hand to express pride in having such a unique culture. It is the Kraton and notably the Sultan himself that are the benchmark for Yogyakarta's culture (Y02, Y03, Y06, and Y07). On the other hand, such single focus bears the threat of making yet popular culture static instead of having a living culture which changes and adapts to societal changes—as the urban culture has done in the past when adapting things from other ethnic groups, e.g. in construction. At the same time not all experts are satisfied with the duality of the Sultan as cultural as well as political leader, making him the only focal point (Y02, Y06, and Y07, and Y09). Probably due to the 'specialty' of Yogyakarta hardly any role model for conservation or regeneration was mentioned. Being asked about the urban future the interviewees are aligned on two sides, believing that the special urban features of Yogyakarta are valuable and have to be preserved, but recognizing the rapid change the urban area is undergoing. Y06 states that the status as 'Special Region' includes a certain responsibility "*to provide a role model for a good city, also for the rest of Indonesia, as a heritage city*". Y07 is optimistic on the one hand as increasingly not only single monuments and archaeological sites are on the agenda of Indonesian conservation schemes, but sites and cultural landscapes (*saujana*). However her vision for the future is not too optimistic:

> When I have a meeting with the government I always say 'I am hopeless!' Because we are working for Yogyakarta but there are too many challenges, too many things destroyed, even next door. Now they have a new regulation, a law about the outstanding value of Yogya, not that they are talking about it! I will now try one more time for Yogyakarta – if it doesn't work I will give up (Y07).

Both interviewees in Y08 have a more positive scenario for Yogyakarta, referring to Indonesia's economic development and the recovery after the 2008 economic crisis. Y09 wishes for more consideration of local wisdom in planning. Y06 concludes: "*Yogyakarta has a kind of local traditional sustainability concept; it is already there and should be considered in the future, in physical, cultural and livelihood aspects.*" *To him in Yogyakarta* it is most crucial to combine urban development with economic aspects and the cosmological planning. Summing it up, interviews and questionnaires show a high level of appreciation of the historic centre. To both groups this should be considered in future urban development, but urban realities do not (fully) meet these requests.

Yogyakarta Summary

- Conservation legislation has changed profoundly since the end of colonial times, the '*saujana*' approach introduced in the 2003 charter is coming closer to the Indonesian/Javanese understanding of interdependencies between cultural, natural and intangible heritage.

- Conservation in reality is challenging, due to the large number of formal and informal actors involved on the different administrative levels, and resulting in the loss of heritage fabric.
- Yogyakarta's urban layout of utmost importance, following a cosmological planning with the Kraton as centrepoint. However, the urban area and its fabric are prone to change due to urban growth, densification and growing traffic burdens, among others.
- Adapting historic buildings or parts of them to present needs and preferences—to a certain extent—is part of the Javanese culture. In Yogyakarta constructions in Javanese, Dutch and Chinese style as well as mixed forms can be found, each requiring distinct conservation schemes.
- Indonesian and foremost Javanese culture is of high significance and still very popular and appreciated, however, in reality it is much less protected by legal schemes than by the Kraton/Sultan as the point of reference for Yogyakarta's culture.
- The Sultan is the main actor and point of reference for Yogyakartanese culture, taking care of the maintenance of local culture, but at the same time also determining urban development and politics.

## 5.3 Case Study III: Recife, Brazil

Like in the other case studies the research methods applied in Recife comprised field visits, surveys of literature and documents in relation to urban planning and heritage in Recife, expert interviews and a questionnaire-based investigation.

Overall 15 expert interviews were conducted in autumn 2013, with experts from urban planning, regional development, practice and academia. The in-depth interviews took between 30 min and 1 h 45 min and were recorded, transcribed and afterwards evaluated with MAXQDA software. As the interviews were done in Portuguese they were translated to English afterwards. Recife interview partners do not appear with their full names but are encoded from R01 to R15 in the text. A detailed list with names and functions is given in Annex II. Three out of the fifteen interviews were carried out in group form with two people at the same time following the interviewees' preferences. In this case an 'a' or 'b' is added to the interview code to indicate the respondent.

In total six interviews have been carried out with representatives of Recife Municipality. R03 is the head of the Directorate for the Preservation of Cultural Heritage of Recife, while interviewee R01 is working for the same Directorate in the field of tangible heritage preservation. Both participants in interview R02

(R02a/b) are working in the branch of immaterial heritage conservation of the Directorate. R12 is the manager of the *Centro de Formação, Pesquisa e Memória Cultural-Casa do Carnaval,* a section of the *Secretaria de Cultura por meio da Gerência de Preservação do Patrimônio Cultural Imaterial* (Cultural Secretariat, by means of the Directorate for the Preservation of Immaterial cultural Heritage, Recife Municipality). The *Casa do Carnaval* serves as hub for studies, research and documentation of the city's culture, comprising, e.g. a large archive, the promotion of project support, or the conduction of courses, seminars as well as the organization of thematic exhibitions.

Interviewee R14 is the head of the *Museu da Cidade do Recife*, located in Forte das Cinco Pontas (Museum of the City of Recife, located in a historic fortification built by the Dutch). Like the *Casa do Carnaval* the museum is associated with Recife's *Secretaria da Cultura* as well. Interview R08 was conducted with the head (R08a) and an expert (R08b) of the *Instituto da Cidade do Recife Engenheiro Pelópidas Silveira (CPS).* The mission of this municipal authority is to support the implementation of Recife's *Plano Diretor* (master plan) as well as major strategic urban planning projects and to coordinate the rehabilitation of central areas.

Five interviews were carried out on the level of Pernambuco state. In interview R11 two managers of *AD Diper*, the Economic Development Agency of the Federal State, were consulted. Overall four interviews were conducted with staff of *Fundarpe*, the Foundation for the Historic and Artistic Heritage of Pernambuco. The relatively large number of interviews represents the different branches and responsibilities the Foundation has. Interviewee R05 is the director of *Fundarpe* himself, interviewee R04 is the lawyer, responsible for safeguarding all kinds of monuments. Interviewee R09 is the coordinator for immaterial heritage at *Fundarpe* and R10 is working in the same field as expert and historian.

Interviewees R06 and R13 both have previously worked for different planning authorities in Recife and Pernambuco (R06 for Recife urban planning authority, he is the former president of the *Instituto da Cidade do Recife Engenheiro Pelópidas Silveira,* and R13 for FIDEM, additionally she was part of the team that elaborated the initial plan for the centre revitalization) and are now working as consultant and architect.

Finally R07 and R15 are both university professors and planners, in parallel both are affiliated to the *Centro de Estudos Avançados da Conservação Integrada*—CECI (Centre for Advanced Studies on Integrated Conservation). R07 is professor for urban conservation and architectural and planning history, while R15 holds a professorship for conservation of urban spaces and parks.

Like in the other case studies questionnaires were handed to university students from different study programmes related to planning, geography, and social sciences. Overall 81 questionnaires were collected. In the case of Recife questionnaires were distributed in two different universities, due to the fact that during the field research period the Federal University of Pernambuco (UFPE) was on an unscheduled study vacation. Therefore UFPE staff took the questionnaires to the Catholic University of Pernambuco. In Recife, questionnaires were filled mostly by Master students, in addition some PhD and Bachelor students responded as well.

**Table 5.4** Period the questionnaire respondents are already residing in Recife ($n = 81$)

| Grown up in Kathmandu | For more than 5 years | Between 2 and 5 years | Between 6 months and 2 years | Less than 6 months | No answer |
|---|---|---|---|---|---|
| 65.4 % | 19.8 % | 4.9 % | 2.5 % | 3.7 % | 3.7 % |

Around two third of the respondents grew up in Recife itself, another almost 20 per cent lives in Recife for more than 5 years already (cf. Table 5.4). All respondents were residing in Recife when filling the questionnaire. The age distribution ranged from 17 to 59, with an average age of 26.7 years.

The 'centre area' as mentioned in the questionnaire was defined as the quarters of Bairro do Recife (southern part up to Praça Tiradentes in the north), Santo Antônio and São José (northern part up to Forte das Cinco Pontas in the south). Figure 5.36 provides a more detailed overview on the research area.

### 5.3.1 A Brief Country and Planning History of Brazil

The Federal Republic of Brazil (*República Federativa do Brasil*) is Latin America's largest country. On a global scale it is number five in both, total population and size. Brazil ranks number 79 out of 187 in the HDI index (UNDP 2014). As an upper middle income country with a considerable economic power is a member of the G20 and considered as one of the BRICS states (Brazil, Russia, India, China and South Africa).

Before the landing of the Portuguese in 1500 Brazil was inhabited by numerous tribes which were defeated by the invaders. Brazil remained colony until 1808, despite some battles with the Dutch and the French in sixteenth and seventeenth century. Sugarcane became the main good of exportation, while numerous slaves of African origin were imported for its production. During Napoleonic times the Portuguese Royal Court fled to Brazil. After the King's return and some years of struggles in 1822 the Empire of Brazil was founded, with the former Portuguese crown prince as Emperor Dom Pedro I. During the nineteenth century coffee production replaced sugar cane as main cash crop. In 1889 monarchy was overthrown in a military coup backed by wealthy coffee magnates, one year after Brazil abolished slavery—the last country worldwide.

Throughout the twentieth century Brazil witnessed phases of governmental instability, military coups and changing governments. Transition to democracy from end of the 1970s onwards paralleled a profound economic transformation and ongoing urbanization. In 2003 the first transition from a democratically elected president to another in more than 40 years took place when Luiz Inácio da Silva, known as Lula, was inaugurated as president. Since then his Workers' Party (*Partido dos Trabalhadores*—PT) is making up the president. In 2011 Dilma Rousseff became his successor, increasingly confronted with protests against corruption, environmental exploitation and social inequalities exacerbated by 2014 FIFA World Cup and 2016 Olympic Games.

Latin America's urbanization ranges back as to pre-colonial times. However, the appearance of its cities is shaped by Spanish and Portuguese urban models that were implemented in Latin America since the continent's colonization (Heineberg 2006). Building tradition follows a strict rectangular layout with the main square in the centre. All main representative, administrative as well as religious buildings and functions are located around this 'plaza mayor', surrounded by upper-class housing. The layout was however stricter in the Spanish sphere, while the cities in the lusophone hemisphere followed the characteristic chessboard layout of the city to a lesser extent (Bähr and Mertins 1995). Outstanding examples of Brazilian colonial cities are, e.g. Salvador da Bahia or Ouro Preto in Minas Gerais (Nobre 2002; Zancheti and Gabriel 2011; Peres Torelly et al. 2014). The colonial city was characterized by a centre-periphery gradient, with the upper-classes near the centre, and lower classes in the urban fringe. Until today this concentric urban layout is still visible in most Brazilian colonial cities (Struck 2008).

Colonial centres lasted almost untouched until the early twentieth century, when Haussmann's Parisian role model was implemented in many urban cores like Rio de Janeiro or Recife. As a consequence much of the historic fabric was destroyed (Kamarid and Leupert 2009; Proença Leite and Peixoto 2009). Since the 1950s then Brazilian Modernity shaped architecture and urban planning, essentially marked by Lucio Costa's and Oscar Niemeyer's iconic plan for Brasilia (Lacarrieu 2013). Brazil's urbanization rate is as high as 85.7 % (prediction for 2015, United Nations-Department of Economic and Social Affairs 2014). Urban growth took place in two main phases, with the first one during the 1930s and 1940s, enabled by modern means of transportation. Between 1945 and 1961, Brazil experienced rapid growth, its major cities almost doubled in size, undergoing radical transformations. During this time the country adopted modernism as key style, with the new capital Brasilia as archetype (Lara 2011). The second phase of urbanization is mainly characterized by peripheral development. This urban sprawl is still ongoing and shaping the urban outline (Bähr and Mertins 1995). The main boom in urban growth and migration to the centres took place between the 1950s and the 1970s. Since then apartment houses became popular (Aragão and Maennig 2013), the starting point of Brazilian Gated Communities, the so-called *Condominios Fechados*. The most prominent examples are Alphaville in the São Paulo metropolitan region and the privatized urban quarter of Barra da Tijuca, in Rio de Janeiro (Lara 2011; Villaça 1998; Coy 2006; Coy and Pöhler 2002). The rapid increase of metropolitan population also led to a growth of the informal sector, including the construction of spontaneous (or informal) settlements, many of them in vacant areas of the centre or peripheral areas (Coy 2006; Coy and Pöhler 2002).

At the same time also formal and large-scale housing projects in urban peripheries took place, mostly gated communities, middle-class housing and major social housing projects for low-income groups (Fernández-Maldonado et al. 2014), fragmenting the urban layout. This trend was accompanied by commercial and business developments, the expansion of road infrastructures and the emergence of shopping malls. As a result, the gap between the spaces of wealth and those of marginality grew, upper-classes preferably settled in suburban luxurious high-rise complexes (Herzog 2013; Coy 2010; Deffner 2006). Since the 1990s macro-political and economic shifts have led to massive spatial extension of the built-up areas (Fernández-Maldonado et al. 2014).

Inner-city regeneration in Brazil was often understood in the context of urban poor, including investments, infrastructure and social interventions in low-income neighbourhoods (Baker 2006) and fragmented and degraded urban cores (cf. Coy 2007, in São Paulo; Kamarid and Leupert 2009), often resulting in processes of gentrification (Proença Leite 2007, 2013). Other regeneration projects were carried out in the also degraded cores of colonial cities, with the aim to also restore the fabric. One of the role models for regeneration of historic cities in Brazil is Salvador da Bahia. The colonial core faced decay and degradation in the 1980s and underwent large-scale regenerations schemes after the declaration as UNESCO World Heritage in 1985 (Nobre 2002; Rothfuß 2007a). Since the 1980s, urban regeneration strategies of many Brazilian States, in particular in the Northeast, are focusing on the promotion of cultural tourism as a way to support economic development. The case of Salvador followed this idea and has much influence on the urban regeneration of many North–East cities like Recife and São Luiz (Nobre 2002; Zancheti and Gabriel 2011; Rothfuß 2007b).

Throughout the past decades a number of innovative approaches in housing and urban development have been implemented (Cunha et al. 2013; Athayde 2011) such as the participatory budgeting, for which the city of Porto Alegre became famous (Novy and Leubolt 2005). Social housing is a major concern since the PT presidency. In 2009, the 'My House, My Life' Program (*Minha Casa, Minha Vida*), was introduced with the aim to provide housing for low-income families. Overall two million housing units shall be funded during the different program phases (Aragão and Maennig 2013; Marques and Rodrigues 2013).

Brazil's urbanization rate still growing, with the current trend in urbanization marked by a growth of second-tier cities like Recife (Serra et al. 2004; Dowall 2006). Currently it is the agglomerations in the country's north and central west which are growing the most (Da Mata et al. 2005). Actual urban discourses are dealing with citizenship rights, marginalization, the right to the city and informality (Roy 2009). Fernandes (2006) names speculative land markets, clientelistic political systems, elitist urban planning practices and exclusionary legal as major concerns, while to Altvater (2003) the high level of informality is among the main challenges Brazilian cities have to deal with.

## *5.3.2 Introduction to Recife*

Recife, located in the Northwest of Brazil, is the countries' fifth largest urban agglomeration and capital of the State of Pernambuco. After being founded in 1537 it became one of the most prosperous cities in the Americas during Colonial times

**Fig. 5.34** Aerial photo of Recife, with the centre area and surrounding high-rises in the middle background

(Zancheti 2005). The reasons behind are its strategic location almost at the easternmost point of the continent and the natural protection by reefs along the coastline. The city's name itself is said to go back to the Portuguese word for reef. After being founded by the Portuguese in 1537 the city witnessed some decades of Dutch reign which have altered the urban outline profoundly. After the restoration of Portuguese reign Recife itself remained the harbour and workers location while the upper-classes settled in the hilly terrains of Olinda, Recife's twin city in the north. Today, the Portuguese and Dutch colonial history is still visible in the urban outline, in particular in the centre, while the periphery has been developed in more recent times. Due to its location near the equator the climate is a tropical monsoon one. Recife is located amidst patches of Atlantic rainforest (*Mata Atlântica*), along the coast some large mangrove areas are still preserved within the urban area. The two main rivers of Capibaribe and Beberibe flow through the city and shape its outline with many islands, connected by bridges, which have given the city the byname 'Brazilian Venice' (Fig. 5.34).

Recife's urban layout is dominated by the coast and the urban rivers. The main quarters are located along the coastline on different peninsula, linked by bridges. The most historic quarters are the Bairro do Recife (or *Recife Antigo*, Old Recife) Santo Antônio and São José, accommodating emblematic urban places and buildings (Antunes Cavalcante 2006). Recife has 3.7 million residents in its metropolitan region which spans 14 districts, has a total land area of 276,143 hectares (United Nations—Department of Economic and Social Affairs 2014). Only the metropolises of São Paulo and Rio de Janeiro surpass Recife in terms of housing density (Rubens de Menezes and Figueira de Souza 2014). However, it is still considered as Brazilian 'periphery' (Campos 2002).

The city has traditionally been the destination of large inflows of migrants from the arid rural areas of the Northeast. After the end of slavery in 1888 many former slaves moved from the sugarcane plantations to Recife and settled in so-called *mocambos* (local expression for *favela* or squatter settlement, another word used in Recife is *morro*) in the mangrove or swampy areas along the urban rivers. In the beginning of twentieth century the local government decided to upgrade these areas. Responses from urban authorities ranged from relocation, social housing programmes to governmental upgrading including sanitation, drainage, electrification and road-building programmes—depending on the spirit of the time and the government in charge (Koster 2012; Sé Carneiro 2010). From the 1920s onwards many efforts were taken to expand and modernize the city, including the demolition of many informal settlements as well as formal but decayed or outmoded inner-city buildings (Barros Filho and Gama Monteiro 2011). From the end of the 1930s to the mid-1970s, slum evictions significantly increased, forcing many poor families to the hills west of the city (Barros Filho and Gama Monteiro 2011; Sé Carneiro 2010).

Since the 1980s Recife grew rapidly, with the most accelerated growth numbers in its peripheral districts (Bitoun 2007). Throughout the 1970s and 1980s many large-scale social housing units were constructed, mostly in remote peri-urban areas (Kirsch-Soriano da Silva 2010; Bähr and Mertins 1988). Between mid of the 1960s and end of the 1980s almost 80,000 housing units were built in the RMR, corresponding to almost 30 % of all legal households in the city in 1987. Many poor families were relocated from central areas of the city to its periphery (Barros Filho and Gama Monteiro 2011; Kirsch-Soriano da Silva 2010).

Recife has a solid tradition in designing instruments for urban land tenure. To provide access to affordable land for the city's growing poor population, in 1987 (revision 1995) the city passed a law on innovative land titling known as the 'Plan of Regularization of Zones of Special Interest' (*Plano de Regularização de Zonas Especiais de Interesse Social*, PREZEIS). This pioneering program gives favela residents security of tenure and the right to receive government services including infrastructure. Furthermore it allows them to participate in decision-making at the neighbourhood and city levels. So far, local government designated 66 areas as areas of particular social interest (*Zona Especiais de Interesse Social*—ZEIS), corresponding to 85 % of the favelas in Recife (Serra et al. 2004; Campos 2002; de Souza 2001a, b; Miranda and Moraes 2007; Bitoun 2006; Aigner and Leite 2006).

The end of the 1980s and the 1990s were then marked by the economic/financial crisis of the country that was especially severe in the Northeast. It had enormous impact on the urban economy of Recife, due to intense de-industrialization followed by a radical change in the economic base of the region: the decline of the sugar cane production (Koster 2012). Different large-scale urban development projects were launched at that time induced by an administrative reform and ongoing decentralization (Freitas Cavalcanti 1997), among them some for people living in vulnerable informal settlements (Bitoun 2007). Acknowledging the poor infrastructure—mid of the 1990s still only about 24 % of Recife's population was connected to the public sewer system (Nance and Ortolano 2007)—the improvement of the urban infrastructure was another aim, e.g. in the *Prometrópole* project (main phase 1996–1999). Its emphasis on private property and autonomous citizenship can be classified as neoliberal although the PT, that is in charge of both municipality and federal government is giving it a 'socialist twist' (Koster and Nuijten 2012).

As a reaction to the economic crisis the years after 2000 are marked by a radical change in the role of the State: It withdrew almost completely from the production activities. Recife's economy was decided to change towards service sector, giving another push to the building sector. The government stimulus includes incentives for people to buy a car and their own home. Consequently, trunk roads are built and numerous housing projects aimed at the working and lower middle classes are launched, mostly in peri-urban areas (Furtado et al. 2014). Within only few years, services became the wealthiest sector in the local economy, bringing with it the development of commerce and industry branches (Zancheti 2005). The strong economic growth came hand in hand with urban expansion. Among the major urban development projects was the *Projeto Capibaribe Melhor* (Project 'Better Capibaribe' 2007–2013), supported by the World Bank, on urban development and social inclusion of the low-income population in peripheral riverine areas (MC Consulting and Empresa de Urbanização do Recife (URB) 2006).

Today Recife is undergoing a substantial urban and peri-urban development, south, north and west of the centre. (Furtado et al. 2014). British consultancy PricewaterhouseCoopers even predicts that Recife will be one of the 100 richest cities in the world in 2020 (Rubens de Menezes and Figueira de Souza 2014). South of Recife, in the Suape area, the largest dockyard in the southern hemisphere is constructed—within a protected area comprising of natural and cultural heritage (*Parque Metropolitano Armando Holanda Cavalcanti,* this area is said to be the spot of the first discovery of Brazilian territory, as stated by interviewee R04). In total more than 100 ventures are already running, among them a petrol refinery and three petrochemical plants, adding up to an investment level of currently US$17 billion. To link Suape with the north and east of the RMR the capacity of the inadequate road network is planned to be improved with the '*Arco Viário Metropolitano*'. This traffic project includes several complementary schemes and is bypassing Recife from South to North, with the intention to provide road access to new suburbs and in particular trade and industry, including the stadium for the 2014 FIFA World Cup and the foreseen surrounding facilities (Furtado et al. 2014).

The expansion of the road network includes the 'Via *Mangue*' (mangrove street), a new North–South shortcut leading partly through former protected mangrove area, going hand in hand with the new giant shopping centre '*RioMar*' which opened 2012. In addition to the road network, another major undertaking is the navigability of the Capibaribe River to disburden the almost collapsing public transportation.

Additional business options are expected and fostered by urban authorities, potentially going along with more legal concessions for environmentally fragile or protected areas (Furtado et al. 2014). Planning legislation has been weakened to allow for such large-scale developments, even such in environmentally or culturally protected areas, harming environment and local identity. The population was either not aware of such legislative changes; or they may be accepted based on the argument of job creation and economic benefits (Furtado et al. 2014). Overall, this is contributing to the environmental crisis that Recife is facing. Since the 2000s multiplied environmental degradation is going hand in hand with suburbanization and expansion into socially and environmentally vulnerable areas—exacerbated by lack of suitable plans and strategies to handle that (dos Santos 2013). Today, Recife's urban landscape is a very disperse one, with modern middle-class neighbourhoods and highly protected upper-class condominiums next to deprived squatter settlements (Koster and Nuijten 2012).

The impressive number of another 28 high-impact projects is awaiting approval by the urban Development Council (*Conselho de Desenvolvimento Urbano*, CDU). To Furtado et al. (2014) this is indicating a lack of long-term visions for the urban development but rather a preference for individual projects. Such kind of 'central suburbs', high-rise condominiums constructed in central places freed from prior structures are the latest urban trend. In recent urban development projects Recife is following PPP approaches, not only including the construction itself but social programs for the population in the process of resettlement (Koster and Nuijten 2012).

Recife is still growing fast due to the actual regional development policy that has encouraged rapid growth, going along with steeply rising property values. However, Recife's infrastructure does not keep pace with its economic growth. As a consequence, urban accessibility and mobility have become major problems. The low priority on public transport results in a lack of coordinated traffic flow, lack of bus lanes, problems with traffic jams (Furtado et al. 2014; Rubens de Menezes and Figueira de Souza 2014; de Almeida Souza and Bitoun 2015; de Andrade and Alves Maia 2009), accelerated by the increasing number of commuters from and to the fringe areas.

Uncontrolled growth, poverty, social inequality with a large share of the population living in poverty and a chaotic urban mobility are probably the most pressing urban problems (cf. de Andrade and Alves Maia 2009). As typical for Brazilian cities, there is a high variety of morphological patterns (do Eirado Amorim et al. 2014), indicating Recife's high social disparities (Furtado et al. 2014; Rubens de Menezes and Figueira de Souza 2014).

While the region's economic development is promising and market competitive, the city has one of the highest rates of poverty and inequality in Brazil. Many of its residents still live without basic services. It is estimated that around half of the population lives in one of the often very densely populated 450 *favelas* (Serra et al. 2004; Furtado et al. 2014). Furthermore the quite flat topography is making the city highly vulnerable to flooding and rising sea level, likely to be exacerbated by climate change—today's drainage and other infrastructure is already not able to cope with periods of high rainfall. The low-income population is most vulnerable because they live on low-lying land or on hilltops with inadequate slope stabilization (Furtado et al. 2014).

### 5.3.3 The 'Historic Urban Landscape' of Recife

After its foundation Recife became the capital of one of the Portuguese *Capitanias*, the administrative entities during colonial times. Over the next century Recife itself developed as the harbour location while Olinda—a hilly spot around six kilometres to the north—became the upper-class residence area. From 1630 to 1654 the Brazilian northeast came under Dutch rule. It was during that period when Recife itself prospered and developed, due to the abilities of the Dutch to handle the swampy and flat land traversed by rivers. During his reign 1637–44, Governor João Maurício de Nassau (Johann Moritz von Nassau-Siegen) initiated the very first urbanization plan for Recife, including street layout and bridge. Famous paintings by Dutch artist Frans Post till today give testimony of Nassau's Recife. The layout was "*a harmonious and well-organized ensemble of right-angled streets and proper-sized building blocks, and a canal with two central public squares as its backbone*" (Van Oers 2014: 61). During his reign the first Jewish community on the American continent settled in Recife, after the return of the Portuguese they migrated to northern America (where some of them became founding members of New York's Jewish community). Until today the Dutch reign is part of Recife's collective memory, often considered as its 'golden age', as, e.g. apparent in one of the city's nicknames, '*Cidade Maurícia*'. It was during that time when the first bridge was constructed to link the islands; its current successor on the same site is called after Nassau *Ponte Maurício de Nassau* (Maia Alvez 2009).

With the return of the Portuguese in 1654, representative and housing functions returned to Olinda, while Recife remained harbour and trade area. For the next almost two centuries residential construction in Recife was rather undefined and following different styles, but less representative than in Olinda. In contrast, it was during that time when most of the numerous churches and convents in Recife's centre were built, in particular in Bairro do Recife on the central island, and the quarters of Santo Antônio and São José on the headland behind (Duarte 2011; Maia Alvez 2009; Van Oers 2014). While most churches stem from eighteenth century, the majority of palace, cultural and administrative buildings was built only in the nineteenth century, e.g. the Alfândega (customs) building in Bairro do Recife or the Teatro de Santa Isabel in Santo Antônio (Duarte 2011).

The development of Recife area was dominated by the harbour on the one hand (Lubambo 1991) and the sugarcane production on the other. In the seventeenth century Recife harbour used to be the largest in the Americas (de Albuquerque Lapa and Almeida de Melo 2007), a transshipment centre for mainly sugarcane and slaves. Many sugarcane plantations (*engenhos*) later became urban quarters, their names still persist, e.g. Madalena or Apipucos (Bezerra Cavalcanti 1998). For the sugarcane production numerous African slaves were brought to Recife area. In 1830 there were still as many as 16 *engenhos* functioning in Recife (Maia Alvez 2009). In the same year the municipal chamber (*Câmara Municipal do Recife*) passed a law to give the city a more uniform appearance, including regulations for facades of new houses and street alignments. Another goal of the governor at that time, the Conde da Boa Vista, was to develop the cultural life by building new theatres (Moura Filha 2000).

After the abolishment of slavery many former slaves moved to Recife and inhabited marginal urban areas, mostly close to the harbour area or riverbanks. In parallel droughts in the urban hinterland and the industrialization of sugar industry led to an inflow of rural population. In the course of the economic decline also the harbour area declined (Maia Alvez 2009), resulting in a large-scale renewal scheme. As a consequence after 1909 a major demodulation of the centre took place, focussing mainly on Bairro do Recife. The urban reform sought to improve the harbour area, based on Haussmann's Paris role model. Plans followed mainly European role models at that time, including a longing for improved hygiene, better road access and representative architecture (de Albuquerque Lapa and Almeida de Melo 2007; Teixeira 2012).The new urban design included main avenues in a radial-concentric network, meant to link the harbour to the sugarcane industry in the hinterland and to prepare the city for new and modern economic activities. French-style residential houses were constructed along the major avenues, turning it into the 'Brazilian Paris'. Expropriations on a large-scale paved the project's way (Maia Alvez 2009; Moura Filha 2000; Pontual 2007). In 1915 the opening of a canal made the peninsula an island, linked to the rest of town by bridges.

The quarters of Santo Antônio and São José were overworked only 30 years after the Bairro do Recife. In particular Santo Antônio (cf. Fig. 5.35, an actual photo of the quarter and the Buarque de Macedo bridge) was regarded as dirty, polluted and unhealthy at that time, therefore it received more attention than São José (Loretto 2008). After the renovation measures the quarter became a centre of luxury commerce in the 1940s, while the Bairro do Recife decayed and turned somewhat to a red-light and slum area (Barreto Lira and Pontual 2006). Industrial growth in the second half of twentieth century triggered urban growth and transformation towards a regional metropolis, as a result many uses and functions moved out of the centre areas towards periphery. The centre area decayed, empty plots

**Fig. 5.35** Ponte Buarque de Macedo and Bairro Santo Antônio

were invaded, and low-standard housing dominated the area. Since the 1980s then various upgrading activities in the centre quarters took place (Prefeitura da Cidade do Recife 1986a, b, 1987a, b):

> The Bairro do Recife is the place where the city was formed, is its most important historical site, and has been the target of urban renovation actions since the 1980s. Today it performs the functions of a business and decision centre, mainly those directed at information technology and governmental institutions (Barreto Lira and Pontual 2006: 28).

Today the urban patterns of the three quarters of São José, Santo Antônio and Bairro do Recife are somewhat different. While Bairro do Recife and Santo Antônio have been largely transformed in early twentieth century, this is much less true for São José, which is currently witnessing a phase of upgrading and transformation. The quarters also differ a lot in terms of predominant uses and functions.

> It is necessary to bear in mind that Recife/Mauritsstad was not designed as a single entity. It was, instead, developed in stages with different sections separately designed and knitted together to form an urban complex. The strength of the design lies exactly in this feature: it is a patchwork that maintains an overall sense of harmony, continuity and functional organization (Van Oers 2014: 63).

Recife also comprises of a number of protected gardens and green areas, including patches of mangrove, Atlantic rainforest as well as gardens or parks. Recife's history of urban gardens started with Campo das Princesas in 1872, and Praça Visconde de Mauá around 1888, introducing elements of English Gardens. Another eight projects were established in 1924/25 only, mixing French and British elements for artistic, hygienic and recreational functions. Aspects of art, botany and education were brought together in the concept of the modern garden as done by famous landscape architect Roberto Burle Marx in Recife of the 1930s, bringing together native species, sculptures and the intention of respecting nature In Recife, the creation of public gardens resulted in a variety of landscapist approaches, dating from the last decades of the nineteenth century until the end of the 1930s. So far, the Recife gardens were hardly tackled in contemporary literature about the urban history. Seven projects from Burle Marx were realized between 1935 and 1937, initiating his career. Projects mixed traditional landscape design elements and native species with new constructions, e.g. Brazilian species and water surfaces with granite benches and sculptures on Brazilian topics (de Figueirôa Silva 2010).

Beside the built heritage Recife is rich of culture, in particular literature, handicraft, dance and music. Recife's culture and particularly carnival is dominated by Frevo and Maracatu music and dancing, which have a more than hundred-years tradition in the region (Cassoli et al. 2007). After a rediscovery of the state's cultural richness in the 1990s (Freitas Cavalcanti 1997) a variety of programmes have been launched to promote and revive different cultural activities. The city is also rich in literature, including popular folk novels (Literatura de Cordel) but also famous authors like Gilberto Freyre, one of the intellectuals and poets that wrote on Recife in the period between 1920–1960, sharing a somewhat conservationist or nostalgic view (cf. Freye's 1934 publication *Guia prático histórico e sentimental da cidade do Recife*—Practical, historic and nostalgic guide on Recife City) (Duarte 2011).

The urban outline is characterized by its canals, bridges and the water. This unique feature in Recife is pointed out in various expert interviews (R03, R04, R07 and R15). Until today Nassau's plan is seen as reference for urban planning (interview R07). In general remnants of the Dutch colonial times are seen much more positive than the ones from Portuguese times, mostly because of Mauricio de Nassau's contribution to urban planning: "*The Dutch came and changed things to the better, they tried to upgrade the situation, whereas the Portuguese basically tried to make the most benefits from their colony*" (R03). Nevertheless different interviewees clearly stated their fear that this amalgamation of water, green and built-up areas is the most exceptional and historic asset of Recife, which should be preserved for the future. However, in various interviews major doubts concerning the future were raised, due to the highly dynamic actual developments as well as due to past mistakes, e.g. by not considering vulnerable groups in urban planning (e.g. R15) or the preference of Bairro do Recife over regeneration in upgrading projects (e.g. R08a/b, R13).

## 5.3.4 Inner City Patterns and Dynamics

Today, the Bairro do Recife is a densely built quarter, because of its relatively small area and relatively tall buildings. The architecture ranges from seventeenth to twenty-first century, grounded in street patterns from early twentieth century based on the Haussmann model (cf. Fig. 5.38/12–13). The road network is a regular one, with streets in North–South direction, crossed by minor roads in east-west direction. The main streets now converge to the central square of Marco Zero (Barreto Lira and Pontual 2006). There sideways are stone-paved, each street with an individual design, from traditional to modern (Córdula Filho and Corrêa de Araújo 2002). The square is one of the major landmarks of town, surrounded by some of the finest architecture (cf. Fig. 5.38/13). To its north, few years ago a complex for local handicraft and culinary has been built to upgrade the touristic value of the area. Former harbour buildings to its south are currently converted for a reuse in Recife carnival as part of a programme on the revitalization of the former harbour areas in Bairro do Recife and Santo Antônio (*Projeto de Revitalização de Áreas Portuárias,* initiated 2008).

Housing typology is somewhat heterogeneous, showing the different urban development phases. The surrounding of Marco Zero and the facades towards the main roads in the southern part of Bairro do Recife are dominated by three- to five-storey rectangular or trapezoid buildings with representative facades, built in the course of the 1909 renovation. R06 describes them as a "*wedding cake of decadent architecture*". In contrast the buildings along the smaller roads and towards north are rectangular, smaller and with less storeys, e.g. around Praça do Arsenal, which is most lively during lunch time, when people from nearby offices

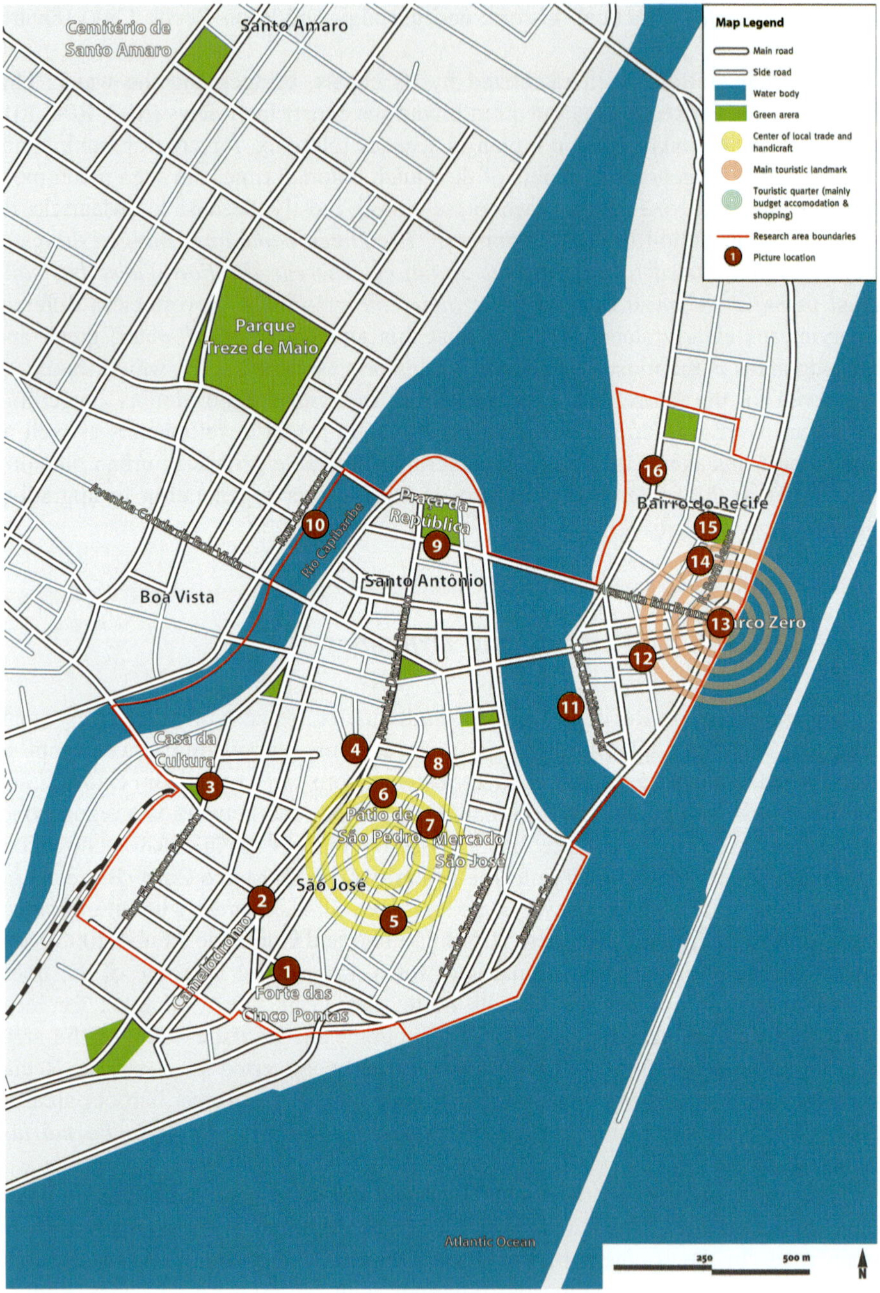


Fig. 5.36 Map of Recife centre area with research area boundaries

**Fig. 5.37** Streetscapes in Recife centre area, 1–8

**Fig. 5.38** Streetscapes in Recife centre area, 9–16

drop by and during events (Arsenal Square, cf. Fig. 5.38/15). The same is true for the very south of the Bairro where two-storey houses in *sobrado* style (usually two-storey mansion with courtyard) are still predominant. Almost at the southernmost point some decade ago a posh shopping has been established within the premises of the former mint, the Paço Alfândega, the former customs duty office from eighteenth century (cf. Fig. 5.38/11). The Rua do Bom Jesus near Marco Zero still comprises of a number of houses in a 'Dutch-like' style, memory of the former Dutch colonial period (cf. Fig. 5.38/14). Here also the former synagogue is placed. The vision of '*Polo Bom Jesus*' (Bom Jesus Pole, one out of five areas of intervention in early regeneration) as starting point of renewing Bairro do Recife was initiated and followed up under the mayors Vasconcelos and Cavalcanti in the late 1980s. R07 elaborates that while Rua da Moeda with its bars and restaurants was a certain success, the second focal area of Polo Alfândega is yet not completed.

One block north of Marco Zero the housing style and buildings' functions change, dominated by industrial and former port building, plus a historic fortification (Fort Brum) and a large favela (Favela do Pilar). The Atlantic seafront is built up with warehouses and harbour facilities while in the northern part there are several high-rise buildings with administrative functions (cf. Fig. 5.38/16), such as the municipality. The residential houses in Bairro do Recife often have economic uses in ground floor and housing facilities in upper ones. Maia Alvez (2009) finds it the most dynamic part of town, while various interviewees are concerned that the housing functions are not well accepted and many buildings are empty (cf. Sect. 5.3.5). Overall, the Bairro do Recife is a very diverse one, home to some of the city's landmarks and iconic buildings, protected by federal law, namely the Madre Deus Church, the Apolo Theatre, the Synagogue of the Americas, the Church of Pilar, Fort Brum, and the Harbourmaster's Cross (*Cruz do Patrão*). Even more buildings and areas are protected on state level (Barreto Lira and Pontual 2006). The quarter layout is "*rather varied and characterized by the spatial overlay of activities*" (Barreto Lira and Pontual 2006: 31). The main uses of the area are still the harbour and industrial one in the northern part, while the southern area is home to a mix of service and business functions. Services are mainly federal, municipal and state government organisms. The main business that has been established within the past decade is the Porto Digital (Digital Port), aiming at establishing an information technology hub in the area. In addition, commercial offices, banks, etc., are located in the Bairro do Recife, in recent years the number of entertainment, leisure and gastronomy increased.

Various interviewees are concerned about the lacking vitality of the Bairro do Recife, as it might easily be torn into an open-air museum without further functions, e.g. R07 is asking "*Who is visiting the centre today? Life and dynamism there is missing, although the vision is a nice one.*" Therefore return of administrative functions to the centre is considered as positive. In the 1990s until around 2000 he perceives it was still there. To R06 the predominance of Bairro do Recife in terms of project and budget is a problem, as "*all is concentrated on the Bairro do Recife, it has to be distributed in a better way*", referring to cultural functions but also respective budgets.

The quarter of Santo Antônio, north of the Bairro do Recife and connected with it by two bridges, also houses administrative and cultural functions. They are concentrated to its very north, where administrative functions and a nineteenth century ensemble composed of the *Palácio da Justiça* (court), *Teatro de Santa Isabel* (Saint Isabel Theatre), *Praça da República*, (Republic Square cf. Fig. 5.38/9) and the *Palácio do Campo das Princesas* (the former governors and today's Federal State government seat and the park) are located, opposite of the quarter of Boa Vista with its historic building ensemble from early nineteenth century (today a part of it is seat of Fundarpe, cf. Fig. 5.38/10). A large part of Bairro Santo Antônio has been remodelled in the 1940s (de Andrade Pereira 2009), in particular the road network and some major buildings which have been constructed during or after this decade. While the north of the quarter is characterized by single buildings, the urban appearance changes towards the southern part, with its rows of houses. The predominant functions change likewise, from cultural and administrative functions on the north to shopping and housing functions in the south. Here, at the transition between Santo Antônio and São José many churches have been preserved, like the Basilica and Convent of *Nossa Senhora* do *Carmo* (cf. Fig. 5.37/4) or the Pátio de São Pedro with the Church of São Pedro dos Clérigos, a favourite event area of the city (cf. Fig. 5.37/6). When Recife was built every community constructed a church of its own, consecrated to different saints, due to the many religious orders in town, and donations from rich individuals or professional groups. Today this high number is resulting in maintenance problems (interview R04), due to available budget and the decreasing importance of catholic religion in daily life.

Handicraft from Recife and Pernambuco is sold in Casa da Cultura (cf. Fig. 5.37/3) in the northwestern part of the quarter. Each cell in this former prison is now housing a small shop. The building is property of Fundarpe, who however has problems with the costly maintenance and with abuses such as drug trafficking (R05).

The shopping functions continue and accumulate towards south, to the quarter of São José (cf. Fig. 5.37/8 of Largo do Livramento, a square with shopping functions heading southwards to São José). Here still many small houses are preserved and/or renovated. The main landmarks of the quarter are the fortification (Forte das Cinco Pontas), Mercado São José (cf. Fig. 5.37/7), the oldest Brazilian premanufactured iron building (Nascimento 2005), declared national monument by IPHAN, numerous churches as well as cay and storage buildings. São José is a foremost residential area and also known for its commercial activities, in particular the sale of regional goods, including all kinds of foods in and around Mercado São José. In front of the building mostly fruit and flower vendors offer their products, in the surrounding streets then any good of daily supply, clothes or household devices can be purchased (cf. Fig. 5.37/5). Vivid street life is dominated by formal and informal trade, with a huge number of shops and street vendors. To reduce the number of informal vendors the municipality has provided them with sales premises along Avenida Dantas Barreto, the so-called *Camelódromo* (cf. Fig. 5.37/2, *cameló* is the nickname for mobile vendors). The outline of São José seems more chaotic than the other ones, due to the fact that it has always been the place of the informal vendors and economic activities, that is visible in the urban layout, it is the *Cidade dos Mascates* (R03), dating back to colonial times when Portuguese traders (named *mascates*) settled here. Until today economic functions outweigh housing. R03 regrets that the quarter is suffering de-characterization due to degradation and many of illegal transformations like added storeys or changed facades. Most alterations are done without any permission; as a consequence the public hand only gets to know once it is done (R03).

Until recently the major intervention stems from a plan made in the 1940s, resulting in the destruction of a number of blocks to improve the traffic situation. Its later implementation caused major changes in urban logics and layout of São José. Recently the quarter has become object of a large-scale renovation. Like in most Brazilian cities a large share of the residential construction is done in form of high-rise buildings. Two examples are the pair of tower blocks built in known as the 'Twin Towers', built on a very prominent seaward parcel in Santo Antônio (cf. Fig. 5.37/1 and Fig. 5.35 on the bottom left) and the nearby Projeto Novo Recife ('New Recife' Project), a complex of 13 towers for housing, hotels, business, cultural and leisure facilities in São José. These projects increase pressure on urban land prices, accelerating excessive verticalisation and building density (Furtado et al. 2014). The towers of 36–45 storeys height will be constructed in an area surrounded by monuments protected on national level (Sabino do Nascimento 2014). In the course of the Projeto Novo Recife the quarter's appearance will change, particularly along the waterfront (cf. Fig. 5.39), as stated in various expert interviews. Protest movements opposed against Novo Recife in the 'Occupy Estelita' campaign, particularly against the demolition of the ancient harbour buildings along Cais (cays) José Estelita southwest of the fortification. Nevertheless, the demolition of Cais José Estelita has begun in the end of 2014 (Truffi 2014; Furtado et al. 2014).

**Fig. 5.39** Waterfront development in Recife, near to the historic centre

R08a/b find the Santo Antônio and Sao José quarters comparably more complicated to deal with and call it an '*Aréa complicadissima*', due to the multitude of actors and functions. Parts of the fabric of these quarters are even older than the Bairro do Recife, but they are mixed with commercial and industrial areas and *ZEIS*, including a major one just aside). Various projects have been carried out, however "*It is an area under transition but no one know where to*" (R08b) as "*The area has a lot of actors and goals, it is super dynamic*" (R08a).

One factor which is probably fostering the differentness of the central quarters is their geographical division by rivers. As a consequence different uses and functions emerged; in addition local projects, policies and the state of conservation differ profoundly. The Bairro do Recife is receiving most attention, since the 1980s various projects have been launched to bring back various functions and to upgrade the fabric. In addition it has become a major event location and attraction for urban tourists. Some of the major representative buildings such as the governor's palace or the court are located in Santo Antônio, which is mainly characterized by large individual buildings in the northern part and blocks of five- to six-storey houses along thoroughfares in the southern part. Urban structures in São José then are very different; particularly between Mercado São José and Forte das Cinco Pontas where mostly one- to three-storied buildings are located. Many of them serve as stores with vending rooms in the ground floor and storages in the upper floor(s). All three quarters have a high fluctuation during the course of the day and the week, the comparably low number of inhabitants implicates very little movements during night time (except during events in the Bairro do Recife or Patio São Pedro) and over weekends or holidays when shops and municipal buildings are closed.

### 5.3.5 Perceived Drivers of Change in the Urban Landscape

Being asked about main impacts on the urban centre, the majority of experts mentioned the rapid economic development of the region. Projects like the newly constructed huge Suape harbour area south of Recife and related establishment of large industrial zones have put pressure on the centre, e.g. by major traffic infrastructure constructions, rising housing demands and growing traffic congestion. However, the perception of the impacts and potentials differed. While in interview R11 a clear positive attitude towards current developments was predominant (most likely due to their professional background in the federal state's economic development authority), interviewees R04 were more reluctant. R01 sees *this "moment of rapid urban transformation*", that people dislike and thus try to preserve their heritage. In comparison to nearby Olinda which is "*frozen in time*" she finds consequences for Recife more visible and faster. Interviewees R08a/b are as well concerned about the rapid urban growth which is putting high pressures on both, historic as well as green areas (R08).

Recife is internationalizing, due to the large-scale economic projects in the area, as mentioned during various interviews. Some understood this as a chance (R11), others as a threat to local culture. R04 and R05 find the ongoing internationalization rooted in the growing number of international companies and nonlocal labour employed in the likewise growing economic sector. As a consequence they consider it very important for Fundarpe to maintain traditions—a major aspect of the foundation's work. In interview R11, however, mostly the economic potentials were raised, which have "*changed the reality of the municipal area, people there will have a higher quality of life*", as they do have more and different job opportunities.

The cultural change different interviewees are witnessing as a consequence of the internationalization is only positively connoted in interview R11, while most of the other interviewees are more sceptical. To R11 the city should be adapted to the needs of people and classes moving to Recife. In relation to the huge number of incoming labour force R04 is concerned that "*people give up their own culture; they take over the one of the people coming to the area or international one and forget their own*". Also modern lifestyles sometimes do not fit traditional houses that were foreseen for large joint families. Particularly buildings declared monument cannot be converted easily to modern functions (R04).

The tourist sector development is pushed by Recife municipality and the federal state, e.g. events like carnival or the World Cup. In the perception of R12 it heads to a higher grade of valorisation of culture and other things in Recife. To him cultural tourism also helping in dissemination of appreciating local popular culture. Nearby Olinda is not perceived as a potential disadvantage that might take away tourists and tourism-generated income. Different interviewees rather see disadvantages in the World Heritage status of Olinda, as Recife allows for more development. However, a process to analyze the potential of Bairro do Recife to become UNESCO World Heritage is ongoing (R01).

The FIFA World Cup 2014 was also mentioned by various interviewees as a major influence on urban development, including in particular the Bairro do Recife, the location of 'FIFA Fan Fest'. In addition, R07 and R15 name the construction of the Arena Pernambuco as one example for the misguided urban planning. R15 is particularly pessimistic about all the constructions for the World Cup, which to her often are neither legally approved nor matching ecological or sociocultural perspectives: "*the Copa is finalising our future*". R03 is concerned that the "*the FIFA Fan Fest has nothing to do with the local culture*", as it will not allow the traditional food and drinks of Recife, as only the huge sponsors will be permitted to sell their products. R10 on the other hand is more optimistic and sees the event as an opportunity to present the Recife and North-eastern culture to the outside world—if permitted by the organizing committee.

Only recently—emerging with protests against the World Cup—processes of social mobilization emerged, as R03 stated. R01 and R04 witness a growing number of public initiatives to preserve single buildings, as, e.g. happening with Cais José Estelita (cf. Sect. 5.3.4) and the Edifício Caiçara, a municipal heritage building in late neo-colonial style constructed in 1940 (Araújo 2012), located in one of the most expensive and posh beachfront locations south of the centre. It gained a sad notoriety after it was partly—and illegally—destroyed in September 2013, likely for profit maximization by replacing it with an apartment building. In this case the destruction has led to certain protests, on the spot, in the newspapers and in the growing number of (online) activist groups concerned with the preservation of Recife's tangible and intangible heritage assets. In July 2015, the local court has in the first instance decided on the reconstruction (Diario de Pernambuco 2015).

The real estate market is very influential in Recife. To R08b it is both, "*the biggest driver and the biggest problem*", having a strong impact on the politics. Public–private projects are even taking place within ZEPH zones(*Zonas Especiais de Preservação do Patrimônio Histórico-Cultural*—Special Zones for the Preservation of historic-cultural Heritage, included in the urban masterplan), including monuments under federal law. In particular the former harbour areas are affected. Here historic buildings are destroyed and replaced under the supervision of the public hand (R01). The private market has a huge influence on politics and the development in the centre, as stated in various interviews, e.g. when R08b states that the actual governor has completely entrepreneurial visions. R08a/b names the example of the harbour area reconstruction around Marco Zero. Here the old harbour and storage buildings are supposed to be renovated, but according to R08b "*this is no renovation, it is a reconstruction. However, the project was permitted by the Recife Harbour authorities, so there is nothing to do against.*"

As a consequence, large-scale modifications of the centre's urban fabric are perceived by various interviewees. R04 and R06 are particularly lacking good use concepts for heritage buildings. According to R03 the centre is suffering de-characterization due to many illegal transformations like added storeys and changed façades. Interviewees R04, R06, R08 and R15 stated a lack of sanitation and proper traffic infrastructure, in particular of a better public transportation system. In addition, urban growth is impacting on the heritage, the centre is densifiying while parts of the historic buildings there are still not inhabited (R06). A growing number of high-rise apartment buildings are even built directly aside the heritage buildings, impacting on its appearance (R01, R04, R06 and R07). The PPP-project Novo Recife was mentioned several times in this context, the largest project within the historic centre. R04 states "*when they do as they want you will be able to see Recife from Africa*", expressing his deep discontent. To R07, this project and particularly the already existing twin towers at the Cais de Santa Rita are "*the spearhead for the takeover of Bairro São José*". R13 is more in favour of it; to her "*this will contribute on a large-scale to the development of the area*" while to R07 "*Recife Novo is completely ignoring the neighbourhood, the landscape, the surrounding, the urban layout and the quarters around*". R04 and R06 are concerned about the trend to purchase heritage buildings with municipal or state budget without suitable use concepts, as, e.g. done in case of the Patio São Pedro:

> The use had to be found after purchasing the buildings. Therefore today around Patio São Pedro you find many public authorities but even the bars and restaurants around are in buildings of municipal ownership. This does work for museums, but not for other things (R04).

The lack of housing functions in the centre is hindering the renewal process, according to different interviewees (R04, R06, and R13). While the Bairro do Recife was made attractive for companies and for business, there was no suitable strategy for bringing back the housing functions (R06). The major reasons that discourage potential inhabitants are the lack of parking lots—resulting in core removal of historic buildings and conversion to garages—and the large number of festivals, resulting in waste and noise pollution. The lack of inhabitants is also hindering the implementation of community-driven initiatives (e.g. the *morar no centro*—'living in the centre' initiative R06 was engaged in) as the number of potential local participants is too small. The same argument was raised in interview R08, adding that initiatives to promote housing areas failed because in parallel policies for establishing touristic activities in the centre were implemented. The *Secretaria do Commercio* transformed the area into a compound for festivals and events, causing noise and other disturbance, so that in the end people already living there moved away again. Overall, the Bairro do Recife has only around 600 inhabitants, Santo Antônio—where less housing facilities are provided—has less than 300 while São José has almost 9,000 inhabitants (Prefeitura da Cidade do Recife 2012).

Recent municipal initiatives comprise the creation of a pedestrian area as well as traffic calming in selected areas of the Bairro do Recife to reduce transit. Interviewees R08a/b are however not too convinced about the success. R13, who was involved in development of the early revitalization schemes, states that "*today the Bairro do Recife already is more solid, there are people, offices and activities*", there are more people permanent in the quarter, but "*what is still pending is a mixed use*". In addition she perceives security problems, particularly at night (matching questionnaire answers), as well as a lack of education, housing and healthcare facilities. R03 is lacking public infrastructure, in particular the public transportation system was mentioned several times, e.g. by R04 who describes the Recife Metro as "*only palliative, it solves the symptoms but not the problem itself*".

Despite the relatively large number of projects and initiatives in the centre of Recife R07 is concerned as to him those projects do not have a holistic vision considering the impacts on the city as whole, and are lacking an integrated vision. To him, projects like Novo Recife or the Arena Pernambuco for the FIFA World Cup 2014 are isolated projects, having not much in common except being huge projects of interest for the managers and entrepreneurs—"*This does not make a city, this makes a Frankenstein!*" (R07). Recife still has a very outdated and old-fashioned vision of urban planning, without considering housing, leisure and work functions together, without mixed use (R08, R13).

R03, R07, R08a and R08b are missing political stability which is the precondition for long-term planning. Interviewees R04, R07, R08a and R08b complain that after each of the many elections the composition of parties and thus power change. As a consequence every 2 years some government or mayor wants to change things, including the policies in the historic centre. As R03 poses there is a "*need for political commitment for any preservation*" as she sees good plans in theory but difficulties in their implementation and a lack of interest on the decision-makers side to do necessary adaptations in conservation policies. To counteract the unstable situation R08a and R08b in their work life try to form strategic alliances with other partners from science, administration and practice, so that the politicians cannot easily change plans. Other interviewees like R03, R11 and R13 are also missing a continuity of policies, on urban as well as on state level which R01 finds "*existing on the paper but lacking in action*".

In addition, different interviewees mentioned the challenging division of responsibilities, particularly on municipal level. One example mentioned was the shift of DPPC from *Secretaria do Planejamento*, to *Secretaria da Cultura* when the PT entered the municipal government (R03, R08). As a consequence, R03 finds their work more complicated now, due to separation of budgets and less involvement in urban planning processes. R07 shares this opinion, to him the division of *Secretaria de Planejamento* into small pieces and sectors resulted in a lack of integration and visions. As a consequence "*there are no more projects dealing with the whole town, today's projects are isolated ones, like Novo Recife, the Arena and the Twin Towers*" (R07).

While all interviewees stressed the good cooperation among the different authorities on a personal level, they however criticized the complex legal framework on different levels. As R09 states "*it is not a problem of the teams, they get along well, but a problem of the legislation*". R01 is concerned that the excellent cooperation among the different authorities concerned with heritage (IPHAN, FUNDARPE & PPC) could easily change as soon as different persons are in charge, as the good partnership is not based on anything else than personal relations. To R01, R10 and R12 the level of administration and bureaucracy is too high, time-consuming and lacking staff to implement all conservation initiatives. In addition R03 is lacking mutual understanding of different actor groups and authorities, e.g. between decision-makers, public, and architects. She concludes that although there are ways to preserve heritage and maintain the centre the city is disordered, not all institutions are working well.

The probably biggest challenge in urban conservation is the noncongruent legislation at the different levels, resulting in often diffuse situations. To R05 the heritage legislation itself is worthy of improvement. R04 is missing qualified staff and the interest of the municipal officials: "*The public administration is the biggest violator of the cultural heritage*" (R04). And R08b states "*The biggest problem in conservation of heritage is mismanagement.*" While R03 acknowledges that urban conservation policies and plans are good in theory she finds the execution difficult and rather improvable. She is mainly concerned about the lack of interest the owners of monuments show in preserving their houses, and about too little public financial support mechanisms. Therefore, R03 sees a need to update municipal legislation and outdated conservation strategies. To R03 and R05 the present laws are too unspecific because decision-makers do not have interest in modifying them.

R14 finds the municipal supervision of heritage objects rather weak, therefore she and her students are doing surveys of different objects like the Burle Marx parks and inform the municipal authorities in case of problems: "*We are the police that takes care of the gardens*". R01 does not even find the public hand democratic in terms of their urban planning and conservation approaches: "*if a decision-maker wants to demolish a building he will do so, they are not related to the heritage*". Again, the Edifício Caiçara is given as an example. R04 himself is involved in the process, to him it shows the difficulties of contemporary heritage conservation in Recife and beyond, as the heritage status would enable the building owner to be compensated, leaving a big dent in the municipal or state budget. He also claims that the municipality did not want to touch the issue and just threw the problem to the federal state level.

R03 finds the public participation processes that have become part of Brazil's urban planning over the past decades, sometimes frustrating. According to her people often feel more linked to their own community instead to other urban areas or the city as a whole. She finds this grounded in the social disparities of Brazilian society. The huge social disparities in the city as a whole were stressed by various interviewees (R01, R11, R12). To R01 the improvement of the social situation of many vulnerable groups is of utmost importance, as 50 % of Recifenses still live in Favelas. Any change to her is closely linked to education, as education which to her is fundamental to improve their situation. Such need for education was mentioned as crucial point for an enhanced awareness of heritage, its values and benefits, as well as for improved urban development strategies considering heritage (R02, R03, R04, R05, R06, R12, R14 and R15). To counteract the ongoing loss of Brazilian culture and to promote local cultural traditions like Frevo, Fundarpe has already launched some projects with schools and in selected Favelas (R03) but "*there is a lack of heritage education, this does only exist in university, but nowhere else, there it is seen as of low value, as folkloristic only*" (R12). R15 would like to have "*heritage education also in the Prefeitura to make them aware of adequate measures*" while R06 asks for a general "*education for citizenship*" to raise awareness and respect for people and the urban environment.

To R12 it is of utmost importance to educate people not to see immaterial heritage separately from material one. Recife is losing skilled technicians and artists (R04, R05, and R12). As a result craftsmen skilled in renovating the historic buildings are increasingly scarce. Many of them do not know any more how to carry out adequate measures in historic buildings.

Budget constraints are another important aspect affecting development and conservation in the centre. The budget allocated for conservation is limited as stated by R02, with the exception of carnival on which in her opinion comparably more public and private budget is spent. Various interviewees are more dissatisfied with the ways the public budget is distributed than with the overall amount. R03, e.g. is complaining about the difficulties arising from the separated budgets of cultural and urban planning departments which are bothersome for joint conservation activities, R04 and R05 are mentioning the challenging distribution of the Fundarpe overall budget among tangible and intangible heritage activities. To R04 most is spent on events and too little on buildings. R03 poses "*Brazil is paradox, there is lots of investment, but only little in culture, heritage and health, this is lacking*". Finally R14 is worried about the huge number of cultural foundations that are absorbing a large portion of the ministerial budget for culture, enabled by Brazil's legal system.

As a consequence much of the ministry budget for culture remains with the biggest banks that have their own cultural foundations, e.g. Santander is having an '*Instituto Cultural*', like many other banks, lowering the public budget for other institutions.

Another issue in the centre area is the informal market which is difficult to control. In interview R08 a lack of management and policies for social as well as economic activities, especially concerning the informal market, was claimed. Another recent development is the growing number of Chinese entering into the commercial sector particularly in São José. Here much of the electronic sector but also clothing stores are owned by Chinese who then partly employ local vendors. As a consequence many apartments in surrounding condominiums are rented out to Chinese, what the local population often perceives as disturbing due to different customs and habits.

Various interviewees state a lack of awareness concerning different kinds of heritage, instead of having a holistic view: "*There are things the Federal State is not interested in*" (R04). R06 is witnessing a preference to preserve baroque and eclecticism architecture, but not architecture of the twentieth century which is almost not protected. To him it is the most important architecture to protect as this is the typical and unique Brazilian style. R15 is more concerned about the natural assets which she finds underrepresented in urban planning: "*Recife has everything: Burle Marx gardens, water, and different kinds of environmental reserves like mangrove and Mata Atlântica. It is all there but not recognized*" (R15). To R14 "*Recife has already lost a lot, and the process is continuing*" as cultural heritage is not maintained adequately. The same is true for the intangible heritage. Pernambuco itself is playing an important role for the whole of Brazil, a lot of historic movements are rooted here, and therefore some of the heritage is of national value, protected by IPHAN (R04). However, R12 himself witnesses a lack of communication and discussion particularly on the Afro-Brazilian cultural assets, which to him are too often seen 'only' as roots of Brazilian carnival, without a critical reflection.

### *5.3.6 Assessment of Urban and Heritage Policies*

In Brazil, legislation and policies can be decreed on three levels: the national one, the Federal State level and on municipal level. This is true for any legal documents on urban planning as well as conservation. This chapter presents the main policies, programmes and statutes on national level; on Federal State and urban level it will focus on Pernambuco respectively Recife Municipality.

In 1973 Brazilian military government installed by law nine metropolitan regions (*Regiões Metropolitanas* (RM)) around the major urban agglomerations of the country, among them Recife. The main idea behind this centralist top-down measure was to instal mechanisms of centralized coordination and macro-planning, by means of new forms of intermunicipal governance (Lacerda and Ribeiro 2014;

Lefèvre 2009; Maricato 2011) and own planning and management institutions, like the Condepe-Fidem (*Agência Estadual de Planejamento e Pesquisa de Pernambuco*—State Agency for Planning and Research of Pernambuco) in Recife RM (Maricato 2011; Klink 2009).

During the past decades a number of innovative urban policies addressing problems of social housing, urban regeneration or stakeholder inclusion have been installed. Since around 2 decades socially underprivileged groups are increasingly organized in social movements claiming for urban reforms, in particular the provision of urban infrastructure in poorer quarters and the construction of housing facilities and option to purchase parcels. The major result in terms of policies was the *Estatuto da Cidade* (Statute of the City), passed in 2001, and the establishment of the Ministry for Urban Affairs in 2003. Although very promising attempts in theory, practice has shown limited outcomes, e.g. in terms of still limited participatory processes (Kirsch-Soriano da Silva 2010). The *Estatuto da Cidade* reinforced planning and mandates that Brazilian municipalities over 20,000 people have to issue a master plan at least every five years. In addition, it provides legal support for municipalities to promote land tenure, and legitimates different new legal instruments for urban areas (Serra et al. 2004). Also the importance of historic sites and their artistic and cultural value is laid down in the *Estatuto* (Barreto Lira and Pontual 2006).

Another main outcome are participatory programmes opening new channels of resource allocation. The most popular is the Participatory Budgeting (*orcamento participativo*), implemented by the Workers' Party counteracting 'old' clientelist politics and enabling local municipalities to engage with slum dwellers as full citizens (Koster 2012). Brazil's current development model however has negatively impacted on the sustainability of its agglomerations in two ways: the preference for individual rather than public transport, and significant changes in urban land use and occupation, heavily influenced by the interests of civil construction and the private building industry (Furtado et al. 2014).

Barreto Lira and Pontual (2006) stress that many instruments incorporated in Brazilian urban legislation were adapted from other countries such as France, USA, Italy and Portugal, e.g. the *Outorga Onerosa do Direito de Construir* (Selling the Right to Build), which is rooted in French Plafond Legal de Densité. Public–private partnerships for urban development and regeneration, which are very common nowadays in Brazil, originated in the US and European countries. For instance some instruments contained in the *Estatuto da Cidade* are quite similar to those used in the rehabilitation of the historical centre of Lisbon since the 1970s (Seixas 2014).

Urban regeneration became one of the main components of Brazil's national strategies for development in the 1980s and 1990s (Barreto Lira and Pontual 2006). Before, the main actors in conservation processes were intellectuals, often organized in professional alliances. The same actors remained active until the early twentieth century, including even modernist architects who were worried about vanishing Brazilian roots due to the loss of traditional buildings. Brazil's conservation movement was elitist at that time, restricted to artists and intellectuals (Adams 2002).

Conservation in Brazil is laid down in the national constitution. It decrees the preservation of *Patrimônio Cultural Brasileiro*' (Brazilian Cultural Heritage), material and immaterial items that relate to the identity, events or the memories of the different groups that compose the Brazilian society (Borba 1994). Compared with other countries Brazil pioneered in conservation legislation, as already article 148 of the 1934 Constitution mentioned about the protection of national heritage. Consequently a *Brazilian legislation on conservation* was enacted as early as 1937, dealing with the preservation of historic and artistic national heritage (*Patrimônio Histórico e Artístico Nacional*). Beside the formal protection it also tackles restauration aspects. According to R03 the 1937 law, enacted during the dictatorship of Getúlio Vargas, is still the basis of contemporary heritage legal framework, on all administrative levels. The process of declaring e.g. a building cultural heritage was done in a top-down approach with hardly any consideration of the population's preferences. For a long time therefore cultural heritage followed a quite conservative definition, inscribing buildings of the colonial upper-class or churches (Leal 1977; Proença Leite and Peixoto 2009). Different from other countries like, e.g. Nepal, protection of private-owned premises is possible, the process to protect a good, whether tangible or intangible can be initiated by anyone, by writing to the responsible authority.

The 1964 Venice charter (see Sect. 3.2) pushed a new policy of heritage in Brazil, focusing now on ensembles and historic landscapes. This fostered the need to look for new forms of conserving urban environments, encompassing cultural values. To do so strategic partnerships with other administrative entities became necessary, what was not successful in all cases. The shifting conservation paradigm matches a shift in the understanding of heritage, moving away from seeing culture as something purely static and historic. From the 1970s on cultural and natural heritage protection became a more systemic approach (Milet 1988). The next advances followed with the 1973 establishment of the *Programa Integrado de Reconstrução das Cidades Históricas*—PCH (Integrated Programme for the Reconstruction of Historic Cities), focussing on supporting and establishing touristic activities, launched in the countries Northeast (Adams 2002). In 1977 Brazil signed the UNESCO World Heritage convention. Three years later the Historic Town of Ouro Preto, image of Brazil's golden age in the eighteenth century, became the first world heritage site in Brazil. In the same year the Brazilian ICOMOS committee and ABRACOR—*Associação Brasileira de Conservadores e Restauradores de Bens Culturais* (Brazilian Association of Conservators-Restorers of Cultural Goods), a Brazilian NGO seeking to promote and disseminate adequate conservation techniques and strategies, were founded.

Articles 215 and 216 of the 1988 constitution, drafted as a reaction to the period of military dictatorship, deal with culture and are of relevance for conservation. Among others it obliges the federal states to do an inventory and listing of heritage, but according to interviewee R04 until today no regulation has been passed on how to do this, the articles are quite unspecific, therefore all federal states and municipalities "*simply do it the way they want*".

Below the national level, other means of protection exist on state or municipal level, depending on its value for the nation, the state or the municipality. Subsequently there are three authorities in charge for heritage,

- IPHAN (*Instituto do Patrimônio Histórico e Artístico Nacional*—National Historic and Artistic Heritage Institute) on national level,
- Fundarpe (*Fundação do Patrimônio Histórico e Artístico de Pernambuco*—Foundation for the Historic and Artistic Heritage of Pernambuco), on federal state level and
- DPPC (*Direitoria de Preservação do Patrimônio Cultural*—Directorate for the Preservation of Cultural Heritage, Recife Municipality), on the municipal level.

Their responsibilities are to a large extent comparable but differ in scale, seeking to protect the cultural (tangible and intangible) heritage on municipal, state or national level. Any heritage on national level automatically also becoming a heritage on the federal state and the municipal level. If something is declared heritage on federal state level it is becoming heritage on municipal level automatically (R04).

IPHAN's predecessor SPHAN (*Serviço do Patrimônio Histórico e Artístico Nacional*—National Historic and Artistic Heritage Service) was founded in 1937 to save monuments of exceptional national value. From this time on the organization guided or realized conservation and restauration measures benefiting 90 % of the monuments inscribed (Leal 1977). In 1970 the organization changed its name to IPHAN. Adams (2002) claims that IPHAN limited itself to administrative processes of identification or registration of heritage, without assuming tasks like multiplication or education on heritage values.

IPHAN Pernambuco began to conserve churches and old fortifications, under the law from 1937, after 1968 the protection of the urban ensemble of Olinda marked a change towards a more diverse understanding. While protection on national level was established in 1937, legal protection in Pernambuco State itself only became possible after 1979 (Prefeitura da Cidade do Recife 1981). The Bairro do Recife then became national heritage in 1998 because of its 'unique expression of 'Paris in Brazil', on municipal level it already had been listed as heritage since 1980. Surprisingly the natural assets were not considered as a reason (Cadena de Melo Filho 2012; Pontual 2007; Barreto Lira and Pontual 2006).

On the level of Pernambuco State conservation is the duty of Fundarpe, founded in 1973. Initially foreseen to protect the cultural heritage as defined during military rule, it has turned its portfolio over time. Today the majority of its fund is spent on culture, events and festivals (Buril 2013b). Its budget spent on protecting the state's cultural heritage was as high as 979,000 R$ (corresponding to around 275,000 €) in 2012, compared to the immense amount of 33.2 million R$ (corresponding to around 9.1 million €) spent on the Funcultura programme (Buril 2013b). Funcultura is focussing on intangible assets, but however also comprising museums and theatres (Buril 2013a). In the years 2012/13 more than 280 projects have received funding, among them also such on capacitation and cultural research (Herold 2013).

Fundarpe is a special case as comparable foundations in other federal states usually deal either with culture or with built heritage, but not with both at the same time. R04 and R14 consider this as good on the one hand as this way tangible and intangible heritage is seen as intrinsically linked. On the other hand it has negative aspects, mostly due to budget constraints which may privilege one over the other. In Recife municipality the focus of Fundarpe still is on tangible heritage, so far no intangible heritage has been inscribed within its borders (Cadena de Melo Filho 2012)—while R05 finds it responsible for operationalizing culture and cultural politics. In addition, Fundarpe is supervising 18 heritage properties owned by the federal state, e.g. the Casa da Cultura and various museums.

Recife Municipality itself has executed various conservation as well as regeneration plans and programs since the end of the 1970s. The initial point has been the Bairro do Recife; subsequently projects in neighbouring quarters were established. Legal frames to be mentioned are the 1978 *Plano de Preservação dos Sítios Históricos* (Preservation Plan for Historic Sites) of the Metropolitan Region of Recife (PPSH/ RMR), and the 1979 State Law n° 13.957, regulating the PPSH and classifying eight subareas as a historical ensemble because of its artistic hand historic values (Prefeitura da Cidade do Recife 1981). In his work, the PPSH followed the international paradigms set by the Venice Charter (1963), Quito Norms (1967) and the Amsterdam Declaration (1975, cf. Sect. 3.2) (de Andrade Pereira 2009).

While most inscriptions on the national level are churches, palaces or fortifications, the variety on state level is broader. On state level, e.g. the ancient Zeppelin tower (1930–38 Recife served as destination of Zeppelin connections from Europe to Latin America) is protected, just as the ancient prison (Casa da Cultura) or a cemetery (Borba 1994). In 2013 the number of tangible cultural heritage sites in Pernambuco state inscribed on state level increased to 62 (29 out of them in Recife itself, out of which 13 are located in Bairro do Recife, Santo Antônio and São José), with another 124 in the process of being inscribed (Buril 2013b, c; Cadena de Melo Filho 2012) The most common instrument for protecting cultural heritage is the inscription as heritage site, for reasons defined by scientists. Cultural and natural aspects are treated individually, in other words, the natural elements that shape Recife and its cultural heritage sites are not looked at together but rather separately, following rather traditional conservation approaches (Cadena de Melo Filho 2012).

The state law contributed to the implementation of a municipal preservation law and the establishment of *ZEPHs* in 1991. Overall 33 ZEPHs were established in Recife Metropolitan Region, covering 56 areas identified under PPSH, among them the quarters of Bairro do Recife, Santo Antônio and São José (de Andrade Pereira 2009). Aside the protection of historic assets and fabric the ZEPHs also aim at harmonizing old and new, allowing for change and development within certain limits. Within the ZEPHs there are two different categories of preservation, 'environmental protection' (*zona de preservação ambiental SPA*) and 'rigorous protection' (*zona de preservação rigorosa SPR*). Within the overall protected area of the two Bairros of São José and Santo Antônio both categories can be found. While in the SPR almost no changes are allowed, in SPA certain modifications are permitted. For any modification the approval of the municipality and DPPC as the responsible authority is needed—or as R01 poses: '*in theory*', as in reality she has witnessed cases where areas were altered without permit. The same categories of conservation areas and single monuments on municipal and state level applies to the Bairro do Recife, where in addition objects protected on national level can be found.

Municipal Law n° 16.290 for the Historical Site of the Bairro do Recife was decreed in 1997. Amongst others it mentions the instruments applied in the Bairro do Recife: public–private partnerships for urban development; selling the right to build; and transfer of development rights (Barreto Lira and Pontual 2006). In the same year the category of *Imóveis Especiais de Preservação* (IEP—special properties of preservation) was added on municipal level, to protect single buildings outside ZEPHs which are of value for urban history. However, as pointed out in interview R08, although the ZEPHs are a very welcome instrument there is a high need to revisit and reevaluate these plans to adapt them to actual challenges in urban development.

Much of the information on material as well as immaterial heritage is accessible in online databases of the responsible authorities, based on an obligation from the national Ministry of Culture that all states and municipalities have to establish a system of "cultural information" (R02). Recife's online database contains a cultural cadastre where all artists of Recife are listed, grouped in different thematic areas. In the five years of its existence, already around 10,000 artists got listed and were awarded with the title "*trabalhadores da cultura*" (cultural workers). The inscription is decided by a committee in which enlisted people are represented and have the right to co-determine criteria for future inscriptions.

When it comes to intangible heritage in Recife IPHAN is the responsible authority for any protection status, as the state of Pernambuco does not yet have a law but only a *decreto* (decree) to protect intangible values. On the national level the 2000 *Decreto n° 3.55 institui o Registro de Bens Culturais de Natureza Imaterial* (Decree for Registration of Cultural Goods of Immaterial Nature) is the relevelant legal document. R09 however mentions that Pernambuco is in phase of developing an own law, Fundarpe is part of the developers. Intangible assets do have a growing importance as stressed by various interviewees (among them R01, R02, R05, R09, R10 and R12), only in 2013 four more assets were proposed to IPHAN on federal state level, among them two different forms of Maracatu, a dance and performance of Afro-Indigenous origin performed particularly during carnival. Capoeira is already inscribed since 2007. Particularly music and different forms of dance are important traditions in Brazil's Northeast, as a consequence many initiatives to support these traditions have been set up in the whole region and notably Recife as one of the regional urban centres.

In 2002 Pernambuco passed a law to establish the *Patrimônio Vivo de Pernambuco* (Pernambuco Living Heritage), which enables the federal state to protect individuals, or groups that have and share a certain artistic knowledge, among them three related to Frevo. The inscription process is time-consuming as it requires inventories and documentations (in which Fundarpe is involved). To R10 it is nevertheless a very important process as to protect it is *necessary* "*to see the reality of the people*", at the same time he finds it "*a necessary process because there are so many, it gives visibility and helps because many are disappearing*". In 2012 the idea emerged to start with two inventories per year, what R10 would support as he perceives that immaterial cultural goods are disappearing, especially in Recife Metropolitan Region.

One could say the fragmented urban landscape that shapes many Brazilian cities is mirrored in the ambitions but somehow scattered legislation on different levels. Successful urban development and conservation depends on the interaction between these levels and agencies. Various interviewees mentioned difficulties between the different municipal authorities that share responsibilities for the centre, their goals and the divided budgets (e.g. R03, R04, R05, R08a/b). The different levels of heritage policies are among the main concerns of many interviewees. Although all of them agreed that in practise the work with the colleagues from other institutions works well, they were concerned that such well-established cooperation schemes are mostly based on mutual understandings of individuals and less on well-defined cooperation schemes and policies (concern of R03). The different levels of protection as well as a focus on single buildings (depicting the initial and partly still predominant preferences of conservation efforts) potentially harm concerted action and a (cultural) landscape perspective (Duarte 2011). In addition, and despite the many heritage levels and categories there is yet none on cultural landscapes (R15).

In practice, there are not only different levels but also different kinds of (physical) protected objects, each one under different laws or decrees and executing authorities, ranging from zones (of environmental protection, of cultural heritage

**Table 5.5** Cultural heritage categories of IPHAN and Fundarpe (*Source* IPHAN 2014; Fundarpe 2015)

| IPHAN (national level) | Fundarpe (Pernambuco State level) |
|---|---|
| • Cultural goods of archaeological, ethnographic or scenic value (*Bens culturais em função do valor arqueológico, etnográfico, e paisagístico*)<br>• Cultural goods of historic value (*Bens culturais em função do seu valor histórico*)<br>• Cultural goods belonging to fine arts (*Bens culturais em função do seu valor artístico particular*)<br>• (Cultural goods belonging to applied arts *Bens culturais em função do seu valor artístico aplicado*) | • Mobile heritage (*Bens móveis*)<br>• Buildings and individual monuments (*Edifícios e monumentos isolados*)<br>• Ensembles and historic sites (*Conjuntos urbanos e sítios históricos*)<br>• Natural monuments, sites and landscapes (*Monumentos, sítios e paisagens naturais*)<br>• Cities, settlements and villages (*Cidades, vilas e povoados*) |

protection, etc.) to single buildings or parts of them. The potential difficulties between the three levels become clearer when considering that, e.g. IPHAN has four different heritage categories while Fundarpe currently has five ones (and R04 would even like to add a sixth one on industrial heritage), as described in Table 5.5.

Different experts (R03, R04, R15) mention problems with the law compliance, as laws and regulations are changed too often, and the laws (on different levels) are not always coherent, e.g. in what to preserve exactly. There are differences between governmental and federal state laws concerning the things and values to preserve, for R04 this is making his work very complicated as it requires lots of negotiation. The respective laws supplement each other, following a top-down hierarchy, most important is the constitution, in particular laws 215 and 216/216a from constitution 1988 (R04). Various interviewees state the need to consolidate legislation, as there is a multitude of laws, decrees, and norms. To R08a/b it has to be found out who is using which ones, no one knows this today as there are too many, sometimes even contradictory. While R04 finds the heritage inscription process itself is working quite well, he is lacking supervision of the inscribed objects as he perceives ongoing alteration and a lack of awareness: "*today the situation is a very serious one, the most efficient measures/penalties are the ones that affect the people's money, that deal with fines*". While the legislation is permitting expropriation of heritage buildings in reality this is hardly done, as the owner would have to be recompensed.

### *5.3.7 Phases of Urban Renewal in Recife*

To refurbish Recife various projects for land use and zoning were elaborated and implemented throughout the first decades of the twentieth century, starting with the Haussmann-like modernization of Bairro do Recife 1909–1915 (see Sect. 4.1). The underlying idea was to construct a new entrance gate to Recife ("*porta de entrada da cidade*", R07). Over the next 30–40 years the neighbouring quarters of Santo Antônio, São José and Boa Vista followed, however, to a lesser extent. Not much changed in terms of uses and functions, until the 1960s, when the centre area decayed while the middle and upper-classes preferred to settle in peri-urban quarters. As a consequence, many functions left the centre. Residential functions were still in place; however the inhabitants changed towards medium to low-income groups.

During the military regime 1968–1973 Brazil faced a rapid economic growth, the "*milagre brasileiro*" (Brazilian miracle). At that time also strategies to develop Recife were developed to counteract the lack of a strong local economic base after the decline of sugarcane plantation and sugar production. The city nevertheless grew, however, without proper planning (as stated in interview R07). In 1973 the first integrated development plan for the metropolitan region (*Plano Integrado de Desenvolvimento da Região Metropolitana do Recife*) was elaborated, in which the preservation of the 'urban ambience' (*ambiente urbano*) was called a priority.

**Fig. 5.40** Main phases of urban regeneration in Recife Centre

Subsequently three years later the plan for preserving historic sites of the metropolitan region (*Plano de Preservação do Sítio Históricos da Região Metropolitana do Recife—PPSH*) followed. R13 describes these years as a difficult phase for the city, due to the rapid changes induced by capital inflow; therefore the concepts of cultural and natural heritage were introduced and kept since then. Beside the impacts on the buildings also social impacts on the centre became obvious, in particular on the informal sector and dwellers, which were forced out (cf. Fig. 5.40 on the main phases of urban regeneration from this time on).

In the early 1980s planners and managers jointly tried to improve the Bairro's negative image, following European role models (R07). To improve the situation in 1986 the Plan for the Rehabilitation of Bairro do Recife was launched, aiming at an improvement of urban infrastructure and considering the local context. It was during that phase when the Prefeitura do Recife decided to move back to the centre, in a new high-rise building at the riverfront of Cais do Apolo. Attempts to (re)turn Bairro do Recife into a business district did not succeed at that time, instead it became a commercial site and almost all residents left the area. By 1991 only 566 residents were left (de Albuquerque Lapa and Almeida de Melo 2007). In this phase almost exclusive attention was paid to Bairro do Recife, leaving the surrounding quarters aside. What R13 explicitly liked about the plan was that it was done with the contribution of people working or living in the Bairro do Recife, e.g. people working there at night in prostitution and other informal things (*trabalhadores da noite*). For her it was '*an interesting exercise*', as, e.g. the project tried to figure out the necessities of the local people, like prostitutes raising children in the Bairro do Recife. Mostly there was a need for infrastructure improvement to prepare the Bairro do Recife also for a day-to-day use during daytime and not only being used at night any more. As a consequence administrative and public uses came back to the area, a result of the 1980s rehabilitation project. Subsequently, in particular the improvement of public transportation and open spaces were addressed, e.g. around

Rua do Bom Jesus. To coordinate the rehabilitation projects the Bairro do Recife Technical Agency was established:

> Nevertheless, without the necessary financial and political autonomy to ensure projects in the area are carried out in full, the Bairro do Recife Technical Agency remained a mere consultative body and, in 2005, its activities were ended. Such fragility, therefore, became aggravated because of the non-continuity of policies between successive administrations (de Albuquerque Lapa and Almeida de Melo 2007: 39f).

The 1993 Revitalization Plan for the Bairro do Recife then aimed at the preservation of its cultural heritage, on bringing back economic activities and increasing tourism. This Plan was more entrepreneurial than the 1986 one, which included participatory processes with actor groups. The 1993 one focussed more on private initiatives and revitalization/revalorization processes (R07). It proposed three different zones of intervention, and five poles of interest (riverside, Alfândega, Rua do Bom Jesus, Favela do Pilar and the mole itself). The plan was realized in a public–private partnership form (de Albuquerque Lapa and Almeida de Melo 2007) and is based on European and North-American experiences, e.g. Boston, Barcelona or Lisbon. Besides physical measures the plan also aimed at changing the Bairro's image, from periphery to a cultural centre which is attractive for investors (Pontual 2007), comparably a more economy-driven approach (Vasconcelos de Medeiros 2001). Van Oers (2014) finds the 1993 plan fitting it initial aims, although not all potentials are fully tapped yet. He emphasizes on the tidiness of the streets in Bairro do Recife and the cultural events that now take place, as well as on the bars and restaurants around Rua do Bom Jesus that are mainly used by middle- and upper-class. Proença Leite (2006, 2007, 2013) on the other hand argues that uses and the counter-uses of different central spaces result in socio-spatial changes and gentrification, as the regeneration triggers a certain elitist shift of the centre.

In this phase also the quarters of Santo Antônio and São José received more attention. Mid of the 1990s the municipality established the Calçadão dos Mascates (better known as *camelódromo*, see Fig. 5.37/2) and the Shopping popular Santa Rita at the Cais de Santa Rita in Bairro São José (see Fig. 5.37/5). Overall they comprise 2,700 boxes for informal vendors who previously were selling their goods in the streets (Antunes Cavalcante 2006). The Plano de Açāo do Bairro de São José pilot project focussed on public spaces, e.g. by installing boxes for the market waste around Mercado São José, and the creation of parking areas.

In 1997 the municipality launched a restoration program again in the Barrio do Recife, based on the 1993 approach. It encouraged private investment to restore heritage properties for recreation, services, and residential functions. The project area was divided into three sectors, focussing either on controlled intervention, renovation or conservation (Pontual 2007; Rojas 2002; Proença Leite 2006). Throughout the next years a number of initiatives at least partly dealing with heritage aspects were launched in the central quarters, as shown in Table 5.6.

The last phase of urban regeneration emerged around the year 2005 with the *Plano de Ação do Bairro do Recife* (Action Plan for Bairro do Recife). Following up on the already quite entrepreneurial foci of the projects listed above (R06) the

**Table 5.6** Main goals and focal areas of past and ongoing regeneration projects (Antunes Cavalcante 2006; de Albuquerque Lapa and Almeida de Melo 2007; Pontual 2007; Vieira 2007; Prefeitura da Cidade do Recife 2015)

| Project | Main goal(s) and focal areas | Actors involved |
|---|---|---|
| Reviver Recife Centro (Revive Recife Centre) launched 1998 | • Mainly economic upgrading<br>• Revitalisation of the main trade and shopping roads in the centre area by attracting consumers, as well as upgrading security and cleanness<br>• Among others implemented in selected locations of Santo Antônio and São José | • Municipality, e.g. DPPU, and private sector (Recife board of storeowners) |
| Porto Digital (Digital Harbour) Launched 2000 | • Economic and technological upgrading of the former harbour area in Bairro do Recife by attracting digital technologies<br>• Contributed to the restoration of distinct buildings and street sections and is one of the large-scale projects in the area | • Consortium with municipality, federal state and local enterprises |
| Morar no Centro (Living in the Centre) Launched 2000 | • Seeks to bring back housing functions use to the underused centre area, also to allow better use and conservation of buildings and public spaces.<br>• Project area covers São José | • Caixa Econômica Federal (Federal Savings Bank) and the municipality |
| Programa Monumenta (Monument Program) Launched 2003 | • Programmes covered under the umbrella of comprise the revitalization of the Cais da Alfândega—completed in 2003—of churches, private houses and central streets in Bairro do Recife | • Private consortium with lead of the federal state government |
| Complexo Turistico Cultural Recife/Olinda (Tourist-Cultural Complex Recife/ Olinda) Launched 2003 | • Improvement of touristic facilities by means of culture<br>• Valorisation and international visibility of tangible and intangible cultural heritage in Recife and Olinda, as well as the requalification of Recife centre<br>• In Recife centre the project area comprises Marco Zero, Alfândega, Forte do Brum, Forte das Cinco Pontas, Bairro São José, Praça da República, and Casa da Cultura. | • Actor consortium consists of city and state level authorities as well as private consultancies |

economic sector was significantly involved in its elaboration. The growing number of high-rise condominiums in the course of the conversion of former harbour areas in Santo Antônio and São José (e.g. Novo Recife or the twin towers) has to be seen in this context as well. In the same year the *Programa de Reabilitação de Áreas*

*Urbanas Centrais* (Programme to Rehabilitate Central Urban Areas) of the Ministério das Cidades declared—among others—Bairro do Recife as well as parts of Santo Antônio and São José areas of integrated regeneration. Different from the *Plano de Ação* it is fostering a "democratic" use of the inner city.

As mentioned above a large number of urban development and upgrading projects are carried out in the centre area, which has become the target of renovation actions since the 1980s (Barreto Lira and Pontual 2006; Vieira 2007). Recife nowadays is the city with the highest number of urban development plans in Brazil. A multiplicity of urban actors and stakeholder groups is actively involved in planning processes on the different levels, sometimes conforming, more often not conforming as they are aiming at different goals. Already under the last PT mayor (2009–12) and particularly with today's administration and mayor R07 witnesses a lack of public authority and involvement, the absence of a legal system for public open space and a prevalent and exclusive interest in the real estate market. R06 is lacking policies concerning housing and social aspects, while there are heritage policies in place in the centre. Although administrative functions are being brought back to the centre, e.g. many functions from Pernambuco State and Recife Municipality, housing functions are still missing, particularly in Bairro do Recife. In the surrounding quarters basically high-rise upper-class condominiums along the coastline are planned or already constructed while lower income classes are much less considered. Despite programmes focussing on social aspects it is foremost the large-scale public-private-partnerships in the former harbour areas, focusing on middle and upper-class exclusively, which are the determining factors for urban development.

### 5.3.8 *Perception of Inner-City Processes*

Like in the other case study cities, a questionnaire were used to investigate on values and feelings related to the historic centre of Recife, to processes perceived there and to intangible assets. A vast majority of the 81 respondents attach positive or very positive attributes to the area (cf. Fig. 5.41). More than 80 % find it fascinating or even very fascinating, while even more describe it as beautiful, with an absolute majority (52.0 %) finding it very beautiful. Answers for the area's overall value are even more explicit, with 93.4 % finding the area valuable or even very valuable (61.1 %).

However, the respondents also raised some critical points, e.g. on the centre's cleanliness, as 75.3 % of the respondents find the area dirty or even very dirty, which is linked to the fact that 48.0 % describe the centre as run-down or very run-down. The impression concerning the centre's vitality, its current shape, the upgrading process and its residents are ambiguous, as here positive and negative answers are almost balanced. A potential reason could be the heterogeneity of the area with its different quarters that have different functions and states of conservation, as also indicated by the respondents (68.0 % find the centre

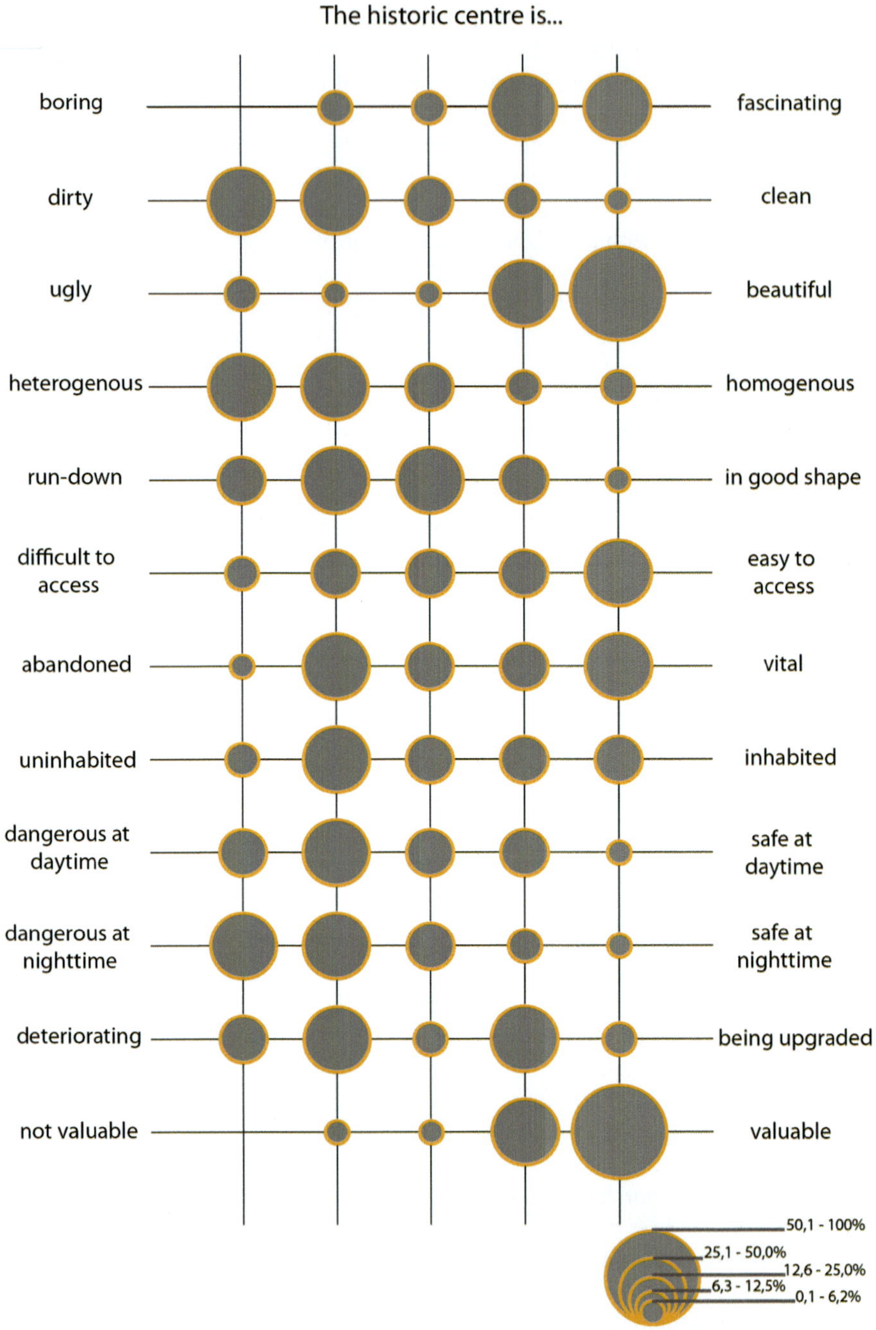


**Fig. 5.41** Responses to the question: What is your feeling towards the historic centre of Recife? ($n$ = 81)

heterogeneous/very heterogeneous). Accessibility seems to be given for most of the respondents (48.7 % find the centre easy or very easy to access), while a vast majority states that the area is dangerous, particularly at night time (59.1 % state the centre is dangerous/very dangerous at daytime, and 76.7 % at night time).

One can conclude, that there is a clear compliance about the overall value, beauty and fascination of the area, while the vitality, the shape and the current state of upgrading are seen more critical and at variance. The most critical parts are the area's security and its dirtiness.

These answers are backed by the results obtained on a question related to the processes perceived in the centre (Fig. 5.42). Although the area is (still) easy to access, the growing number of vehicles and the lack of parking lots are perceived as major problem (84.8 % of the respondents agree or definitely agree here). The same issues were raised in various expert interviews (cf. Sect. 5.3.5) and in an open question in the questionnaire on suggestions for project to upgrade the historic areas of Recife, where traffic infrastructure (namely improved accessibility by public transportation, pedestrian areas and better parking facilities) were named 13 times.

The answers concerning gentrification processes and the eviction of user groups are less explicit, as the majority of respondents agreed only partly, however, still more than 40 % agree or completely agree that such processes are going on. The replacement of old buildings was perceived only partly, here consent and dissent has almost the same amount of answers, while 48.0 % witness that traditional uses of the buildings get lost. Even 79.2 % agree or definitely agree that people prefer to live in modern buildings instead of the traditional ones, potentially linked to the fact that to 45.6 % planning does not follow local traditions but international trends.

The most explicit responses were obtained on questions regarding the users of the centre and individual preferences for or against it. Here, 80.0 % disagreed or even completely disagreed with the hypothesis that only tourists and not locals are frequenting Recife's centre. The impressive number of 88.8 % of the respondents don't find the area outmoded and instead prefer more modern parts of town, out of them even 53.8 % completely disagreed. Both questions reveal that the overall impression of the centre is a positive one, it is found valuable, beautiful and not outmoded. However, the respondents also gave clear statements on problems they perceive, including traffic, eviction of user groups and the loss of buildings as well as functions.

Again this is matching the topics of some suggested upgrading projects (answers on the question "If you had the chance to suggest some project to upgrade the historic center, what and where would it be?, as overall nine times improved housing facilities—out of that four times social housing in the centre—was named. Beside housing and traffic infrastructure, other answers were related to promoting activities and uses, to enhancing the area's security and cleanliness and to the renovation of historic buildings. The majority of responses dealt with the historic centre in general, if more explicit proposals were made they mostly concerned areas in Bairro do Recife and São José. Improving the area's security is mentioned in ten proposals, mostly by bringing back uses to the area. Overall 17 times proposals were made related to activities in the centre, mostly on promoting cultural activities

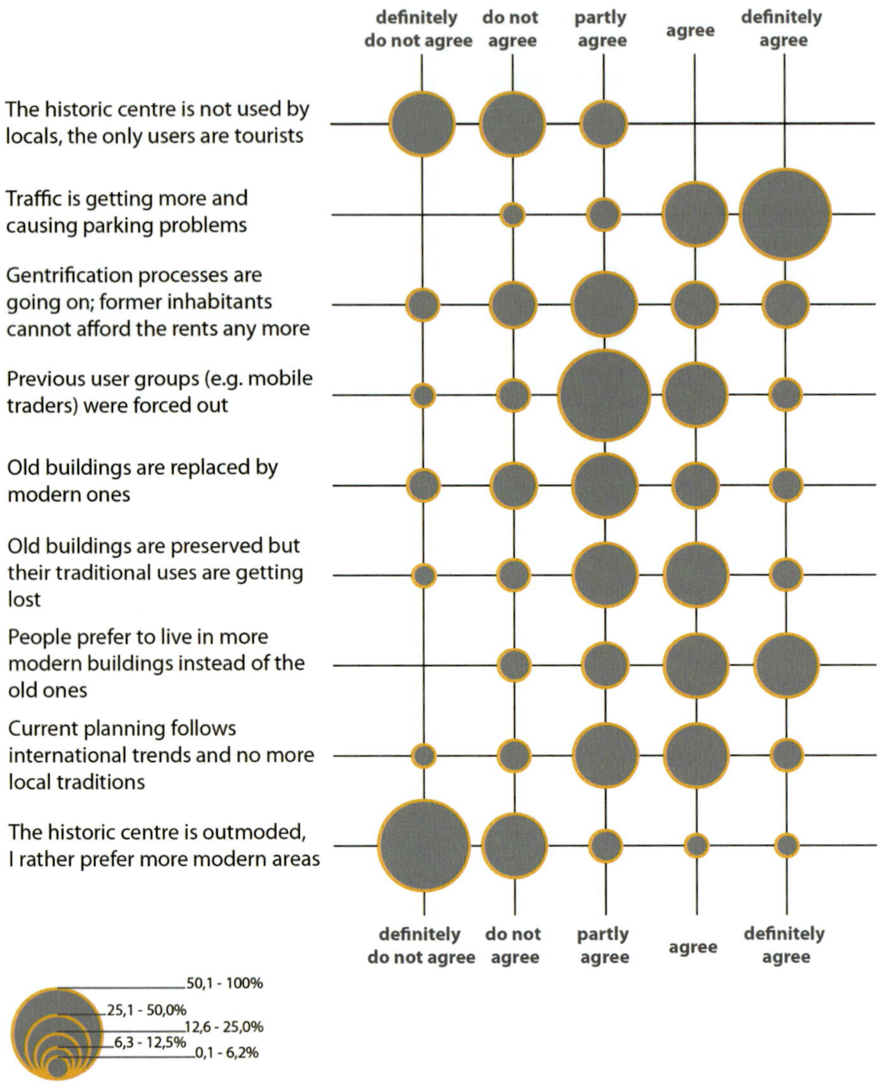


**Fig. 5.42** Responses to the question: Which processes do you perceive in the historic centre of Recife? ($n$ = 81)

in Bairro do Recife, sometimes linking such activities to accessibility, e.g. in one proposal to have "*more cultural events that cherish our traditions, and allocation of public spaces for leisure activities of local population during weekends*" or another on the "*integration of Recife Metro in the regeneration of the vacant harbour facilities to create a cultural centre*".

### 5.3.9 *Importance of Tangible and Intangible Assets*

After having given their opinion on ongoing processes in Recife centre the questionnaire respondents were requested to assess values attributed to the area (cf. Fig. 5.43). The commercial activities are not regarded as very important. Most

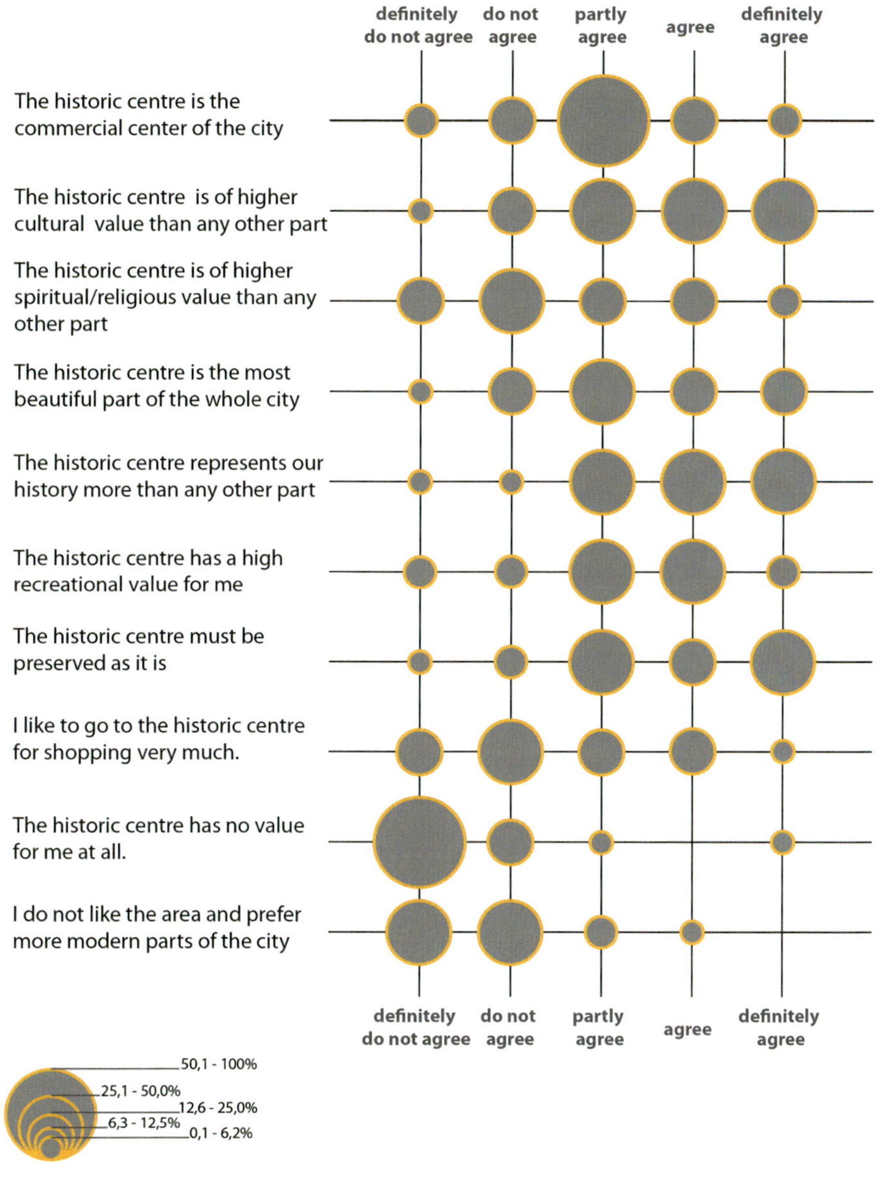


**Fig. 5.43** Responses to the question: Which values do you attribute to the historic centre of Recife? ($n = 81$)

respondents (51.3 %) only partly agreed that the centre is also the urban commercial centre; as a consequence 51.3 % dislike or completely dislike shopping there, although particularly the quarter of São José is a popular shopping area. However, it is mostly known for daily needs and low-budget items, while the large shopping centres are located in other quarters. Overall 58.4 % do not ascribe the area a higher spiritual or religious value, despite the high concentration of churches. A potential reason may be the decreasing importance of Catholic Church in Brazil. On the other hand overall 51.9 % find the centre of higher cultural value (answers for 'agree' and 'definitely agree' summed) than other parts of the city, 50.0 % agree or definitely agree it represents their history more than any other part.

For 39.7 % the centre has a high recreational value, while 39.7 % only partly agree to this, matching previous answers on security issues and the proposals for more cultural activities in the area. To 39.5 % it is even the most beautiful part of town, while 42.0 % only partly agree here, maybe due to ongoing degradation and problems with the cleanliness as stated before. However, 60.0 % want the centre to be preserved as it is, even 78 % completely disagree (another 17.5 % disagree) that the centre is not valuable for the individual respondent. 91.3 % do like the area and don't prefer more modern parts of town. Overall, the responses are very much matching the statements given during the expert interviews, where the need to improve the centre's infrastructure, uses and housing facilities were stressed very often. At the same time all experts agreed that the centre does have an outstanding value in both, built fabric but also cultural aspects, as many cultural facilities are located or taking place here. However, it seems that public cultural events are only taking place in certain occasions, while the permanent cultural facilities that have been implemented throughout the past years, e.g. the *Paço do Frevo* (Frevo Museum) or facilities to shop local handicraft are (yet) less popular.

Local handicraft coming from Recife and Pernambuco however is strongly promoted and advertised in Recife. In 2008 Pernambuco launched the *Programa de Artesanato de Pernambuco* (Pernambucan Handicraft Program). This may indicate a primarily economic view on handicraft and local culture, as the responsible authority AD Diper is foremost concerned with the state's economic development, instead of cultural or heritage issues. The initiatives main goal is to "*promote handicraft and to stimulate the producers to produce and sell their goods*" (R11).

The artisans are part of the more than 2,000 local producers of 'cultural goods', supported by Fundarpe under the FUNCULTURA umbrella in more than 2,300 initiatives only in 2013 (R04, R05). R05 however laments that the number of *feiras* or traditional markets is decreasing, as "*in Recife the culture of shopping centres is more prominent today, you will find handicraft and other non-industrialized products only in the traditional markets*". To protect local knowledge like handicraft but also the local poetry R05 describes that Fundarpe visits the *feiras* and contracts them for practice and teach their knowledge, "*this makes a difference, it is a question of valorisation, and these people have never thought that what they do could be paid from official side*".

Besides handicraft music and dances are among the intangible cultural goods mentioned most often. Carnival is considered of utmost importance by the intangible heritage are dealing with is music, dance and events related to carnival. Overall there are three cultural cycles in Recife, (*carnavalesco, junino, natalino*), each with distinct rituals, music, and performances (R02, R12). R01, R10 and R12 mention, e.g. municipal trainings for transmitting the Frevo knowledge, which has been declared Intangible World Heritage UNESCO (see Sect. 3.5). During such trainings people look for potential future musicians, and new music arrangements for the coming parades. Special dance trainings are offered for children but also for teenagers and grown-ups to teach them Frevo and maybe find new dancers. Other trainings deal with costume design, for event planning and organization. Some of these trainings are done in the *Casa do Carnaval* (House of Carnival), which is located at Pátio de São Pedro. The building belongs to municipality and comprises space for different exhibitions, an auditorium for presentations, a gallery for the flags of the different carnival associations, and the documentation and research centre. Its collection comprises books, journals, photos, films and more than 400 carnival musical scores. In addition the municipality (by means of the *Casa do Carnaval)* promotes contests, including awards of which some have a history of already more than 80 years, gives incentives to found new associations to participate in these contests and facilitates the selection of competition topics and contest evaluation schemes (R12). R12, the head of the *Casa do Carnaval* states, "*we see ourselves as multipliers, trying to preserve the immaterial heritage*". In this context he adds:

> The urban history of Recife is also the history of Frevo, it is in our blood. But it is also important to say that Frevo is not a relic, it is a living heritage. It is important to explain also to the people that it should not be frozen, also not by all the documentation, there has to be room for development (R12).

Fundarpe is also doing a survey on Maracatu and other local/regional music styles with the goal to put them on the List of *Patrimônio Cultural Nacional* (national cultural heritage), as elaborated by interviewees R01 and R05. *The Patrimônio Vivo de Pernambuco* (Living Heritage of Pernambuco) is another initiative to promote intangible heritage, by supporting groups and individuals that are preserving old cultural and handicraft traditions. The overall 33 groups/persons are receiving a monthly scholarship, what R09 calls "*an activity of valorization*". Although "*Pernambuco is a source of culture*" she is concerned that appreciation for intangible values is diminishing, as, e.g. Frevo music is listened to only during carnival, while it used to be heard all over the year. R10 lacks the protection and appreciation of popular culture (*cultura popular*); to him too much is spent on single artists or groups but not on initiatives to support the culture as a whole.

Intangible heritage like music, dance and handicraft is also seen and promoted as tourist attraction, as, e.g. stated by R11 and R12. Both, state and municipality have launched different projects and incentives to foster cultural tourism, e.g. by trainings or the provision of showrooms. However, to both, experts and questionnaire respondents, immaterial heritage is of personal importance. Figure 5.44 gives the answers on which intangible assets (given list) are considered as typical and important. Here, the carnival of Recife and Olinda is found important (91.3 %), out of which 70.0 % find very important. This value is only surpassed by results questionnaire respondents, and in fact a major part of what authorities working on

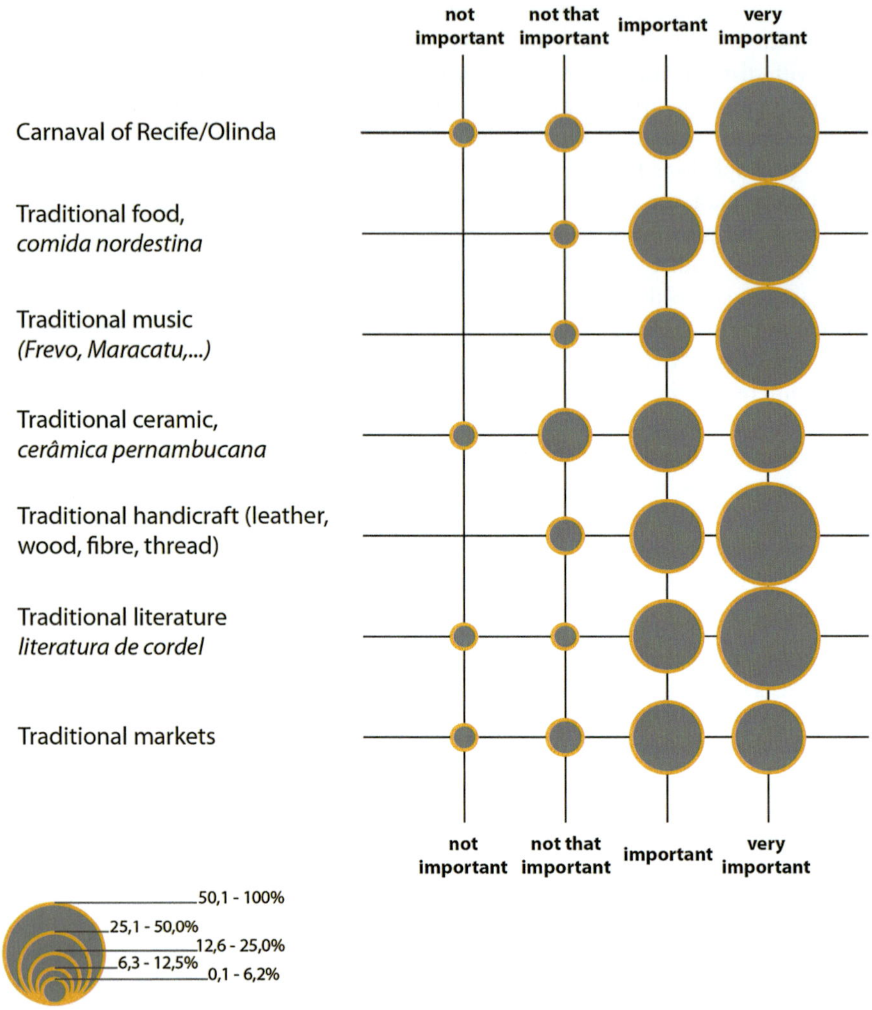


**Fig. 5.44** Answers to the question: Which intangible things or activities do you consider as typical for Recife area, how important is it to preserve them? ($n$ = 81)

obtained for traditional music, which 98.0 % (75.0 % respectively) regard as important or very important. Also typical food from the Brazilian Northeast (68.8 % very important, 27.5 % important) is highly appreciated, just like handicrafts. While traditional handicraft made from leather, wood or fibre is appreciated by overall 93.7 % (58.2 % state 'very important'), the typical regional ceramics obtained lower results with 83.8 % (36.3 % respectively). The regional popular literature sold in small booklets with xylographs (*Literatura de Cordel*) achieved results of 95.0 % out of which 61.3 % find it 'very important'. The comparatively lowest importance is attached to the local markets, although still the impressive number of 43.0 % considers them important and 45.6 % even very important (together 88.6 %).

The appreciation of intangible values is even higher than the one of the built heritage in the urban centre, in particular the values the interviewees attach to music and carnival. This data leads to the conclusion that overall there still is a high appreciation of such intangible assets.

Many of the intangible assets listed in Fig. 5.44 are linked to the case study area, particularly music and carnival, which is performed, promoted and supported—though not exclusively—in the centre. There are distinct places for distinct rituals, e.g. the inauguration of carnival on Marco Zero. All kinds of local food and handicraft can be found in and around the traditional market of São Pedro—and increasingly in other newly built locations in the centre.

The comparably high level of appreciation of intangible heritage is also witnessed by various experts, e.g. R02 thinks that the urban population appreciates immaterial heritage more than the material one. At the same time R06 is more critical about the common treatment of immaterial heritage, to him "*the vision of immaterial heritage in Recife is a very folkloristic one*" K14 is lacking awareness regarding the importance of the country's history in all its facets, including both, tangible and intangible urban heritage. This is not matching the questionnaire results, where a large majority of the respondents state they are interested (48.8 %) or even very interested (32.5 %) in the urban history (cf. Fig. 73, Annex IV). A potential explanation could be an interest in only certain aspects of history or a more passive interest without getting engaged in any of the activities. For instance different interviewees mentioned Maurício de Nassau and his plans for the urban outline as point of reference for Recife's history, among them R07 and R15. The Dutch colonial reign is a phase in the urban history which is cited very often, while the Afro-Brazilian background of most intangible heritage items like the different music styles is causing less attention (R02, R04).

While welcoming the protection of immaterial heritage, different interviewees are concerned about potential and real budget constraints, which in the end favour the financing of immaterial assets, as stated, e.g. by R04, R05 and R14: "*Carnival is absorbing a huge portion of the budget, politicians prefer to spend the budget on events*" (R14). Nevertheless most experts are positive about the growing concerns on intangible heritage and also its growing linkage to the tangible one, as R12 poses it: "*There has been a change in the Prefeitura thinking: the immaterial heritage is no longer seen as isolated from the material one.*" He gives the example of events in the Bairro do Recife, where the protected buildings are taken care of in event planning, e.g. by picking adequate decoration during carnival and other events, not to affect the old buildings. Place and size of decoration is chosen carefully not to destroy or affect anything. Contemporary conservation projects do not only concern the buildings but also the lifestyle of the local communities (R04). R09, however, still wishes for more integration of material and immaterial heritage, "*this is not a problem of the teams; they get along well, but a problem of the legislation*".

Overall the attachment to urban tangible and intangible heritage and the centre are high in groups, experts and questionnaire respondents. Two more questions asked to the student peer group dealt with (predefined) distinct historic places in the urban area and (not predefined) places the students would choose for sightseeing and leisure in case a friend from outside would visit them The given answers are very congruent, emphasizing the importance of the historic centre, its history and cultural heritage.

The top four answers concerning the value of certain places for remembering Recife's history are the Bairro do Recife itself, two more spots within the Bairro (Marco Zero and Rua do Bom Jesus with the former Synagogue) and the bridges connecting the central quarters (Fig. 5.45). For each location ranked among the top eight more than 90.0 % of the respondents found the place important or even very important. The reasons behind were mostly "*history*" and "*tradition*". Bairro do Recife is regarded, e.g. *as* "*place where the city was born*", while Marco Zero is the "*symbol of our city*" and "*meeting point of the city*". The bridges are, e.g. "*important for the urban identity*" and "*defining the urban character*". Reasons why Rua do Bom Jesus with the former Synagogue are places of remembrance exclusively deal with the inherent historic value as first Jewish community abroad For overall 94.9 % it is an important or very important place of remembrance. Answers given to Casa da Cultura, and Mercado São José varied between the historic importance of the buildings and the importance of local goods that are nowadays sold there while the reasons for the former *engenhos* (sugarcane plantations) either

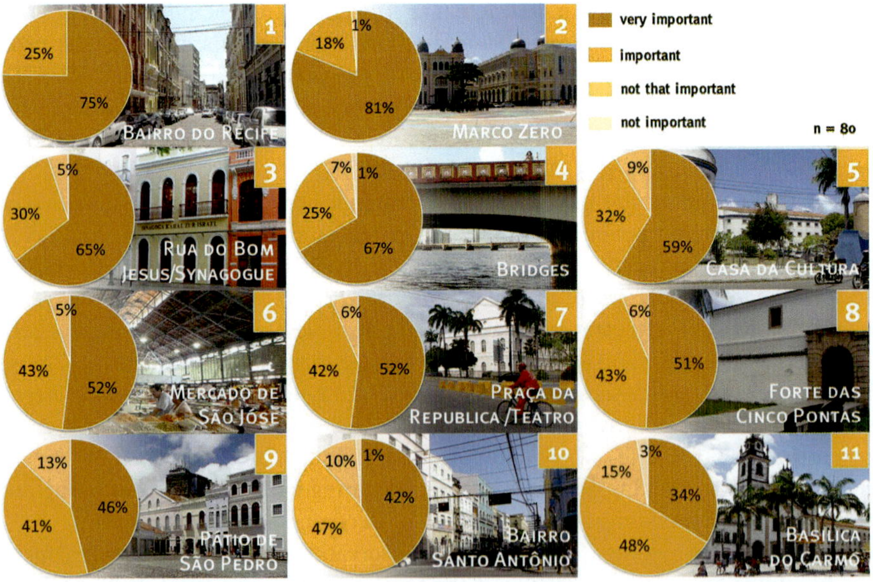


**Fig. 5.45** Answers to the question: Which places/buildings do you consider important as place of remembrance of Recife history?

only mentioned the overall historic value or the economic base of Recife's development and the history of slavery. However, together with the Basilica this is one of the least valued objects on the list, though still achieving more than 80 % of consent on its importance—maybe indicating the diminishing importance of Catholic Church overall as well as indifference towards the large number of churches in the historic centre.

Being asked about places to visit for leisure activities 68 different places were mentioned. Figure 5.46 is giving the top eighteen locations which have been named four times or more. Out of them six are located within the historic centre, and another two are partly. Out of the five locations, the Bairro do Recife and the Marco Zero within that quarter rank number one and two. Mercado São José is ranking number six, Casa da Cultura is ranking number eight and Pátio de São Pedro is no. seventeen. Again, most of the justifications given (free text) deal with the cultural value of the locations, most of the reasons to visit the Bairro do Recife and Marco Zero are related to the urban origin, e.g. "*this is the origin of Recife*", "*to show the history of Recife and its development*", or "*symbolic value of Marco Zero as entrance gate to Recife*". The words used most often are "*history*" and "*origin*", while reasons to visit Mercado São José and Casa da Cultura are mostly related to "*handicraft*" and "*historic building*", e.g. "*to see the market's architecture and its local goods and to show how such a market is functioning*".

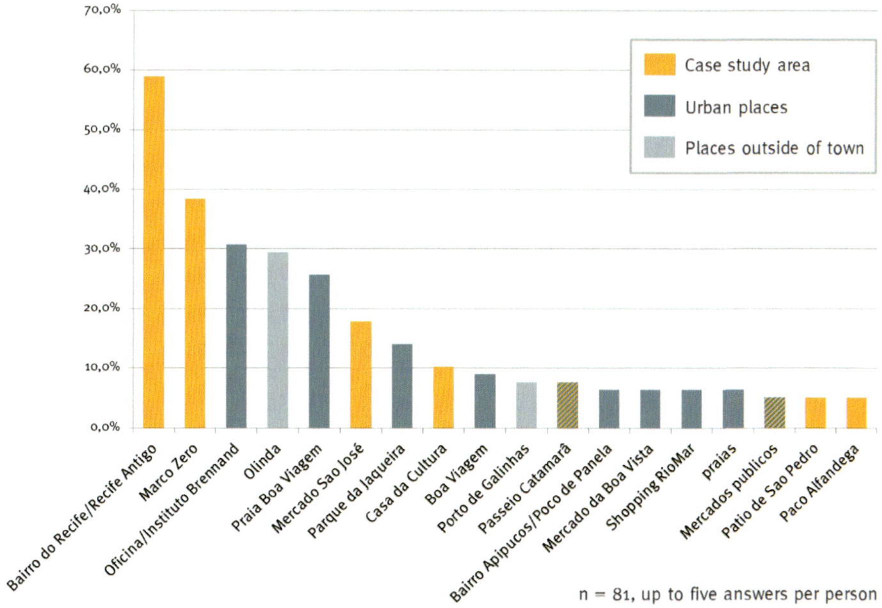


**Fig. 5.46** Answers to the question: Imagine some friends from outside visit you in Recife. Where in and around the town would you take them for sightseeing and leisure?

Overall, the appreciation for intangible assets is very high, including values and memories attached to distinct places within the historic centre. Questionnaire results revealed a comparably high cultural value attached to the centre, which is also regarded as manifestation of the urban history. The importance ascribed to distinct local cultural expressions and activities is particularly high. Authorities have already implemented a comparably large number of policies to particularly support and protect intangible values, particularly related to music, dance and handicraft. One of the underlying reasons to do so is the promotion of tourism, however the main target group is yet local residents.

### 5.3.10 Conclusions on the Role of Heritage in Shaping the Urban Identity of Recife

Summing it up, questionnaire respondents are aware of the cultural and historic value of both, urban tangible and intangible heritage and feel attached to it, in particular to (objects in) the historic centre. Nevertheless, experts and respondents are very aware of challenges to maintain it and named various problems they perceived. The main trigger named in the expert interviews is the rapid urban growth induced by the regional economic development, which is resulting in the conversion of large-scale inner-city areas into high-rise condominiums. In the course of these developments the urban landscape will change fundamentally. Surprisingly, all of these initiatives promote themselves by means of the cultural value of the surrounding. Sabino de Nascimento (2014: 44) talks about a "cultural economy" (*economia cultural*) he finds in Recife, e.g. in the Porto Digital and Projeto Novo Recife, comprising urban planning, social aspects, cultural expressions and activities.

R15 believes that Recife is still offering many qualities already lost in other towns, which are now in the course of being lost due to planning mistakes, including the urban heritage. "*Maybe we change the planning approaches early enough, maybe not, then the discussion will start too late, and we will have to struggle to rebuild what we have lost by that time*". Each quarter has its distinct characteristics. On the other hand may complicate planning and urban regeneration, on the other hand it can be seen as a chance, as to R08a: "*It is an asset to have a town which is not homogeneous*". However most urban regeneration plans were carried out in Bairro do Recife. As a consequence the quarters of São José and Santo Antônio received much less attention until recently.

Various initiatives are already linking material heritage with immaterial one, often including social aspects, like the *Cinema Sâo Luis*, in Boa Vista, one of the oldest cinemas in Recife, restored and owned by Fundarpe. Nowadays it is running again as a cinema, with entrance costs of only 4R$ (compared to up to 30R$ in the popular shopping centre multiplex cinemas) to allow people of less income the entrance (R05). To most experts it was obvious and part of their daily job to think these two aspects together, e.g. R12 states: "*The immaterial heritage must not be seen separately from culture and material heritage*". He also perceives a growing interest in the urban immaterial heritage in general, since the Frevo has become immaterial heritage of UNESCO.

Opinions of the interviewees concerning the urban future differ a lot. R01 is convinced that Recife will become a completely new town with only few local points that preserve the memories of the past, she and R02 are convinced that "*our kids will live in a different city*". R12 on the other hand is optimistic and sure that people will be more aware of their history, "*Recife will be more beautiful than today*". R12 shares the same opinion, stating "*I am optimistic about the future; the new development goes well together with the preservation of old structures and one also has to allow for new developments*". R06 and R11a/b expect a growing living standard and better infrastructure including the traffic. To achieve all this R06 and R08a/b recommends suitable policies for bringing back life to the centre, including mixed uses and social housing programmes—and allowing for change (R07, R14) linking old and new, but preserving the popular culture (R02, R12): as "*many people think that popular culture is a standalone thing, but it is not, it is a part of the urban history*" (R12). To both of them it is crucial to preserve the immaterial heritage, but to allow for change and future development.

Recife Summary

- Comprehensive urban and heritage legislation is in place, on national, state and municipal level, covering both, tangible as well as intangible heritage.
- However, differences between conservation theory and practice come to light, due to discrepancies in legal documents, time-consuming processes and informal arrangements ('*jeitinho brasileiro*') disregarding legal issues and resulting in the loss of heritage fabric.
- Rapid urban and economic change is impacting on the historic centre, where public-private initiatives are fundamentally altering the appearance by building high-rise apartments.
- Intangible heritage is of high significance, potentially at the expense of tangible heritage which is suffering comparably more budget constraints.
- Conservation of intangible heritage is done in a selective way, promoting distinct assets (mostly related to carnival), potentially depreciating continued development and popular culture.
- The level of attachment to the historic centre is high; the challenge is to cope with the diversity of the quarters, social disparities and rapid urban change.

## 5.4 Comparative Analysis of the Case Studies' Findings

The previous chapters have elaborated on the urban history, urban tangible and intangible heritage, urban and heritage policies and regeneration of the centres of Kathmandu, Yogyakarta and Recife. The perception of ongoing processes and their influences on urban patterns was assessed in interviews and questionnaires. Despite the differences in geographic location, cultural background, political and economic situation the three cities share distinct similarities which permit to draw generalizable conclusions. The case studies' findings are analyzed and interpreted subsequently.

### *5.4.1 Urban Patterns and Processes Impacting on the Historic Centres*

Comparing the case study cities first of all reveals a number of comparable impacts on the historic centres, as visualised in Fig. 5.47. The major impacts are to a large extend the same ones, but differing in two ways: the intensity, compared to other impact factors, and the way different impacts are correlated.

When taking a look at the overall urban development it is obvious that all three cities are still expanding and growing, facing rapid suburban growth, accompanied by major road constructions intended to cope with the growing traffic congestions (cf. Rubens de Menezes and Figueira de Souza 2014 in the case of Recife; and Rana and Marwasta 2015, for Kathmandu and Yogyakarta). Urban growth and economic development is also visible in ongoing densification processes of the central urban areas. In all three cities high-rise apartment buildings are constructed. This is even true for Yogyakarta where high-rises so far were mostly hotel buildings as people preferred to live in private houses with only few floors. In Kathmandu high-rise apartment blocks have emerged only within the past few years, as a result of the skyrocketing land prices and lack of available urban lands. In both cities such buildings are yet located outside the core area, while Recife's latest urban development project 'Novo Recife' is located by far more central. Comparing the impacts and the main triggers of urban development, growth and densification are comparably more impacting on the centres of Yogyakarta and Recife, while such densification process is yet less the case in Kathmandu's centre. Here the main challenges are foremost the unplanned urban development, hardly following any regulations, and ongoing processes of marginalization, resulting in the decay of whole quarters.

The urban development is closely linked to inner-city processes, one of the major impact factors for the historic centre's fabric and functions. Nevertheless, the main triggers differ profoundly. In Yogyakarta processes of densification and tertiarisation are increasingly shaping the inner-city area, visible in the growing number of hotel constructions in and nearby the centre, the replacement of small

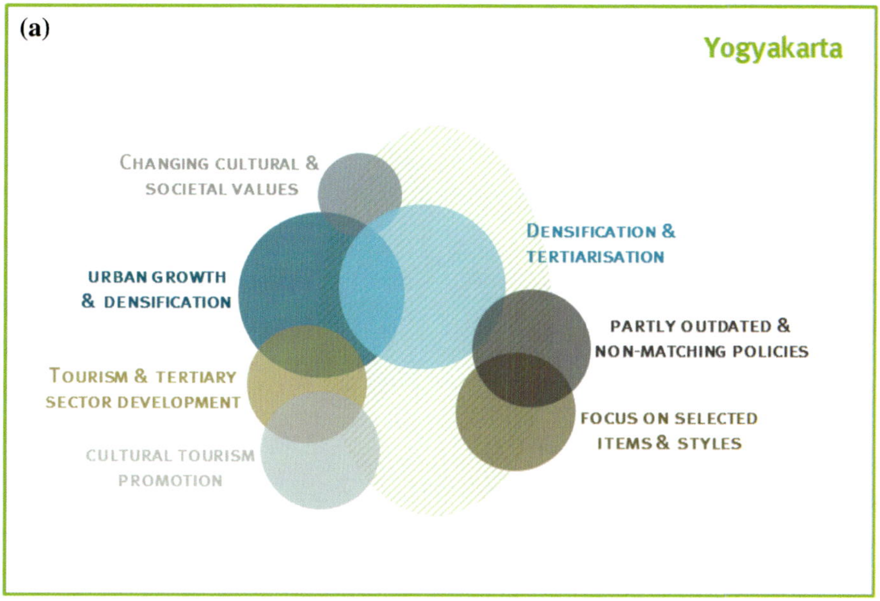


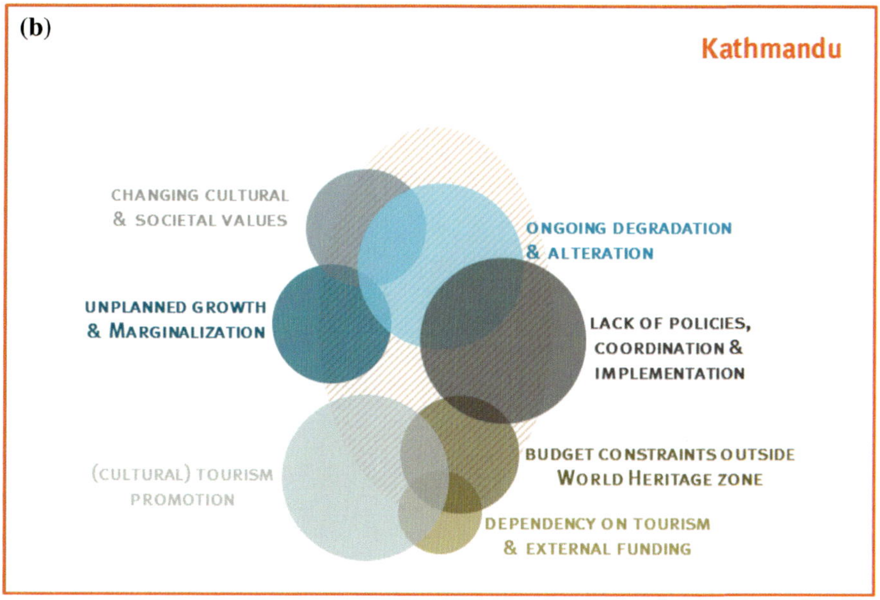


**Fig. 5.47** **a–c** Impacts on the historic centre development in the case study cities

trade by showrooms, and the ongoing conversion of the area west of Malioboro Road for tourism and hotels. Kathmandu's centre is more suffering from ongoing alteration of existing buildings, e.g. by vertical division or added storeys, and the degradation of the building stock which often is poorly maintained.

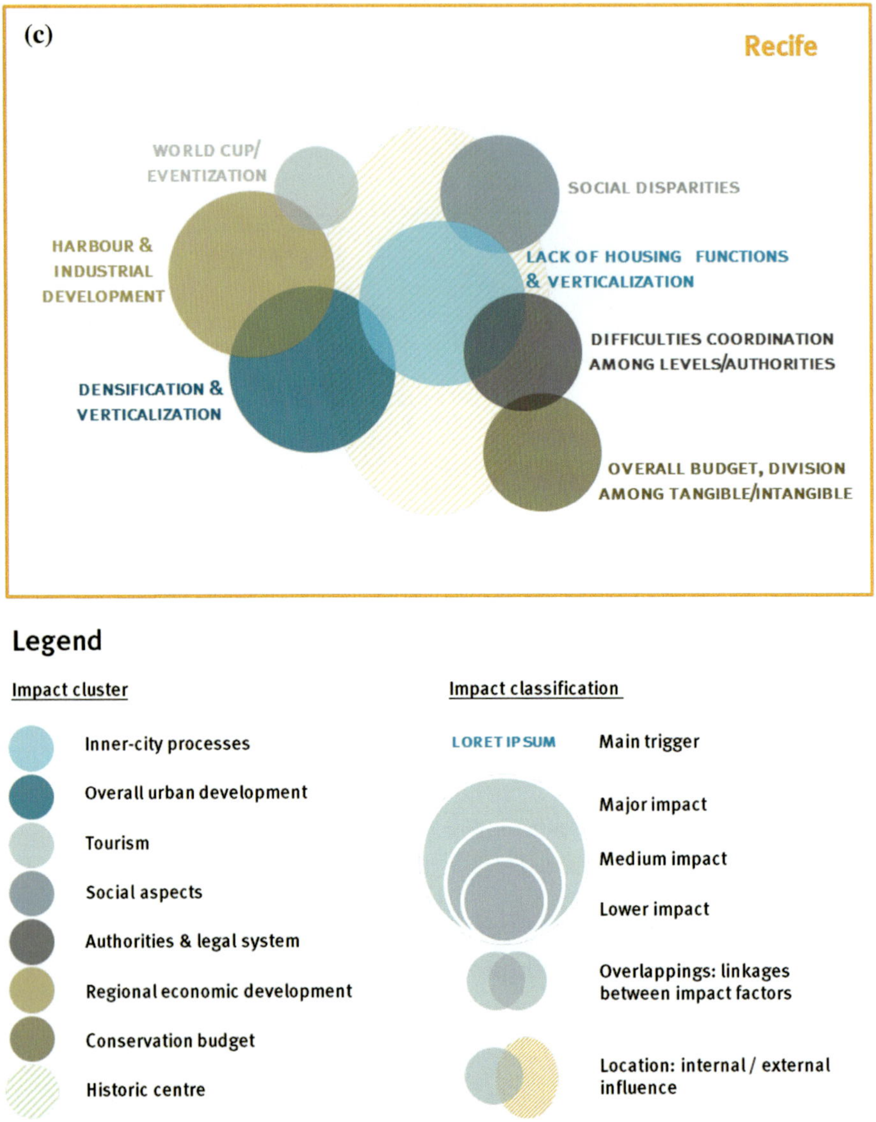


**Fig. 5.47** (continued)

While here the buildings are still inhabited (though often rented out instead of inhabited by the owners), Recife's city centre is partly inhabited at the best. Here the lack of housing functions in the central quarters is an obstacle for development as the lack of inhabitants hinders community-based initiatives. In addition, the trend to insert high-rise apartment buildings in or near the city centre is changing the urban appearance profoundly. However, the inhabitants of these posh apartment buildings will hardly be involved in urban processes as they are part of gated communities. This is showing the huge social disparities of Brazilian society, where the centre is perceived as dangerous, particularly during night time, probably because it is little inhabited and due to the lower social stratum living and working there. Such perception can hinder regeneration processes. In Kathmandu it is foremost the changing cultural values which are apparently impacting on the building stock in the historic centre. Declining numbers of joint families, e.g. lead to the vertical division or replacement of historic houses mentioned above. The maintenance of religious sites and infrastructure, formerly done by local communities, is declining. Societal change is also apparent in Yogyakarta, but to a comparably lesser extent. Here, social contacts and communal spaces for socializing are still essential and a strong foundation for participatory processes (Larasati 2007).

Instead, tourism is comparably more impacting on Yogyakarta's inner-city development. Yogyakarta and likewise Kathmandu are depending very much on tourism as one of the main economic sectors. Particularly cultural tourism is a major source of income in the historic centres. The entry fees charged for Kathmandu's World Heritage sites are a major source for maintaining the urban heritage. Beside the intention to promote the urban centres as tourist destinations, it is also the hotel business which is impacting on these two cities, as whole quarters have been converted for this purpose nearby (five-star hotels) or even within (budget tourism) the centre. In contrast Recife has predominantly developed inner-urban areas for events, bars and restaurants to revive inner-urban spaces (Proença Leite 2013). Unanimously, in all three case study cities tourism was also seen as a chance for heritage conservation—experts believed that once foreign visitors appreciate the historic centre, traditional architecture or intangible assets, then authorities and local people might change their attitude. Y07 poses "*usually we also use foreigners to support us, to tell the governor what we want*". K01 gave a very similar comment as he believes that the conservation of privately owned historic buildings got easier since owners and authorities recognized the tourist appreciation as well as the economic potential.

Tourism development is a focal point of urban development strategies. While Recife is still more a destination for domestic tourism (though trying to change this by promoting the city as a destination in the course of the FIFA World Cup in 2014), Kathmandu is clearly relying on international guests, what is understandable given the country's economic situation and the comparably few inhabitants. Yogyakarta is based between these two strategies, with both, national and international visitors. Different from Kathmandu and Yogyakarta, tourism is not the dominant economic development strategy in Recife. Here, industrial development with the allocation of large industrial areas in the urban fringe which emerged in the course of Suape harbour construction is the predominant economic strategy.

Taking a look at the legal framework (cf. Sect. 5.4.2) and the related conservation budgets reveals basic conformity, again particularly among the two Asian cities. In the case of Yogyakarta it is mostly the partly outdated and separated understanding of heritage in the legal framework which is causing problems, exacerbated by the number of actors in charge for urban planning as well as the different forms of heritage—despite the growing appreciation of the 'saujana' approach. The main challenge in terms of conservation budget is the still existing biassed focus on certain sites and styles. In addition, in both cases the Sultan as cultural and political reference person is very influential, a special case even within Indonesia. Kathmandu has both, a lack of suitable policies to protect those parts of the historic centre which do not belong to the World Heritage, and a lack to implement those policies which are in place. In addition, the coordination among the involved authorities for conservation, urban planning and monitoring of both, is poor. Budget is another problem, as only very limited finances are available to maintain the historic centre; again it is much easier for the World Heritage sites. Furthermore the continuity of budget allocation is not given as it is often depending on foreign financial means. Like the other cities, Recife is dealing with a challenging coordination of authorities on the different administrative levels. Unlike the other cases heritage is regulated by a comprehensive legal framework, however with some inconsistencies between the levels. In terms of budget it is mostly the distribution among tangible and intangible heritage which is challenging, besides the probably ever existing lack of overall budget.

While the main impacts on the case study cities can be grouped in comparable clusters, their interlinkages differ at least partly. Obviously the legal framework and the conservation budget are related, but only in Kathmandu conservation budget shows a very high correlation with tourism and the economic sector, as the urban heritage is one of the main income sources for the centre and its conservation. In Recife and Yogyakarta tourism and economy are more linked to the overall urban development, with the difference that tourism is comparably more important in Yogyakarta's economic strategy. Social aspects as well are inevitably linked to the inner-city processes. In the case of Recife it is mostly the local social disparities (particularly the still comparably large number of marginal population) and the related feeling of insecurity which are hindering more and new usages of the centre. In the cases of Kathmandu and Yogyakarta changing cultural and societal values are triggering a more fundamental change in the overall urban area, which then again is impacting on the centre itself.

In conclusion, the main triggers of change in the historic centres can be clustered in similar way, but they differ in terms of intensity and connectedness. A suitable and sustainable regeneration strategy will have to consider not only the triggers but as well their interlinkages. Particularly a neglect of the interlinkages and dependencies may decide upon the success or failure of urban development and regeneration projects.

Interim conclusion on impacts

- While there are comparable major impacts clusters influencing on the urban centre, they differ in terms of intensity and correlation.
- Impact factors are comparably more 'external' in case of the larger countries that comprise of more cultural and urban hubs (Recife and to a

lower extend Yogyakarta), while influences are more local in the case of the Kathmandu, which is the only major urban agglomeration of Nepal.
- To develop suitable strategies for regenerating historic centres the triggers of urban development as well as their interlinkages have to be taken into account.

## 5.4.2 Findings on Laws and Policies

Any practical policy for heritage and urban development within the historic core area is based on a legal framework. When comparing the legal systems and policies related to the historic urban centres in Kathmandu, Yogyakarta and Recife, three major challenges were detected:

- The content itself, comprising the question what is considered tangible or intangible heritage, what kinds of protective mechanisms are in place and how congruent they are,
- The implementation, dealing with responsible authorities, potential additional actors and their cooperation, and with the execution of protective mechanisms in reality, and
- The adequateness, comparing the challenges mentioned before against the local cultural background.

Global paradigms on heritage conservation and urban renewal have shaped the discourses not only on a global level but have influenced policies and partly even legislation in many countries, as elaborated in Chap. 3. In Nepal, e.g. the UNESCO World Heritage sites are having a higher level of protection by law, and also in practical life these sites are receiving more attention than any other. Likewise in Indonesia and Brazil categories of heritage and the objects protected by law are strongly influenced by the Euro-American understanding that has shaped global discourses for a long time. In all three countries it is mainly the religious and upper-class buildings and sites that are receiving attention and that are guaranteed legal protection while categories for other objects or sites are lacking. In particular policies on an urban scale are lacking, as well as such combining tangible with intangible assets. This is true, e.g. in the urban layout of Yogyakarta which is still reflecting very much its Hindu-Javanese roots and later colonial adaptations—but which is comparably more complicated to be protected than, e.g. the Kraton compound that is under a much more restrictive legislation. Overall, the protection level of cultural heritage is still higher than the one of intangible heritage. Intangible heritage in fact is only considered in legislation in Brazil so far, while there are attempts to mainstream it into Indonesian legislation with the 2003 charter and its saujana approach. In either case, natural and cultural heritage are separated by

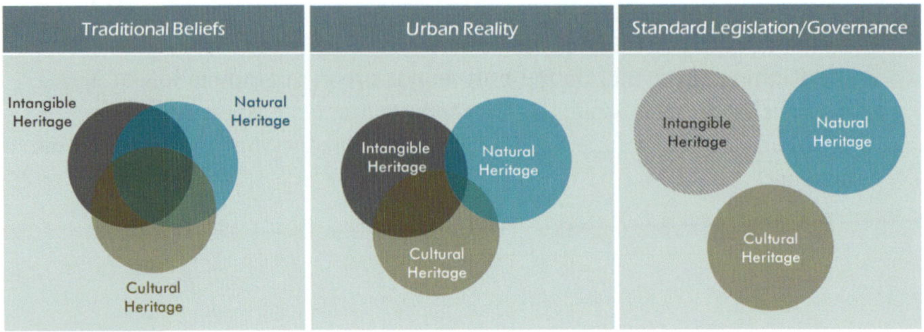


**Fig. 5.48** Abstraction of interlinkages between the heritage types in different contexts

legislation, in addition it is different authorities that are responsible—with a partial exception in the case of Brazil, where ZEPHs can contain natural environment like riverbanks as well. As a consequence, the interlinkage of natural and cultural, tangible and intangible heritage which has become the global conservation paradigm and which is (meant to) reflect traditional belief systems of any society (Albert 2013; von Droste 2012; Scazzosi 2011; Garden 2009), is hardly depicted in legislation (cf. Fig. 5.48). In reality, there are two (natural and cultural in the sense of built) or three (natural, cultural in the sense of built, as well as intangible cultural expressions) different legal frameworks. Even if obliged to cooperate at least in certain occasions (e.g. in case of combined heritage, like built objects surrounded by parks) realities are giving a different picture. Having the results of the empirical study and the close interlinkage of built and intangible heritage in historic urban centres in mind, it is even more astonishing that until quite recently there was hardly any statutory protection in place, in none of the case study areas—not to talk about an interlinkage.

Taking a look at the implementation, the comparably much higher complexity and extent of the Brazilian conservation legislation is striking. To illustrate this: While the UNESCO Database of National Cultural Heritage Laws (2015g) lists four documents in the case of Indonesia and five in Nepal, as many as 100 are documented in the case of Brazil. This, however, has to be seen against the background of how laws and policies are executed. Here the case studies' results correspond most highly in their very critical view of the current governance system. The major burdens can be summarized as:

- The unsuitability or even lack of legal frameworks or conservation policies (e.g. the absence of proper conservation schemes outside the World Heritage zones in Kathmandu or the yet missing consideration of intangible values in Indonesian legislation, despite its importance for society, or the different classification systems for heritage on Brazilian national and federal state level),
- A lack of existing policy implementation or conflicting policies hindering implementation (e.g. confusing responsibilities between heritage authorities and police in Kathmandu, between the Kraton and urban authorities in Yogyakarta),

- Different heritage authorities on different administrative levels and for different categories of heritage, e.g. on urban, federal state and national level in Brazil,
- A legal framework rooted in colonial past or based on international conservation schemes, as the post-independence Indonesian heritage legislation which was taken almost unchanged from Dutch colonial times, or the Nepalese conservation schemes, which only consider the UNESCO World Heritage sites as top category of conservation on national level.

The evolution of authorities and the division of responsibilities among them usually dates back to the state foundation—or may even be rooted in earlier political systems, e.g. the kingdom in Nepal or former constitutions in Brazil. In Indonesia the relevant laws were taken over almost unchanged from the Dutch colonial legislation from 1933. Changes only happen slowly over time when legislation is amended or in case legal power is given to another administrative level, as happened, e.g. in the Indonesian or Brazilian decentralization processes. This is a potential window of opportunity to adapt regional legislation and subsequent policies to changing understanding, local circumstances and cultures.

However, the mills of administration grind slowly. As national law is still binding for any legal document on lower administrative levels, change induced on lower administrative levels works its way up only slowly. This might be explained by the fact that on the national levels relics from earlier administrative systems as well as understandings of heritage are preserved.

Talking about the adequateness of legal systems one has to consider its origins. In the case of the Brazilian conservation system, R02 finds it rooted in US-models, while R06 sees Brazil following a European role model since the 1970s, e.g. in the attempts to preserve ensembles. R07 and R08 as well name European cities as role models for contemporary and future urban conservation and regeneration, like Barcelona in the early 1990s, San Francisco or Amsterdam, because of their culture and history. To interviewee K09 international discourses on heritage and conservation are important because they will be mainstreamed into country or regional levels at least after a while. Talking about conservation in Kathmandu K01 states that "*any process Europe went through some ten years ago is now happening here. Mistakes are repeated instead of learning from mistakes of others.*" K03 distinguishes between the 'Western' understanding of conservation as something which should not touch old fabric, and the Asian understanding where change is part of the conservation process itself:

> It has to be understood against the Hindu-Buddhist background of Nepal and India where time is not linear but cyclic, it comes back, this is represented in the 'rebirthing' of temples, where, based on the lunar calendar, worshipping is done, at many important temples this is done annually at the same date, the worshipping to the god represents the rebirth and renewal of the temple. In this context Nepalese do conservation, it is part of the worshipping process. During any of these ceremonies and ritual conservation processes some things are added. This change aspect is very important (K03).

Incorporating the 'present' into architecture and conservation to reflect contemporary culture, is no contradiction to conservation in the Nepalese and

Indonesian context. Therefore different interviewees (K01, K09 and Y07) are confident that recent paradigm shifts will have a positive impact in the urban future.

The legal systems' adequateness is as well very much depending on how far it matches the local and/or national cultural understanding. There is a broad agreement that "*development should be based on the local culture*" (Y06) but less agreement what this development should look like. Surprisingly, it is Brazil that already has implemented impressive protective schemes for intangible heritage, despite the fact that the origins of the discourses on the consideration of intangibility aspects are rather based in Asia. The legislation for preservation of the living heritage is done after the role model of France (R04), while R06 finds the recently introduced concept of regarding individual persons as heritage a Japanese one—indicating the global importance of Asian concepts on global heritage paradigms. The "*shift from the object-centric to the subject-centric conservation*" that Kwanda (2010: 9) is witnessing in Asia since the end of the twentieth century and that is acknowledging "*the plurality of meanings, functions and values including intangible heritage*" seems to have gone global.

Interim conclusion on laws and policies

- Common problems of conservation legal frameworks are overlaps or contradictions between different legal documents issued on different levels or by different authorities, outdated contents which are no more matching urban realities and/or which have been implemented by replaced governments.
- In addition, the execution of the legal framework in force is often poor due to a high level of informality, a rapid urban change that authorities are not able to cope with, and a lack of operational coordination between the executive authorities.
- Holistic approaches that are considering tangible and intangible heritage are not yet reflected in laws and policies, on the contrary natural, cultural and intangible heritage is yet mostly treated by different laws and administrative bodies—if intangible heritage is protected by law at all.

### 5.4.3 Findings on Urban Regeneration

In the previous chapters urban regeneration measures in Kathmandu, Yogyakarta and Recife have been described in detail. This section now seeks to assess the main similarities and differences, as shown in Fig. 5.49, which itself is based on the figures on urban regeneration phases in the Kathmandu, Yogyakarta and Recife chapters. Basically, regeneration can be grouped into three main phases, starting

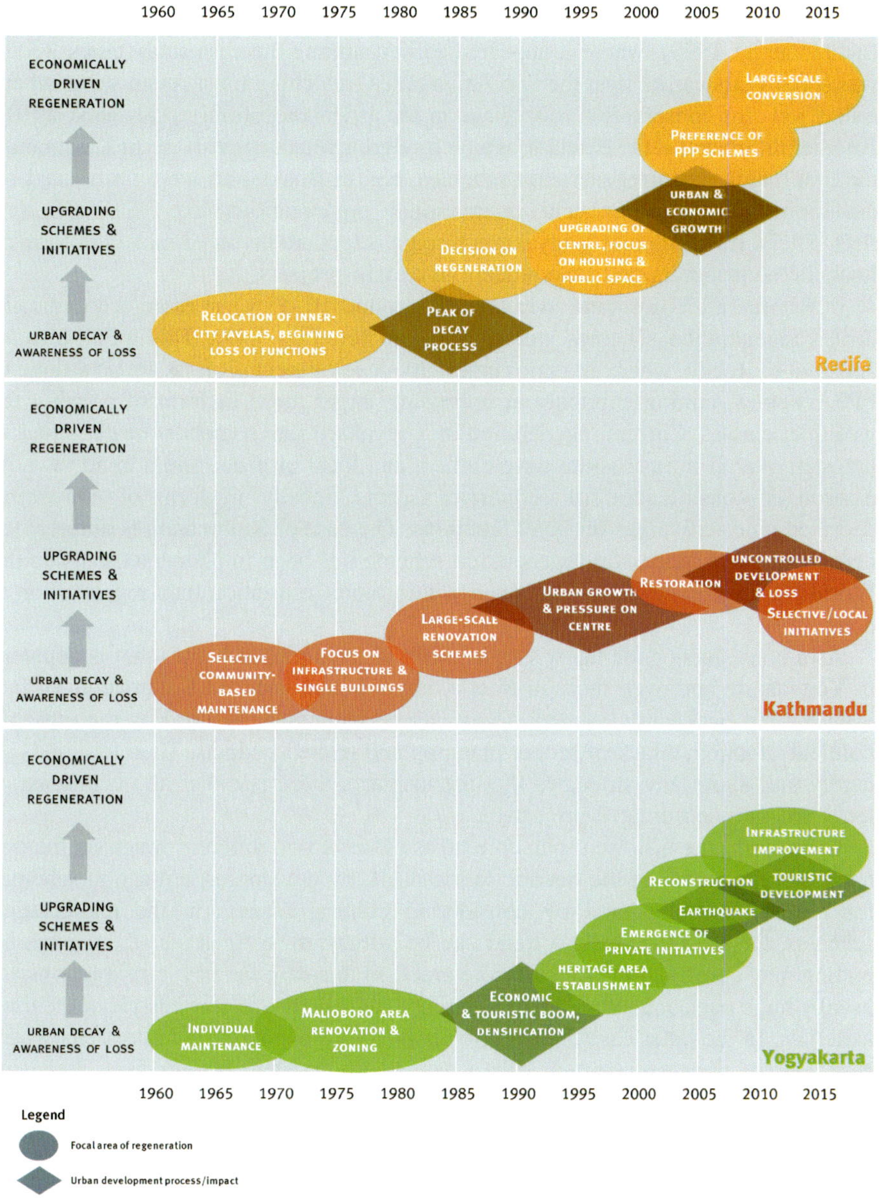


**Fig. 5.49** Comparison of urban regeneration

from processes of inner-city decay, the subsequent loss of functions (though to different extents) and increasing awareness of the ongoing loss. On an average, processes of urban decay and rapid urban change peaked in the 1960s and 70s. As a reaction and due to the growing awareness of ongoing losses upgrading schemes and projects were implemented. The trend is comparable to Western cities (cf. Lichtenberger 1990), where it appeared earlier, among the case study cities Recife was some years earlier than the two Asian cities (matching most Asian states where a first wave of conservation took place in the 1970s, according to Kwanda 2010). Another difference is the extent to which foreign actors are involved: In Kathmandu the first major urban regeneration schemes, e.g. in Bhaktapur, were introduced by development cooperation and international organizations like UNESCO (cf. Sect. 5.1.7). In Recife and Yogyakarta foreign actors are involved mostly in form of financial contributors and stakeholders in certain projects.

In Recife and Yogyakarta over the past around 10–15 years more economically driven regeneration schemes were introduced (cf. Fig. 5.49). This trend is more apparent in Recife, where all major inner-city development projects are now done in PPP schemes, aiming explicitly at upgrading larger areas in form of housing for wealthier classes. Current regeneration in Yogyakarta can somehow be classified as between overall infrastructure development and local projects, and a trend towards a stronger consideration of economic aspects, mostly in form of converting deprived inner-city areas for hotel functions. The case of Kathmandu is different, as here hardly any successful regeneration scheme has been implemented outside the World Heritage core areas—and even within more restoration than regeneration is done.

As a consequence the latest step in urban regeneration schemes that is apparent in Yogyakarta and even more in Recife, is yet non-existent in Kathmandu. In fact even a trend backwards to more decay can be witnessed, induced by the unstable political situation, a lack of proper planning and related budgets. Therefore one can hardly talk about any strategies that are in place, there is rather an occurrence of local and non-coordinated projects.

Upgrading' the historic urban core takes place in two different ways: on the one hand by taking care of the historic fabric itself, by developing utilization schemes for historic buildings and by considering cultural aspects in the regeneration schemes. On the other hand a trend can be witnessed to 'upgrade' the centres by putting new constructions into the core areas. In this case the new constructions are usually huge ones, matching the local context neither in size, nor in style or the used materials. In the latter case the historic urban centre is nothing but the arena where new constructions are erected. These new buildings shall benefit from their surroundings, but are not integrated in the centre. Local examples are the vertical gated communities that are constructed in the former harbour area of Recife. Here, advertisement is done with the beautiful surrounding and the central location, but the access to the area will be a restricted one, contributing to the socio-spatial disparities.

In Yogyakarta, a huge number of hotels are under construction in or in direct vicinity to the historic centre, many of them in kampong areas, altering the quarters' outline profoundly and partly resulting in the eviction of resident population. Most of these people are working in the centre, often in informal trade, contributing to that environment that tourists—for whom the hotels are constructed for—want to visit, a contradiction in itself. Even in Kathmandu such processes emerge; over the past few years malls have become popular shopping facilities for the middle and upper-class. Most of them are constructed south of Kathmandu Durbar Square, along Kanti Path (cf. Fig. 5.7/6 in Sect. 5.1.4). In total, quite a stagnation of urban regeneration can be witnessed in Kathmandu, induced by political instability, urban growth but also a lack of adequate instruments and their enforcement.

Recife on the other hand has implemented a huge number of regeneration projects in the urban core, with a growing tendency towards public-private partnerships and large-scale upgrading for middle and upper-class housing. In Yogyakarta such upgrading trend as part of regeneration schemes can be witnessed as well, in this case mostly for hotel and tourism business. In either case the historic fabric is rather the 'scenery' than the target of the recent regeneration projects. However, in terms of current regeneration schemes the case study cities differ profoundly, the economic differences are very much visible. On the other hand they have in common that increasingly private actors are involved in regeneration schemes, particularly in Recife and Yogyakarta, in Kathmandu such economic upgrading is at least intended, though not yet in place.

In all three cities regeneration is mostly dealing with:

- Conservation of historic fabric in all three core areas,
- Infrastructure, e.g. the improvement of traffic infrastructure in Recife and Yogyakarta, and
- New constructions—mostly hotels in Yogyakarta, apartments in Recife, and residential buildings in Kathmandu.

Values ascribed to certain places and uses are mostly left aside in project planning. Intangible heritage is considered in terms of providing areas for vending handicraft or constructing museums for different cultural expressions, while living vernacular heritage is vanishing. It is surprising that it is exactly the intertwinedness of built and intangible heritage emphasized so much throughout the empirical research (as described in the next chapter), which is left aside.

Here, the outcomes of expert interviews help in the interpretation. Considering their lines of argumentation and perspectives, the interviewees can be divided into two major groups: The first group (and the larger one, due to the targeted selection of interview partners, therefore no conclusions on the representativeness of the peer group size are permitted) is very concerned about the ongoing loss of heritage due to urban and also economic development. These 'maintainers' regard both, tangible and intangible heritage as important assets for both, urban development in a locally suitable and adapted context, and urban identity. Mostly, they are well informed about global heritage discourses and named examples from other countries or cities they regard as 'best' or sometimes 'worst' practise in conservation and/or urban development.

The second group on the other hand has a much more positive attitude towards ongoing changes and economic development, coming along with urban upgrading and new constructions. To them, the 'modernizers' built heritage is important for both, economic and cultural reasons, while intangible heritage is of lesser concern, unless it is, e.g. promoted for touristic purposes and/or regional branding. They themselves regard their attitude as forward-looking and modern, while conservation is partly backward-oriented and should be done only building wise or in selected spots rather than on quarter-scale. In general, the ‘modernizers’ are more internationally oriented than ‘maintainers’ and seek to push ‘their’ city to become more global and modern.

Both groups however agree on allowing for change in historic urban centres, but disagree significantly in the quantity they would allow for. It is likely that the gaps between these two groups are bigger in emerging and developing countries compared to Euro-American ones. As long as influential actor groups still opt for a modernisation that considers urban heritage and the historic urban cores at the best in terms of its economic value (and rather for tourists as for locals), more losses are likely.

It can be concluded, that the holistic view of (urban) heritage as promoted in recent charters and documents (see Sect. 3.3) has not yet found all the way to authorities. As a recommendation, it would make sense to build upon the strong attachment to and recognition of the historic centre that the research peer group has shown. As all of them are students of courses related to sustainable development they are the potential future key actors, as mentioned in Sect. 2.6.

Interim conclusion on urban regeneration schemes

- Urban regeneration schemes are difficult to compare. Although they emerged around the same time, resulting from growing awareness on the loss of historic fabric, later development differs profoundly.
- Economic aspects are a major influence in regeneration; cultural heritage turns to become more important as scenery and setting than being the focal area of intervention itself.
- Aspects of intangible heritage are hardly tackled in regeneration schemes, particularly when it comes to lower social stratum.

### *5.4.4 Findings on Urban Identity and Place Attachment*

The questionnaires handed to a comparable group of respondents reveal some significant conformity, besides some particular features that apply to one city only. To compare the findings for feelings and values attached to the centres and processes observed there, the mean values of the questionnaire results were used and visualized, allowing for a qualitative analysis of main similarities and differences (compilation of semantic differentials as used in the chapters on the case study cities

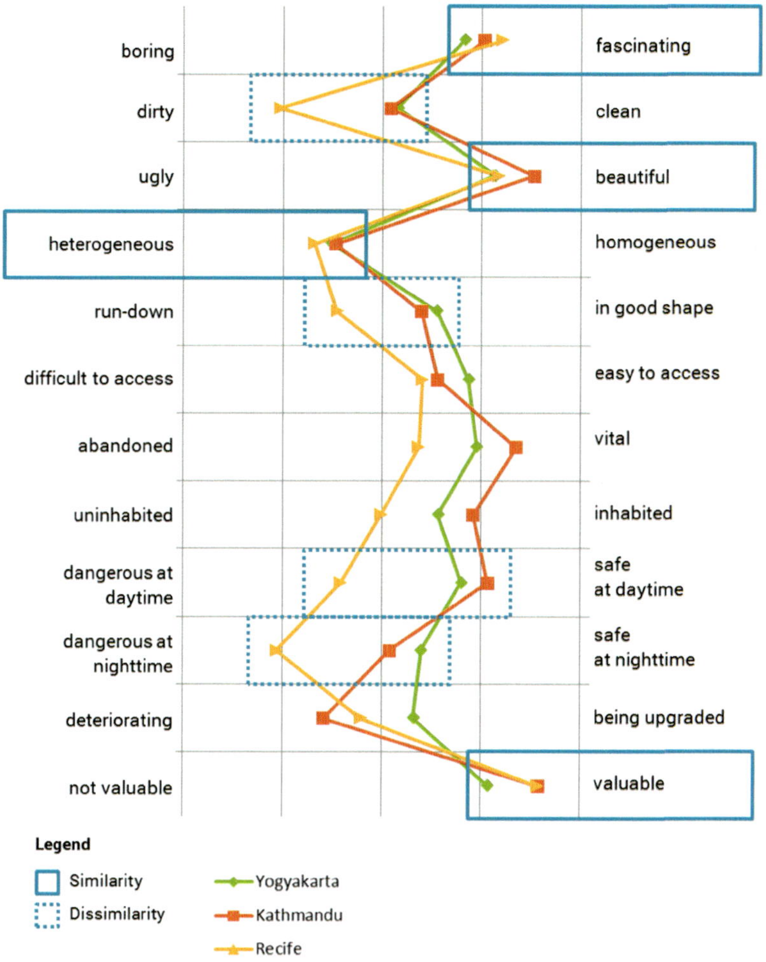


**Fig. 5.50** Comparative analysis of case studies findings on feelings towards the historic centre

and tests for correlations see Annex IV). Figure 5.50 depicts the feelings the peer groups have towards the urban centres, calculated on an average base for each city. Overall, the three peer groups find their centre quite fascinating, beautiful and valuable, indicating very positive associations.

On the other side the groups also agree in finding the centre areas rather heterogeneous, matching the results from survey and mapping that found different clusters of buildings, uses and shapes within the centre areas (see Sects. 5.1.8, 5.2.8, and 5.3.8 on urban patterns). On an average, the results for Kathmandu and Yogyakarta mostly match, with the one exception that the Yogyakartanese find their centre being upgraded, while the Kathmandu and Recife groups rather see processes of deterioration in their cities. Results for Recife are comparably less positive in some points; respondents there find their centre comparably more dirty, run-down, abandoned and dangerous than the other groups, matching problems that experts tackled during the interviews.

The varied results in the three cities can partly be interpreted by differences in cultural backgrounds or comparative differences between the centre area and other urban quarters, what can, e.g. be the reason for the comparably more negative evaluation of cleanliness in the centre itself or the shape of the buildings. Debates on safety are omnipresent in Brazil, resulting in a feeling of insecurity, but not permitting any absolute statement on a higher risk level here than in the other cities. Overall, the respondents in all three cities assessed the historic centre in a quite positive way, allowing for the conclusion that the area is of value. The points rated negatively are probably at least partly reasons for not frequenting the centre and potential points of intervention.

In the next step the perception of processes going on in the historic centre is compared (see Fig. 5.51). Overall, there are again many similarities between the cities. The highest levels of consistency were achieved in the category on tourism where large majorities in all three cities disagree that tourists are the main users of the historic centre instead of locals. Finally, all three peer groups disagree very much that their centre area is outmoded, with the clearest result for Recife. Traffic is seen as very problematic in all three cases, while the perception of gentrification processes, outmigration of previous user groups is higher in Kathmandu than in Recife and Yogyakarta. In Kathmandu the replacement of old buildings or their uses is observed more than in the other cities. Yogyakarta respondents see less preference of people to live in modern buildings than respondents in Kathmandu, while the results in Recife match the ones in Kathmandu. Consequently, Kathmandu respondents' find a high preference for international planning trends instead of following local traditions, while this is less the case in the other cities.

Taking a more detailed look, an analysis of correlations between the different processes (see Annex IV, Spearman correlations between the different process items) reveals a number of linkages which are partly the same in the three cities. In all three peer groups the analysis displayed correlations between replacement of traditional buildings and gentrification processes as well as with a tendency towards modern planning trends instead of traditional ones. This perception and the correlations match very much the statements obtained during expert interviews that also mentioned a bias towards presumably 'modern' planning trends, most pronounced in Kathmandu and Recife.

Subsequently, different values attached to the centre areas are compared in Fig. 5.52. Like in the two preceding figures, results again give a fairly coherent picture of how respondents relate to the historic centres. Matching the results obtained in the question on feelings towards the centre, all three peer groups agree

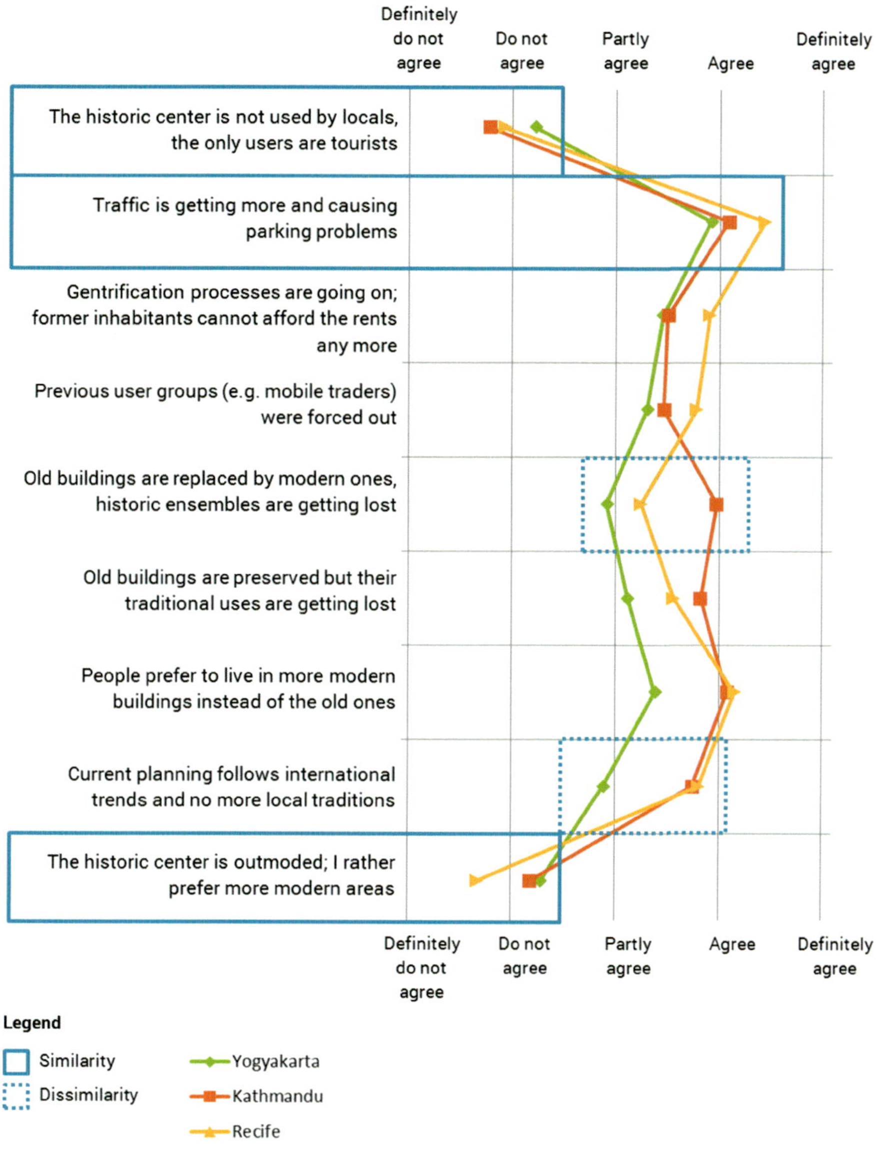


**Fig. 5.51** Comparative analysis of processes perceived in the historic centres of the case study cities

that their centre should be preserved as it is and that it has a value for them. Answers to other items disclose that in all cases the centre is not the preferred shopping or commercial area and that the recreational value is comparably less apparent than the historic one. Cultural and spiritual/religious value was rated highest in Kathmandu and lowest in Recife.

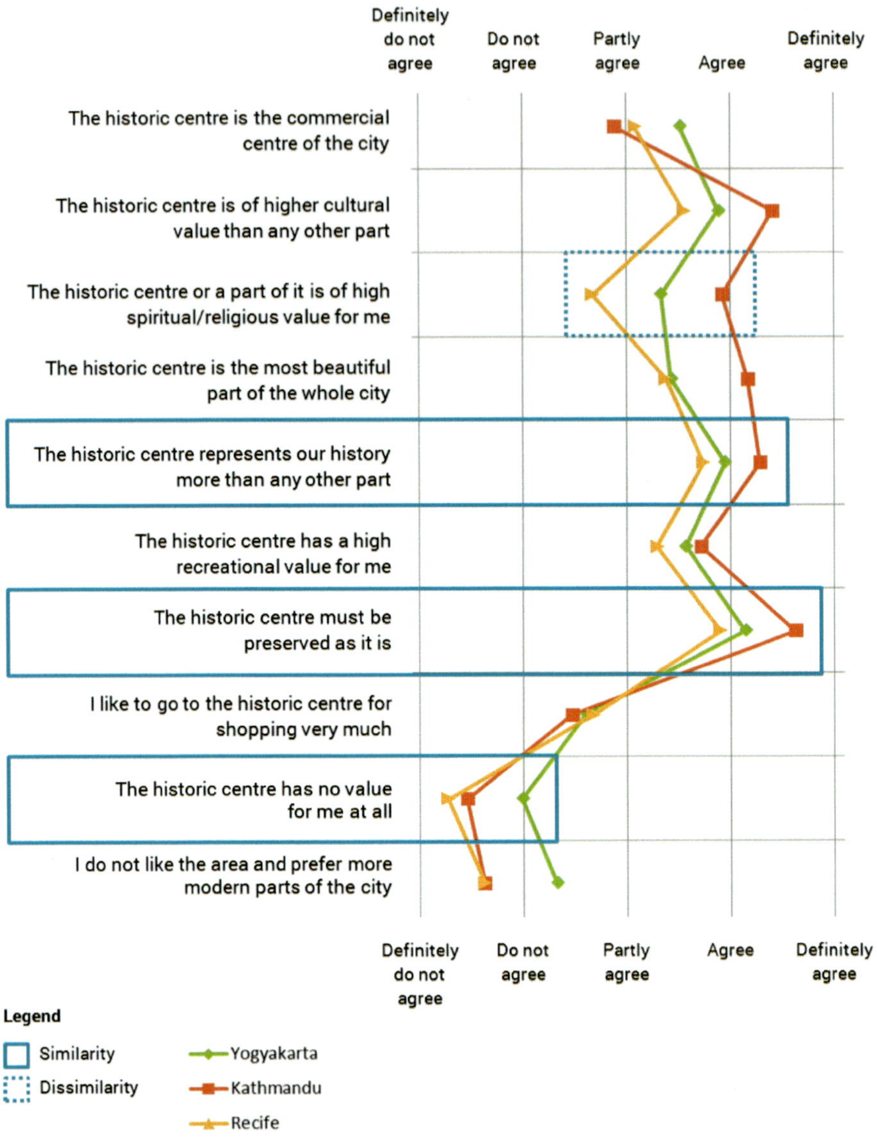


**Fig. 5.52** Comparative analysis of values attributed to the historic centres of the case study cities

Summing up the findings of all three questions (cf. Fig. 5.53), in all three cases the historic centre is found valuable, still used by locals and worth being preserved. Processes and uses partly differ but do not impact on the overall value ascribed to the area.

- In Recife the main issues are safety and cleanness, which are perceived as being quite low. The centre does not have much religious value and is partly uninhabited.

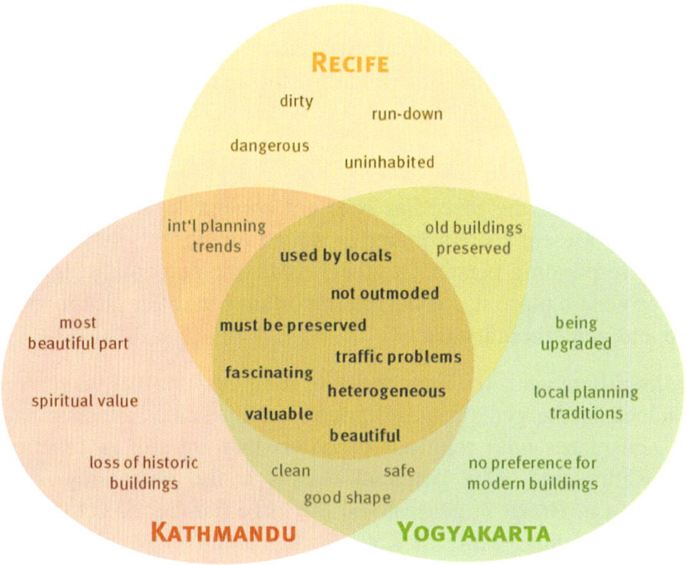


**Fig. 5.53** Summary of questionnaire findings on values and processes found in the historic centres of the case study cities

- By contrast Kathmandu's centre is ascribed a high cultural and religious value and found vital, while the problems detected are the deteriorating buildings and preferences towards replacing them with modern constructions.
- In Yogyakarta the respondents perceived an ongoing upgrading process and comparably less loss of historic buildings and/or their uses. As a consequence they found planning to still follow more local traditions than in the other cities.

Besides the items that can be localized in the centre itself, the questionnaire also investigated on intangible values and certain places of remembrance located in the centres and beyond. In either case the results revealed high appreciation. All intangible values, namely traditional handicraft, traditional craft skills, music, festivals and food are found valuable or even very valuable, in all three cities. As the cultural backgrounds are quite different, the skills themselves cannot be compared directly; however, in general the level of appreciation is very high, matching the results obtained for the historic centre.

The same applies to most of the locations asked for in the question on places of remembrance that mostly were rated as very important or important, no matter which city (cf. Figs. 5.17, 5.32 and 5.45), with comparable mean values and also standard deviation, which is least in first places, higher towards end of scale). Taking a look in more detail, there are two aspects of particular importance.

- First, besides the generally high importance ascribed to the locations asked about, there are certain gradients. Most places ranked in top positions are such ones that are accessible (with the exception of the Kraton in Yogyakarta which is partly accessible but as a museum area) and that are attractive spaces for spending time, permitting different uses, such as leisure, commerce or shopping.

- Second—and backing the first finding—the reasons why these places were found important can be grouped into two main categories: most answers were related to the historic value, with 'history' or 'historic place' mentioned most often in no matter which city. In second place the present contemporary use was mentioned (in case the place has one, not applicable in case of a religious or political function etc.), often linked to a high quality of space perceived by the individual responding person. In other words, the outcome is the same, but the reason for what a place is considered important can differ profoundly—between historic and contemporary use.

One major difference is the importance of religious/spiritual value. Matching the results obtained in the general question concerning values attached to the centre, also here such places with religious functions that are still in function are rated highest in Kathmandu. One potential reason is the decreasing importance of catholic religion in Brazil in general and the large number of colonial churches in Recife and surrounding, while in Kathmandu Hinduism is still practised comparably more. Furthermore the major Hindu and Buddhist religious sites do have religious and historic values going far beyond the urban borders. Pashupati is one of the holiest Hindu temples while Boudha and Swayambhu are two of the major pilgrimage sites for Tibetan Buddhists. The central Mosque in Yogyakarta then again is considered of medium importance, the major reason given is that 'it is just one mosque', with many more in the city. In this case the historic value is acknowledged while the religious one seems equal to other mosque sites.

To take a closer look on underlying reasons for appreciation or non-appreciation of heritage and potential differences in perceiving inner-city processes, the correlation with sex, length of stay in the case study cities and with the respondents' interest in the urban history was tested. Sex had comparably little influence, while the interest in urban history (which itself is correlated as well in case of Kathmandu and Yogyakarta, with each $p < 0.01$ in Spearman-Rho test, but not in Recife) does show more correlations, as shown in more detail in Table 17 in Annex IV. In only two cases the three cities coincided, one is a correlation between the length of stay in town and the value ascribed to the centre, the longer the time of residence the more valuable the centre is found. The second one is the correlation between the interests in urban history, which is negatively correlated with preference for more modern areas. One potential conclusion to draw from this is that arising more interest on urban history could improve the appreciation of the centre area.

The length of stay in town matters very little in Recife, in Yogyakarta it matters to some extent on the assessment of accessibility and overall value as well as on perception of traffic problems and gentrification processes, which probably are perceived more the longer someone is living in a city. Obviously it matters a lot in Kathmandu, where the perception of a comparably large number of items is influenced by the length of stay, such as cultural, recreational or historic value and the wish to preserve the centre as it is. The longer the stay, the more unambiguous is the statement. However, in the majority of cases where significant differences were found, they still differed in the intensity, but not the overall positive or negative evaluation. A potential explanation for the difference between Nepal and the other cities could be the country's multiethnicity and the Newari roots of the urban heritage. In addition, Kathmandu is the only large urban agglomeration of the country, and people coming from more rural areas may perceive the undeniable urban problems in a more negative way than those raised in Kathmandu itself. Obviously, Indonesia is a multiethnic country as well, but here results are different and do not show many correlations with the length of stay. Potential explanations are the comparably better condition of the city itself, and the fact that Yogyakarta is branded as a cultural and educational hub for the whole country.

The questionnaire also investigated on urban planning paradigms that the respective city should follow. Out of a pool of overall 14 proposed answers, selected on base of current debates on sustainable urban development, in each city the protection of cultural and natural heritage was ranked first. The protection of local values and beliefs was ranked comparably lower, highest in Yogyakarta (first third), in top second third in Kathmandu and in lower second third in Recife. Overall rankings for Yogyakarta and Kathmandu differed not very much, while the ranking in Recife in lower first and in second third is quite different, as shown in Fig. 5.54. The paradigms ranked among the last three are again matching and the same in all cities (it would go beyond the scope of this research to go more into detail at this point; however it is astonishing that, e.g. climate change adaptation is found unimportant). Overall, answers given here match some of the problems detected when analyzing preceding questions, like the major traffic problems that were found most severe in Kathmandu, accordingly here the wish to have a better public transportation system is ranked highest. Safety problems were rated worst in Recife, accordingly a safe city is ranked comparably higher than in the other cities, closely linked to a need to create more social equity (as highlighted in the figure, coinciding the expert interviews' statements).

Interpreting the results against the background of other questionnaire and interviews' findings, the top position of heritage protection—although ranking it number one may have been influenced partly by the questionnaire itself, addressing heritage only—matches the overall high or very high value attached to the centre and different places of remembrance, as elaborated on the previous pages. What is more astonishing is the comparably lower ranking of 'protection of local values and beliefs', as specific intangible heritage items were ranked mostly important or even very important. Potentially 'local values and beliefs' is too abstract wording and not very meaningful to the respondents while the intangible items listed were more meaningful and thus easier to grasp. The conclusion therefore could be that despite the decidedly importance of certain intangible assets, 'intangible heritage' may still be a too cloudy concept, although it describes something that is part of daily life.

Being asked about unique features of their city, the experts often mentioned the urban outline itself. All three cities are characterized by a strong linkage to natural assets. Yogyakarta's urban layout is a planned one, half-way between Mount Merapi and the sea, and located along streams originating from the volcano's

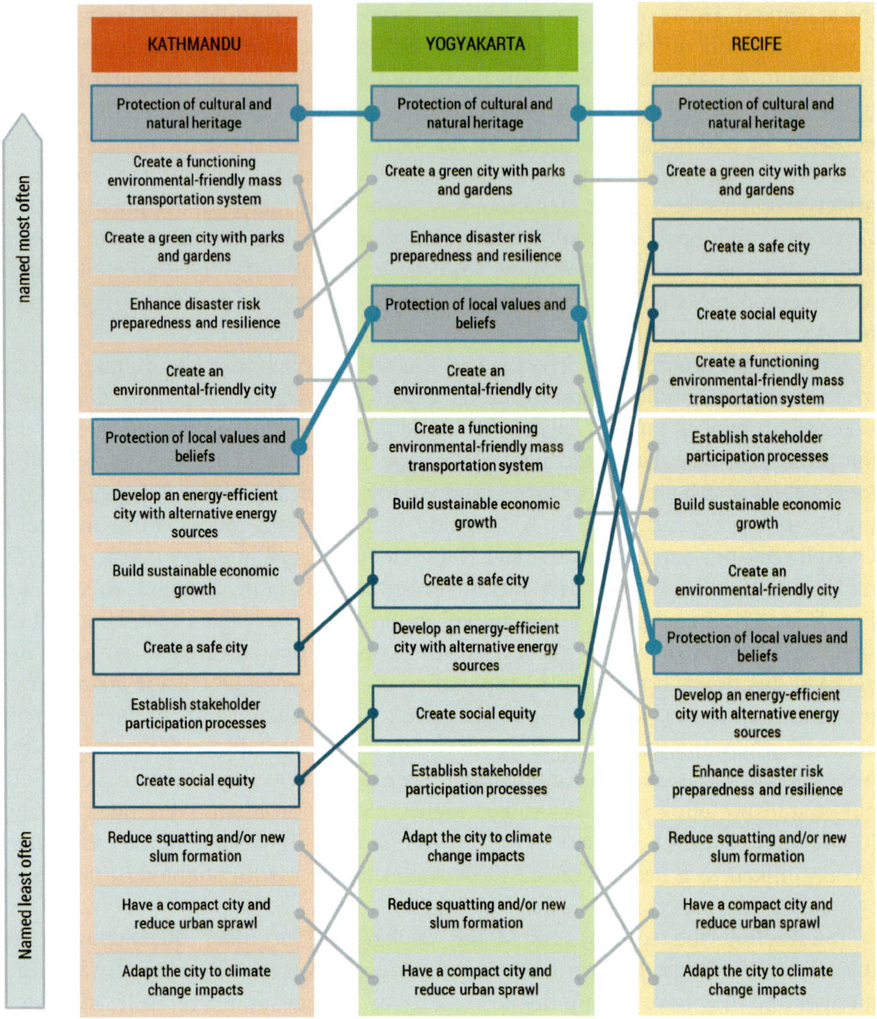


**Fig. 5.54** Ranking of urban planning paradigms their city should follow in the opinion of questionnaire respondents (14 options given, up to five answers per person)

slopes. Values and beliefs attached to these assets are still alive, like in Kathmandu, where mythological origins of Kathmandu Valley date back to an emanation of the Wisdom Buddha Manjushri, who drained the water-filled valley by cutting a notch into the surrounding hills. The lotus flower that blossomed in midst of the valley became the location of Swayambhu Stupa (Department of Archaeology 2011b). Like Swayanbhu certain spots in the surrounding hills are attached to myths and beliefs and became temple locations. In both cities religious, cultural and spiritual values are attached to natural assets and have influenced the urban layout. Even if this religious aspect is not evident in Recife, natural assets have shaped the urban layout in this town as well. Here in particular the interaction between sea, rivers, peninsulas and islands that emerged during Dutch colonial times is considered as unique feature of the city and part of the collective urban identity. In all three cases natural and cultural assets intermingle, built fabric, natural assets and also historic events or religious as well as mythological aspects together define the urban uniqueness.

Summing it up, tangible and intangible heritage does matter in each of the three cities, with astonishing parallels between them. Taking a look at the construction of urban identities within the three case study cities it becomes apparent that in all cases the importance of urban heritage goes far beyond its mere object value.

Interim conclusion on urban identity

- The attachment to the historic urban core is strong. Neither gender nor the length of stay in the respective town does correlate with the appreciation of the historic centre.
- In all three cities the centre is regarded beautiful, fascinating and valuable, and should be preserved as it is.
- The historic centre is of importance, the same applies to distinct buildings and places of remembrance. It is important to consider underlying reasons, as they range from historic value to contemporary (leisure) activities.
- The different forms of expression of intangible heritage are considered of utmost importance and should be preserved as they are part of urban identities. However, and despite the very high appreciation of distinct intangible assets, the overall awareness and recognition of intangible values seems comparably fuzzier than it is in the case of tangible heritage and may result in potential losses.
- There is a strong consensus that actual planning does rather follow international paradigms instead of local traditions which are in danger of getting lost.

### *5.4.5 Conclusions on Case Studies*

While the level of appreciation of tangible as well as intangible heritage in the centre and beyond is very high among questionnaire respondents and experts, current urban development with its rapid change is putting at least part of that heritage at risk. Although suitable planning instruments and policies do exist, they seem rather fuzzy, with time-consuming processes of coordination among the levels and institutions. The attachment to the historic centre is there, the challenge is to cope with the social difficulties, societal change, economic circumstances (development or stagnation) and rapid urban change.

Questions of identity surely are political, decisions on what is protected and what not are based upon value systems.

The Yogyakarta heritage was somehow constructed during the New Order phase to contribute to the establishment of a genuine Indonesian and Javanese identity. The city and particularly its traditions like Batik and the political system with the Sultan court that combated the Dutch rulers made the city and particularly the Kraton a role model and emblematic place for Indonesians.

- Kathmandu and its Durbar Squares are a (purposeful) manifestation of the shared national history that is uniting the different tribes and societal groups.
- In Recife particularly the remnants of the Dutch colonial phase are a glorified symbol of the urban past. This part of the urban history is clearly preferred over the Portuguese colonial times—while in Yogyakarta Dutch architecture is partly regarded as a symbol of the former foreign rulers. Interestingly, efforts to construct a national identity based on 'Brazilian' cultural heritage started in the 1970s, with the preservation and appreciation of post-independency eclecticism architecture as a way to negate the Portuguese colonial times.

However, urban heritage and the historic urban cores are not only political but still something alive and lived, even for comparably young people such as the students asked in this research. The appreciation of the historic centre is high, while at the same time these centres are undergoing rapid changes, altering their outline fundamentally. Legal protection does exist, but is not always suiting the local culture, particularly when it comes to intangible values. Too often, heritage legislation is rather caught up in antiquated concepts which may even range back to colonial times. The coexistence of intangible values rooted in physical spaces in form of built heritage is hardly depicted in the legal systems and even less in project realities—although present in people's perception.

## References

Adams B (2002) Preservação Urbana: Gestão e Resgate de uma História. O patrimônio de Florianópolis, Editora da UFSC, Florianópolis

Adishakti LT (2008) Community Empowerment Program on the revitalization of Kotagede Heritage District, Indonesia Post Earthquake. In: Kidokoro TO, Matsumura S, Shima N (ed) Vulnerable Cities: realities, innovations and strategies, vol 8. cSUR-UT Series: Library for Sustainable Urban Regeneration. Springer, pp 241–256

Adishakti LT, Hadiwinoto S (eds) (2010) Pendidikan Pusaka Indonesia. Panduan Untuk Guru Sekolah Dasar di Daerah Istimewa Yogyakarta, Erfgoed Nederland, BPPI, UGM, UNESCO, Jakarta

Adishakti LT, Tarekat H, Rinandi N, Hadiwinoto S, Zaidulfar EA (eds) (2011) Guidelines for managing post-disaster conservation of heritage buildings. BPPI/Badan Pelestarian Pusaka Indonesia, Jakarta

Agsten K, Hoppe HD (1995) Stadtteilerhaltung in Bandung/Indonesien. In: Grohmann P, Hannemann V, Hoffmann D (eds) Andere Städte, Anderes Leben - Stadtentwicklung, Umweltkrise und Selbsthilfe in Afrika, Asien und Lateinamerika. Verlag für Entwicklungspolitik, Saarbrücken

Ahmad AG (2006) Cultural Heritage of Southeast Asia: preservation for world recognition. Malays Townplan J 4(1):52–62

Aigner M, Leite S (2006) Beteiligungshaushalte für informelle Siedlungen in Brasilien. Das Beispiel Recife. Standort - Zeitschrift für Angewandte. Geographie 30(1):21–27. doi:10.1007/s00548-006-0300-6

Albert M-T (2013) Heritage studies—paradigmatic reflections. In: Albert M-T, Bernecker R, Rudolff B (eds) Understanding heritage: perspectives in heritage studies. de Gruyter, Berlin/Boston, pp 9–17

Altvater E, Mahnkopf B (2003) Die Informalisierung des urbanen Raums. Learning from – Städte von Welt, Phantasmen der Zivilgesellschaft, informelle Organisation. Neue Gesellschaft für Bildende Kunst (NGBK), Berlin

Amatya S (2008) The World Heritage Sites of Kathmandu Valley. History and personal experiences. In: Luger K, Wöhler K (eds) Welterbe und Tourismus: Schützen und Nützen aus einer Perspektive der Nachhaltigkeit. StudienVerlag, Innsbruck, pp 281–304

Anggraini LD (2012) Spatial connectivity of Indonesian Shop Houses. Asian J Environ-Behav Stud 3(10):45–55

Antunes Cavalcante M (2006) Espaços Publicos. Universidade Federal de Pernambuco, Recife, Um ensaio sobre os Bairros de Santo Antônio e São José

Aragão T, Maennig W (2013) Mega sporting events, real estate, and urban social economics—the case of Brazil 2014/2016, vol 47. Hamburg Contemporary Economic Discussions, Hamburg

Araújo L (2012) Defesa da memória da cidade parte da população. Diario de Pernambuco 05 (10):2012

Athayde C (2011) Periferia: Favela, beco, viela Agenda Brasileira. In: Botelho A, Schwarcz LM (eds) Agenda Brasileira. Temas de uma sociedade em mudança. Companhia das Letras, São Paulo, pp 400–407

AusHeritage (2015) Developing heritage trails in historic Yogyakarta and Batavia. http://www.ausheritage.org.au/projects/developing-heritage-trails-in-historic-yogyakarta-and-batavia/. Accessed 15 April 2015

Badan Pusat Statistik (2015a) Number of foreign guests in classified hotel by province, Indonesia 2003–2013. Badan Pusat Statistik. http://www.bps.go.id/linkTabelStatis/view/id/1376. Accessed 16 April 2015

Badan Pusat Statistik (2015b) Percentage of urban population by province, 2010–2035. Badan Pusat Statistik. http://www.bps.go.id/linkTabelStatis/view/id/1276. Accessed 16 April 2015

Badan Pusat Statistik (2015c) Population projection by province, 2010–2035. Badan Pusat Statistik. http://www.bps.go.id/linkTabelStatis/view/id/1274. Accessed 16 April 2015

Bähr J, Mertins G (1988) Einfachhaus-Siedlungen in Groß-Recife/ Nordostbrasilien. Zwei Fallstudien über Zielgruppen, bevölkerungs- und baustrukturelle Veränderungen. Kieler Geographische Schriften 68:193–222

Bähr J, Mertins G (1995) Die lateinamerikanische Groß-Stadt: Verstädterungsprozesse und Stadtstrukturen. Wissenschaftliche Buchgesellschaft, Darmstadt

Baker JL (2006) Integrated urban upgrading for the poor: the experience of Ribeira Azul, Brazil. Working Paper Number 3861. World Bank, New York. doi:10.1596/1813-9450-3861

Barreto Lira F, Pontual V (2006) Is conciliation of the old and new possible? Urbanistic instruments applied to the historical area of the downtown district of Bairro do Recife—Pernambuco. City Time 2(1)(3):27–46

Barros Filho MNM, Gama Monteiro CM (2011) Segmented cities with fuzzy walls: changes in informal settlements as seen through a multiscale analysis. In: Perrone C, Manella G, Tripodi L (eds) Everyday life in the segmented city. Research in urban sociology, vol 11. Emerald, Bingley, pp 143–168

Bezerra Cavalcanti C (1998) O Recife e seus bairros. Camara Municipal do Recife, Recife

Bhattarai K, Conway D (2010) Urban vulnerabilities in the Kathmandu Valley, Nepal: visualizations of human/hazard interactions. J Geograph Inf Syst 2(2):63–84. doi:10.4236/jgis.2010.22012

Bista SK (2000) Rural development in Nepal. An alternative strategy. Udaya Books, Kathmandu

Bitoun J (2006) Territoires du dialogue: les mots de la ville et les défis de la gestion participative a Recife. Géocarrefour 81(3):217–222. doi:10.4000/geocarrefour.1561

Bitoun J (2007) Programa Camaragibe em Defesa da Vida: um novo desenho para a ação habitacional em município da periferia metropolitana do Recife. In: Cardoso AL (ed) Habitação social nas Metrópoles brasileiras: Uma avaliação das políticas habitacionais em Belém, Belo Horizonte, Porto Alegre, Recife, Rio de Janeiro e São Paulo no final do século XX. Coleção Habitare ANTAC, Porto Alegre, pp 436–455

Böhme H, Körte A, Tokya-Seid M (eds) (2003) WOHNEN - BAUEN - PLANEN. Erneuerung historischer Kernstädte in Südostasien und Europa im historisch-architektonischen Vergleich, Wissenschaftliche Buchgesellschaft, Darmstadt

Borba FdB (1994) Pernambuco: Viagem á Estética do Tempo (Manual do Patrimônio Cultural de Pernambuco). FUNDARPE, Recife

Bowen JR (1986) On the political construction of tradition: Gotong Royong in Indonesia. J Asian Stud 45(03):545–561. doi:10.2307/2056530

Brillantes ABJ, Flores HP (2012) Decentralization and good urban governance in Southeast Asia: focus on the Philippines, Indonesia, and Thailand. In: Yap KS, Thuzzar M (eds) Urbanisation in Southeast Asia: issues and impacts. Institute of Southeast Asian Studies, Singapore, pp 293–319

British Red Cross, Nepal Red Cross Society, Groupe Urgence - Réhabilitation - Développement (2014) Urban preparedness—lessons from the Kathmandu Valley. British Red Cross, London

Brown D, Dodman D (2014) Understanding children's risk and agency in urban areas and their implications for childcentred urban disaster risk reduction in Asia: insights from Dhaka, Kathmandu, Manila and Jakarta. Asian Cities Climate Resilience Working Paper Series 6. IIED, London

Bulkeley H, Schroeder H, Janda K, Zhao J, Armstrong A, Chu SY, Ghosh S (2011) The role of institutions, governance, and urban planning for mitigation and adaptation. In: Hoornweg D, Freire M, Lee MJ, Bhada-Tata P, Yuen B (eds) Cities and climate change: responding to an urgent agenda. The International Bank for Reconstruction and Development/The World Bank, Washington DC, pp 125–160. doi:10.1596/978-0-8213-8493-0

Buril B (2013a) Equipamentos culturais da Fundarpe estão entre os menores investimentos. Jornal do Commercio, 14.10.2013

Buril B (2013b) FUNDARPE: reflexos do tempo. Diario de Pernambuco, 16.10.2013

Buril B (2013c) Preservação do patrimônio material inclui tombamento. Jornal do Commercio, 14.10.2013

Campos R (2002) Stadtplanung und Verwaltungsdezentralisierung. Eine Analyse der brasilianischen Realität am Beispiel des PREZEIS-Programms der Stadt Recife/PE. dissertation.de Verlag im Internet

Cassoli C, Falcão LA, Aguiar R (2007) Frevo: 100 anos de folia. Timbro, São Paulo

Chapagain NK (2008) Heritage conservation in Nepal: policies, stakeholders and challenges. Paper presented at the third annual Himalayan policy research conference, Madison, WI (USA)

Colombijn F, Barwegen M (2009) Racial segregation in the (post) Colonial City: the case of Indonesia. Urban Geogr 30(8):838–856. doi:10.2747/0272-3638.30.8.838

Córdula Filho R, Corrêa de Araújo B (2002) Caminhos de Pedra: Calçadas do Bairro do Recife, 2nd edn. Prefeitura, Departamento de Preservação dos Sitios Históricos/Ed. Tanahlot, Recife

Coy M (2006) Gated communities and urban fragmentation in Latin America: the Brazilian experience. GeoJournal 66:121–132

Coy M (2007) Innenstadtentwicklung und Innenstadterneuerung in São Paulo – Akteure, Wahrnehmungen, Interessenkonflikte. In: Rothfuß E, Gamerith W (eds) Stadtwelten in den Americas., vol Passauer Schriften zur Geographie 23. Passavia Universitätsverlag, Passau, pp 57–69

Coy M (2010) Stadtentwicklung und Stadtpolitik. Sozioökonomische Fragmentierung und Beispiele zukunftsorientierter Planung. In: Costa S, Kohlhepp G, Nitschack H, Sangmeister H (eds) Brasilien heute: Geographischer Raum - Politik - Wirtschaft - Kultur. Vervuert, Frankfurt am Main, pp 51–73

Coy M, Pöhler M (2002) Gated communities in Latin American megacities: case studies in Brazil and Argentina. Environ Planning B 29:355–370. doi:10.1068/b2772

Cunha JMPd, Stoco S, Dota EM, Negreiros R, Miranda ZAId (2013) A mobilidade pendular na Macrometrópole Paulista: diferenciação e complementaridade socioespacial. Cadernos Metrópole 30:433–459

Da Mata D, Deichmann U, Henderson JV, Lall SV, Wang HG (2005) Examining the growth patterns of Brazilian Cities. World Bank Policy Research Working Paper 3724. World Bank, New York. ISSN1813-9450, Working Paper Number 3724. doi:10.1596/1813-9450-3724

Dahles H (2001) Tourism, heritage and national culture in Java: Dilemmas of a local community. Curzon, Richmond, Surrey

Dangol P (2007) Elements of Nepalese temple architecture, vol New. Delhi, Adroit

de Albuquerque Lapa T, Almeida de Melo RF (2007) Interventions in historic areas, mobility and urban conservation: the case study of Bairro do Recife District. City Time 2(3–4):37–46

de Almeida Souza MA, Bitoun J (eds) (2015) Recife: transformações na ordem urbana. Território, coesão social e governança democrática. Letra Capital, Rio de Janeiro, Metrópoles

de Andrade MO, Alves Maia ML (2009) The Recife Metro—the impact on urban development after 20 years. Flux 1(75):57–68

de Andrade Pereira JN (2009) Renovar Preservando: os imóveis especiais de preservaçao do Recife. Universidade Federal de Pernambuco, Recife

de Figueirôa Silva A (2010) Jardins do Recife. Uma Historia do Paisagismo no Brasil (1872 – 1937). Companhia Editoria de Pernambuco CEPE, Recife

de Melo Cadena, Filho DR (2012) Para além da fisionomia - identifição da paisagem cultural do centro do Recife Universidade Federal de Pernambuco. UFPE, Recife

de Souza FAM (2001a) The future of informal settlements: lessons in the legalization of disputed urban land in Recife, Brazil. Geoforum 32:483–492

De Souza FAM (2001b) Perceived security of land tenure in Recife, Brazil. Habitat Int 25(2):175–190. doi:10.1016/S0197-3975(00)00023-0

Deffner V (2006) Lebenswelt eines innerstädtischen Marginalviertels in Salvador da Bahia (Brasilien): Umgang mit sozialer und räumlicher Exklusion aus Sicht der armen Bevölkerungsgruppen. Geographica Helvetica 61(1):21–31

Department of Archaeology (2011a) Ancient Monument Preservation Act 2013. Department of Archaeology. http://doa.gov.np/content.php?id=250. Accessed 22 Jan 2015

Department of Archaeology (2011b) Swayambhu. Nepal Government, Ministry of Culture, Tourism and Civil Aviation, Department of Archaeology. http://www.doa.gov.np/content.php?id=266. Accessed 09 Sept 2015

Diario de Pernambuco (2015) Justiça determina reconstrução do Edifício Caiçara. Diario de Pernambuco. http://www.diariodepernambuco.com.br/app/noticia/vida-urbana/2015/07/03/interna_vidaurbana,584659/justica-determina-reconstrucao-do-edificio-caicara.shtml. Accessed 18 Aug 2015

Dixit A, Chettri MR, Dixit KM, Wenju R, Aryal M, Dongol U, Rai DR, Thapa K, Devkota M, Subedi Y (2014) Building urban resilience: assessing urban and peri-urban agriculture in Kathmandu. Nepal, United Nations Environment Programme (UNEP), Nairobi

do Eirado Amorim LM, Barros Filho MNM, Cruz D (2014) Urban texture and space configuration: an essay on integrating socio-spatial analytical techniques. Cities 39:58–67

Donner W (1971) Ein Entwicklungsplan für das Kathmandu-Tal. Nepal bekennt sich zur Regionalplanung. Zeitschrift für Wirtschaftsgeographie 15:22–26

dos Santos OAA (2013) Planejamento Urbano e Problemas Ambientais: Redirecionamentos téorico-metodológicos ao Enfrenamento da Crise Ambiental Recifense. Geoambiente On-line 21:68–84

Dowall DE (2006) Brazil's urban land and housing markets: how well are they working? Working Paper, Institute of Urban and Regional Development, No.2006,08. UC Berkeley, Berkeley
Duarte M (2011) Para além dos mapas…a conservação da paisagem como espírito do lugar. Universidade Federal de Pernambuco, Recife
Earthquakes and Megacities Initiative (2010) Risk-sensitive land use plan—final report. Mainstreaming disaster risk reduction in megacities: a pilot application in Metro Manila and Kathmandu. EMI, Manila
Ellingsen W (2010) Ethnic appropriation of the city. The Territoriality of Culture in Kathmandu, LAP Lambert, Saarbrücken
Ellisa E (2010) Sustainability of spatial context of historical core areas in Indonesian cities. In: Hernández Ibañez S, Tiezzi E (eds) The Sustainable City VI: Urban Regeneration and Sustainability. WIT Press, Southhampton, pp 471-482
Evers H-D (2007) The end of urban involution and the cultural construction of urbanism in Indonesia. Internationales Asienforum 38(1–2):51–65
Evers H-D, Korff R (2003) Southeast Asian Urbanism. Lit, Hamburg
Fernandes E (2006) Updating the declaration of the rights of citizens in Latin America: constructing the "Right to the City" in Brazil. In: UNESCO (ed) International public Debates: urban policies and the right to the city. UNESCO, Paris, pp 40–53
Fernández-Maldonado AM, Romein A, Verkoren O, Pessoa RPP (2014) Polycentric structures in Latin American Metropolitan areas: identifying employment Sub-centres. Reg Stud 48 (12):1954–1971. doi:10.1080/00343404.2013.786827
Firman T (2010) Multi local-government under Indonesia's decentralization reform: the case of Kartamantul (The Greater Yogyakarta). Habitat Int 34:400–405
Ford LR (1993) A model of Indonesian city structure. Geogr Rev 83:374–396
Freitas Cavalcanti JF (1997) Modernização do Estado. Conselho Superior de Orientação Politica e Social de FIESP/CIESP - COPS, Recife
Fundarpe (2015) Patrimônio Cultural - O que é. http://www.cultura.pe.gov.br/pagina/patrimonio-cultural/material/o-que-e/. Accessed 15 Sept 2015
Furtado F, Alcântara E, Bezerra O (2014) Is urban sustainability possible in the face of accelerated property development and major public works? In: Edgerton E, Romice O, Thwaites K (eds) Bridging the boundaries: human experience in the natural and built environment and implications for research, policy, and practice, vol 5. Advances in people-environment studies. Hogrefe, Boston, pp 245–260
Galla A (ed) (2002) Protection of cultural heritage in Southeast Asia. In: Workshop Proceedings, Hanoi, Vietnam, 9–13 April 2001. ICOM
Garden M-CE (2009) The heritagescape. Looking at heritage sites. In: Stig Sørensen ML, Carman J (eds) Heritage studies. Methods and approaches. Routledge, London, New York, pp 270–291
Gautam KC (2015) Lost in transition: rebuilding Nepal from the Maoist Mayhem and Mega Earthquake. Nepa-laya, Kathmandu
Gellner DN, Quigley D (eds) (1999) Contested hierarchies: a collaborative ethnography of caste among the Newars of the Kathmandu Valley, Nepal (Oxford studies in social and cultural anthropology). Claredon Press, Oxford
German Federal Foreign Affairs Office and EMI (2010) Risk-sensitive land use plan: Kathmandu Metropolitan City, Nepal. German Federal Foreign Affairs Office and EMI, Quezon City
GFDRR (2012) Disaster risk management in South Asia: a regional overview. The World Bank, Washington, DC
GoN NTNC (2009) Bagmati action plan (2009–2014). Government of Nepal & National Trust for Nature Conservation, Kathmandu
Government of Nepal (2007) The Interim Constitution of Nepal, 2063 (2007). vol Section 3.17.3 Civil Right Provision. Kathmandu
Government of Nepal, Ministry of Finance, International Economic Cooperation Coordination Division (2013) Foreign Aid in Nepal (FY 2011–12). Government of Nepal, Ministry of

Finance, International Economic Cooperation Coordination Division, Singhadurbar, Kathmandu

Governor General (1931) Monuments Ordinance. Statute 1931 No. 238. http://www.unesco.org/culture/natlaws/media/pdf/indonesie/indonesia_decree_13_06_1931_engl_orof.pdf

Gutschow N (1982) Stadtraum und Ritual der newarischen Städte im Kathmandu-Tal. Eine architekturanthropologische Untersuchung, Kohlhammer, Stuttgart

Gutschow N, Kreutzmann H (2012) Handlung schlägt Plan: Stadtentwicklung im Kathmandu-Tal (Nepal). Geographische Rundschau 64/4(April 2012):42–49

Hadi S (2008) Improving vulnerable urban space in postdisaster in Yogyakarta and Central Java, Indonesia: participatory and comprehensive approach. In: Kidokoro TO, Matsumura S, Shima N (ed) Vulnerable cities: realities, innovations and strategies, vol 8. cSUR-UT Series: Library for Sustainable Urban Regeneration. Springer, pp 225–240

Hampton MP (2003) Entry points for local tourism in developing countries: evidence from Yogyakarta, Indonesia. Geografiska Annaler: Series B, Human Geography 85(2):85–101

Harsono D (2011) To reign for the people: exercising the 'Democratic Monarchy' in Yogyakarta. Institute of Social Studies, The Hague, Netherlands

Heineberg H (2006) Stadtgeographie, 3rd edn. UTB, Paderborn

Helbig K (1949a) Die Südostasiatische Inselwelt. Franckh'sche Verlagshandlung, Stuttgart

Helbig K (1949b) Indonesien. Eine auslandskundliche Übersicht der malaiischen Inselwelt, Franz Mittelbach, Stuttgart

Hengky SH (2015) Beneficial images: Batik handicraft tourism in Yogyakarta, Indonesia. Bus Econ Res 11–23. doi:10.5296/ber.v5i1.6760

Herold V (2013) Funcultura seleciona 288 projetos. Jornal do Commercio 14(09):2013

Herzog LA (2013) Barra da Tijuca: The Political Economy of a Global Suburb in Rio de Janeiro, Brazil. Latin Amer Perspect 40(2):118–134. doi:10.1177/0094582X12467765

Historical Stone Spouts and Source Conservation Association (HSSCA) (2007) National Convention on Stone Spouts. Kathmandu

Hizbaron DR, Coy M, Baiquni M, Sartohadi J, Rijanta S (2013) Is vulnerability 'a mere matter of socio-physical problem' in urban areas? A case study in Bantul District—Indonesia. In: Sartohadi J, Pokharel JR, Nehren U, Fernandez G, Sandholz S (eds) Joint scientific program and one-day seminar on: ecosystem based disaster risk reduction. Faculty of Geography - Universitas Gadjah Mada, Yogyakarta

Höflich de Duque S (2006) „Javanische Geschichten" führten zum Erfolg — ein Beispiel von Entwicklungszusammenarbeit in Yogyakarta. Indonesien. Gruppendynamik und Organisationsberatung 37(1):101–107

Hollé A (1998) La cour des Néwar de Katmandou. The Newar courtyards in Kathmandu. Espace géographique 27 (4):349–358

Hook W (2005) Pedestrianization in Yogyakarta: transforming the Malioboro one step at a time. Sustain Transp 17:16–19

Hutt M, Gellner DN, Michaels A, Rana G, Tandan G (1994) Nepal: a guide to the art and architecture of the Kathmandu Valley. Kiscadale, Gartmore, Stirling

Indonesia UNDP (2006) Data dan Informasi Bencana Indonesia. UNDP Indonesia, Jakarta

Indonesian Network for Heritage Conservation and International Council on Monuments and Sites (2003) Indonesian charter for heritage conservation. ICOMOS Indonesia supported by Ministry of Culture and Tourism, Indonesia

International Federation of Red Cross and Red Crescent Societies (2011) Analysis of legislation related to disaster risk reduction in Nepal. IFRC, Geneva

IPHAN (2014) Livros do Tombo. Instituto do Patrimônio Histórico e Artístico Nacional. http://portal.iphan.gov.br/pagina/detalhes/608. Accessed 15 Sept 2015

Jha PK, Shrestha KK (2013) Climate change and urban water supply: adaptive capacity of local government in Kathmandu City. Nepal Journal of Forest and Livelihood 11(1):62–81

Jigyasu R (2014) Fostering resilience: towards reducing disaster risks to World Heritage. World Heritage 74:4–13

Jogja Walking Association (2015) 7th Jogja international heritage walk. http://jogjaheritagewalk.com/. Accessed 15 April 2015

Jones S, Oven KJ, Manyena B, Aryal K (2014) Governance struggles and policy processes in disaster risk reduction: a case study from Nepal. Geoforum 57:78–90

Joshi J (1997) Planning for sustainable development. Urban Management in Nepal and South Asia, Lajmina Joshi, Kathmandu

Kamarid A, Leupert D (2009) Stadterneuerung am Praça XV de Novembro - Entwicklungskonzept für einen Stadtplatz in Rio de Janeiro. VDM Verlag Dr, Müller, Saarbrücken

Kathmandu UNESCO (2004) Conserve!. Kathmandu Valley World Heritage Site. Potential Areas for Cooperation, UNESCO, Kathmandu Office, Lalitpur

Kathmandu Metropolitan City Office (2015) Kathmandu Metropolitan City Office. http://www.kathmandu.gov.np/. Accessed 23 Sept 2015

Kirsch-Soriano da Silva K (2010) Wohnen im Wandel. Mutationen städtischer Siedlungsstrukturen in Recife/Brasilien, Lit, Wien

Klink JJ (2009) Novas governanças para as áreas metropolitanas. O panorama internacional e as perspectivas para o caso brasileiro. Cadernos Metrópole 22:415–433

Korn W (1977) The traditional architecture of the Kathmandu Valley Bibliotheca Himalayica. Ratna Pustak Bhandar, Kathmandu

Koster M (2012) Mediating and getting 'burnt' in the gap: politics and brokerage in a Recife slum, Brazil. Critique Anthropol 32(4):479–497. doi:10.1177/0308275X12456643

Koster M, Nuijten M (2012) From preamble to post-project frustrations: the shaping of a slum upgrading project in Recife, Brazil. Antipode 1:175–196. doi:0.1111/j.1467-8330.2011.00894.x

Kötter H, Roeder ROG, Junghans KH (eds) (1979) Indonesien. Ländermonographien, Band 11. Horst Erdmann Verlag für Internationalen Kulturaustausch, Tübingen und Basel

Kraas F, Nitschke U (2008) Megaurbanisierung in Asien: Entwicklungsprozesse und Konsequenzen stadträumlicher Reorganisation. Informationen zur Raumentwicklung Themenheft „Raum- und Stadtentwicklung in Asien“ (8-2008):447–456

Kwanda T (2010) Tradition of conservation: redefining authenticity in Javanese architectural conservation, 22–26 June 2010. Paper presented at the Heritage 2010—international conference on heritage and sustainable development, Evora, Portugal

Lacarrieu M (2013) Kulturerbe, Materialität und die Frage nach dem Städtischen. In: Huffschmid A, Wildner K (eds) Stadtforschung aus Lateinamerika: Neue urbane Szenarien: Öffentlichkeit - Territorialität - Imaginarios. Transcript, Bielefeld, pp 419–439

Lacerda N, Ribeiro S (2014) Limites da Gestão Metropolitana e Impasses à Governança Cooperada Intermunicipal no Brasil. Revista EURE 40(121):185–202

Lademacher H (2001) Batavia. In: Gründer H, Johanek P (eds) Kolonialstädte - Europäische Enklaven oder Schmelztiegel der Kulturen? Lit, Münster, pp 95–108

Lama AK, Job H (2014) Protected areas and road development: sustainable development discourses in the Annapurna Conservation Area, Nepal. Die Erde 68(4):229–250. doi:10.3112/erdkunde.2014.04.01

Lara FL (2011) New (sub)urbanism and old inequalities in Brazilian Gated Communities. J Urban Design 16(03):369–380. doi:10.1080/13574809.2011.571160

Larasati D (2007) Towards an integral approach of sustainable housing in Indonesia with an analysis of current practices in Java Technische Universiteit Delft, PhD Thesis

Leal FM (1977) Restauraçao e Conservaçao de Monumentos Brasileiros. Série Patrimônio Cultural, Publicação no 1. Centro de Artes e Comunicação UFPE, Recife

Lefèvre C (2009) Governar as metrópoles: questões, desafios e limitações para a constituição de novos territórios políticos. Cadernos Metrópole 22:299–317

Lichtenberger E (1990) Stadtverfall und Stadterneuerung. VÖAW, Wien

Liechty M (2010) Out here in Kathmandu: modernity on the global periphery. Martin Chautari Press, Kathmandu

Loretto RP (2008) Paraíso & Martírios: histórias de destruição de artefatos urbanos e arquitetônicos no, Recife edn. Universitária da UFPE, Recife

Lubambo CW (1991) Bairro do Recife: entre o Corpo Santo e o Marco Zero. CEPE/Fundação de Cultura Cidade do Recife, Recife

Luger K (2008) Welterbe-Tourismus. Ökonomie, Ökologie und Kultur in weltgesellschaftlicher Verantwortung. In: Luger K, Wöhler K (eds) Welterbe und Tourismus: Schützen und Nützen aus einer Perspektive der Nachhaltigkeit. StudienVerlag, Innsbruck, pp 17–42

Maharatta PS (2012) Community based earthquake vulnerability reduction in traditional settlements of Kathmandu Valley. PhD Thesis, Department of Architecture and Urban Planning, Pulchowk Campus, Tribhuvan University, Kathmandu

Maharjan M (2012) Conflict in world heritage sites of Kathmandu Valley: a case study on the conservation of private houses in three Durbar Squares. Nepal Tour Dev Rev 2(1):87–104

Maia Alvez PR (2009) Valores do Recife. O valor do solo na evolução da cidade. Luci Artes Gráficas Ltda., Recife

Marcillia SR, Ohno R (2012) Importance of social space in self-built and donated post disaster housing after Java Earthquake 2006. Asian J Environ-Behav Stud 3(7):25–34

Maricato E (2011) Metrópoles desgovernadas. Estudos Avançados 25(71):7–22

Marques E, Rodrigues L (2013) O Programa Minha Casa Minha Vida na metrópole paulistana: atendimento habitacional e padrões de segregação. Revista Brasileira de Estudos Urbanos e Regionais 15(2):159–177

MC Consulting, Empresa de Urbanização do Recife (URB) (2006) Brazil—municipal lending program—Recife—project: environmental assessment (vol 1 of 3): Environmental impact assessment—executive summary

Mercer J, Gaillard JC, Crowley K, Shannon R, Alexander B, Day S, Becker J (2012) Culture and disaster risk reduction: lessons and opportunities. Environ Hazards 11(2):74–95. doi:10.1080/17477891.2011.609876

Milet V (1988) A Teimosia das Pedras. Um estudo sobre a preservação do patrimônio ambiental no Brasil, Prefeitura de Olinda, Olinda

Ministry of Culture and Tourism (2003) Compilation of Law and Regulation of the Republic of Indonesia concerning Items of Cultural Property. Republic of Indonesia, Jakarta, Ministry of Culture and Tourism

Minnery J, Argo T, Winarso H, Hau D, Veneracion CC, Forbes D, Childs I (2013) Slum upgrading and urban governance: case studies in three South East Asian cities. Habitat Int 39:162–169. doi:10.1016/j.habitatint.2012.12.002

Miranda L, Moraes D (2007) O Plano de Regularização das Zonas Especiais de Interesse Social (Prezeis) do Recife: democratização da gestão e planejamento participativo. In: Cardoso AL (ed) Habitação social nas Metrópoles brasileiras: Uma avaliação das políticas habitacionais em Belém, Belo Horizonte, Porto Alegre, Recife, Rio de Janeiro e São Paulo no final do século XX. Coleção Habitare ANTAC, Porto Alegre, pp 414–435

ICIMOD, MoEST, UNEP (2007) Kathmandu Valley Environment Outlook. International Centre for Integrated Mountain Development (ICIMOD), Kathmandu

Moura Filha MB (2000) O Cenário da Vida Urbana: a definição de um projeto estético para as cidades brasileiras na virada do século XIX/XX Centro de Tecnologia/Editoria Universitária João Pessoa, João Pessoa

Müller-Böker U (1988) Spatial organization of a caste society: the example of the Newar in the Kathmandu Valley, Nepal. Mountain Res Dev 8(1):23–31. doi:10.2307/3673403

Muzzini E, Aparicio G (2013) Urban growth and spatial transition in Nepal: an initial assessment. World Bank, Washington, DC, Directions in Development. doi:10.1596/978-0-8213-9659-9

Nance E, Ortolano L (2007) Community participation in urban sanitation: experiences in Northeastern Brazil. J Planning Educ Res 26(3):284–300. doi:10.1177/0739456X06295028

Nascimento SR (2005) Mercado de São José: história e cultura popular. Fundação de Cultura Cidade do Recife, Recife

Nehren U, Subedi J, Yanakieva I, Sandholz S, Pokharel J, Chandra Lal A, Pradhan-Salike I, Marfai MA, Sri Hadmoko D, Straub G (2013) Community perception on climate change and climate-related disaster preparedness in Kathmandu Valley, Nepal. J Nat Res Dev 04:35–57

Nitisudarmo S (2009) The role of the informal sector in contributing to the urban landscape in Yogyakarta—Indonesia concerning on the urban heat island issue REAL CORP 2009: CITIES 3.0—smart, sustainable, integrative strategies, concepts and technologies for planning the urban future

Nobre EAC (2002) Urban regeneration experiences in Brazil: historical preservation, tourism development and gentrification in Salvador da Bahia. Urban Design Int 109–124. doi:10.1057/palgrave.udi.9000066

Novy A, Leubolt B (2005) Participatory budgeting in Porto Alegre: social innovation and the dialectical relationship of state and civil society. Urban Stud 42(11):2023–2036. doi:10.1080/00420980500279828

Nyaupane GP, Timothy DJ, Poudel S (2015) Understanding tourists in religious destinations: a social distance perspective. Tour Manag 48:343–353. doi:10.1016/j.tourman.2014.12.009

Obermayr C (2013) Sustainable city management in Surakarta (Indonesia). Dealing with informal settlements an example for 'Good Governance'?, Leopold-Franzens Universität Innsbruck, Innsbruck

Parajuli YK (1986) Bhaktapur Development Project. Experiences in Preservation and Restoration in a Medieval Town (1974–1985). Bhaktapur Development Board (BDP) and Deutsche Gesellschaft für Technische Zusammenarbeit (GTZ), Kathmandu

Peres Torelly LP, Soares A, Menescal M, Bonelli A, Souza C, Martins F (2014) Cultural heritage in the Country of the World Cup. IPHAN, Brasília-DF

Phelps NA, Bunnell T, Miller MA, Taylor J (2014) Urban inter-referencing within and beyond a decentralized Indonesia. Cities 39:37–49

Pontual V (2007) Prácticas urbanísticas em áreas históricas: o bairro de Recife. Revista Bibliográfica de Geografía y Ciencias Sociales, vol XII, nº 752. Universidad de Barcelona, Barcelona

Prabawa TS (2010) The tourism industry under crisis: the struggle of small tourism enterprises in Yogyakarta (Indonesia). Vrije Universiteit Amsterdam, Amsterdam

Prefeitura da Cidade do Recife (1981) Preservação de Sitios Históricos. Prefeitura da Cidade do Recife, Recife

Prefeitura da Cidade do Recife (1986a) Projeto Recife: Seminário de Redirecionamento. URB, Recife

Prefeitura da Cidade do Recife (1986b) Proposta Habitacional para Programa Mutirão da Moradia na Cidade do Recife. URB, Recife

Prefeitura da Cidade do Recife (1987a) Programa Melhorias Urbanas. URB, Recife

Prefeitura da Cidade do Recife (1987b) Projeto Recife: Seminário de Redirecionamento. URB, Recife

Prefeitura da Cidade do Recife (2012) Perfil dos Bairros. Prefeitura da Cidade do Recife. http://www2.recife.pe.gov.br/servico/bairro-do-recife?op=NTI4Mg. Accessed 20 Aug 2015

Prefeitura da Cidade do Recife (2015) Reabilitação do Centro Expandido. Prefeitura da Cidade do Recife - Secretaria de Planejamento Participativo, Obras e Desenvolvimento Urbano e Ambiental. http://www.recife.pe.gov.br/pr/secplanejamento/reabilitacao.php. Accessed 30 March 2015

Proença Leite R (2006) Patrimônio e enobrecimento no Bairro do Recife. Revista CPC 2:17–30. doi:10.1590/S1980-44662006000200003+

Proença Leite R (2007) Contra-usos da cidade: lugares e espaço público na experiência urbana contemporânea. vol 2. ed., ver. e ampl. edn. Ed. da UNICAMP/ Ed. UFS, Campinas/Aracaju

Proença Leite R (2013) Consuming heritage: counter-uses of the city and gentrification. Vibrant Virt Braz Anthropol 10(1):165–189. doi:10.1590/S1809-43412013000100009

Proença Leite R, Peixoto P (2009) Políticas urbanas de patrimonialização e contrarrevanchismo: o Recife Antigo e a Zona Histórica da Cidade do Porto. cadernos metrópole 21 93-104

Puppim de Oliveira JA, Doll CNH, Suwa A (2013) Urban development with climate co-benefits: aligning climate, environmental and other development goals in cities. UNU-IAS Policy Report. United Nations University Institute of Advanced Studies (UNU-IAS), Yokohama

Rana S, Marwasta D (2015) Urbanisation trends in developing countries: comparative study of Yogyakarta City and Kathmandu Valley. J Nat Res Dev 5:29–36. doi:10.5027/jnrd.v5i0.04
Ranjitkar RK (2007) Kathmandu Valley World Heritage Site, Nepal. Heritage Homeowner's Preservation Handbook. Heritage Homeowner's Preservation Manuals for UNESCO World Heritage Sites series. UNESCO, Bangkok/Kathmandu
Rojas E (2002) Urban Heritage Conservation in Latin America and the Carribean. A task for all social actors. Sustainable Development Department Technical Papers Series:25
Röll W (1979) Indonesien. Entwicklungsprobleme einer tropischen Inselwelt. Länderprofile, Klett, Stuttgart
Rothfuß E (2007a) Der Risikodiskurs über das historische Zentrum von Salvador (Brasilien). Eine diskursanalytische Annäherung an den Prozessen der Innenstadtsanierung anhand der Presse zwischen 1967 und 2005. Kieler Geographische Schriften 117:233–255
Rothfuß E (2007b) Tourismus-Gentrification im Pelourinho. Urbane Deformation des historischen Stadtzentrums von Salvador da Bahia (Brasilien). Passauer Schriften zur Geographie 23:41–56
Roy A (2009) The 21st-century metropolis: new geographies of theory. Reg Stud 43(6):819–830. doi:10.1080/00343400701809665
Rubens de Menezes T, Figueira de Souza J (2014) Transportation and urban mobility in mega-events: the case of Recife. Proc Soc Behav Sci 162:218–227. doi:10.1016/j.sbspro.2014.12.202
Rukmana D, Purbadi D (2013) Street vending in Indonesian cities: Their characteristics and activities in Yogyakarta. In: Perera N, Tang W-S (eds) Transforming Asian Cities. Intellectual impasse, Asianizing space and emerging translocalities. Routledge, Milton Park, pp 123–136
Rutz W (1985) Die Städte Indonesiens: Städte und andere nicht-landwirtschaftliche Siedlungen, ihre Entwicklung und gegenwärtige Stellung in Verwaltung und Wirtschaft. Urbanisierung der Erde, vol 4. Borntraeger, Berlin
Sabino do Nascimento A (2014) Projeto Novo Recife: identidade, espaço e gentrificação revista eletrônica e-metropolis:38–46
Salazar NB (2008) "Enough stories!" Asian tourism redefining the roles of Asian tour guides. In: Doquet A, Evrard O (eds) Tourisme, Mobilités et Alterités Contemporaines, vol Civilisations, vol.LVII, n°1–2. Institut de Sociologie, Université libre de Bruxelles, Bruxelles pp 207–222
Salazar NB (2010) The glocalisation of heritage through tourism: balancing standardisation and differentiation. In: Labadi S, Long C (eds) Heritage and globalisation. Routledge, New York, pp 130–147
Salazar NB (2011a) Imagineering cultural heritage for local-to-global audiences. In: Halbertsma M, van Stipriaan A, van Ulzen P (eds) The heritage theatre: globalisation and cultural heritage. Cambridge Scholars, Newcastle upon Tyne, pp 49–72
Salazar NB (2011b) Managing the local-to-global dynamics of world heritage interpretation. In: Bourdeau L, Chassé S (eds) Conference proceedings: world heritage and tourism: managing the global and the local, June 2–4, 2010, Quebec City, Canada. Presse de l'Université Laval, Québec, Canada, pp 1044–1055
Salazar NB (2012) Shifting values and meanings of heritage from cultural appropriation to tourism interpretation and back. In: Lyon S, Wells EC (eds) Global tourism. Cultural heritage and economic encounters, vol 30, Society for Economic Anthropology Monographs. Altamira, Lanham, pp 21–41
Sandholz S (2006) Preservation of urban heritage: potential for urban development. In: Kraas F, Gaese H, Mi Mi Kyi (eds) Megacity Yangon: transformation processes and modern developments, vol 7. Southeast Asian modernities. Lit, Berlin, pp 109–121
Sangachhe SB (2008) Development of building code experience of Nepal In: Subedi JK, Nakamura H (eds) From code to practice challenges for building code implementation and the further direction of housing earthquake safety. Records and outcomes of International Symposium 2008 on Earthquake Safe Housing 28–29 November 2008. UNCRD, Tokyo
Scazzosi L (2011) Limits to transformation in places' identity: theoretical and methodological questions. In: Roca Z, Claval P, Agnew J (eds) Landscapes, identities and development. Ashgate, Surrey, pp 9–24

Sé Carneiro AR (2010) Parque e paisagem: Um olhar sobre o Recife. Ed. Universitária da UFPE Recife
Seixas J (2014) A Cidade na Encruzilhada. Repensar a Cidade e a sua Política, Edições Afrontamento, Porto
Sengupta U, Bhattarai Upadhyaya V (2016) Lost in transition? Emerging forms of residential architecture in Kathmandu. Cities 52:94–102. doi:10.1016/j.cities.2015.11.007
Serra MV, Dowall DE, Motta D, Donovan M (2004) Urban land markets and urban land development: an examination of three Brazilian Cities: Brasília, Curitiba and Recife. IURD Working Paper Series. Institute of Urban and Regional Development, University of California at Berkeley, Berkeley
Shiraishi T (ed) (1990) Reading Southeast Asia. Southeast Asia Program Publications, Ithaca
Shrestha BK (2011) Street typology in Kathmandu and street transformation. Urbani izziv 22 (2):107–121. doi:10.5379/urbani-izziv-en-2011-22-02-004
Shrestha BK (2013a) Implementing the proposed outer ring road in Kathmandu Valley. J Manage Dev Stud 25(1):23–38
Shrestha BK (2013b) Residential neighbourhoods in Kathmandu: key design guidelines. Urbani izziv 24(1):125–143. doi:10.5379/urbani-izziv-en-2013-24-01-003
Siauw GT (2003) Yogyakarta. In: Böhme H, Körte A, Tokya-Seid M (eds) WOHNEN - BAUEN - PLANEN. Erneuerung historischer Kernstädte in Südostasien und Europa im historisch-architektonischen Vergleich, Wissenschaftliche Buchgesellschaft, Darmstadt, pp 131–160
Slusser MS (1982) Nepal Mandala: a cultural study of the Kathmandu Valley (2 vols). Princeton University Press, Princeton
Sörensson E (2008) Making a living in the world of tourism: livelihoods in backpacker tourism in urban Indonesia. Umeå University, Umeå, Sweden
Sosiawan RR (2009) The management of Sultan Land (SG) in Yogyakarta for the accomplishment of legal certainty: a case study in Sleman Regency. Graduate School Press Team, Gadjah Mada University, Yogyakarta
Struck E (2008) Historische, koloniale Stadtstrukturen und moderner Wandel in Lateinamerika. Am Beispiel der Altstädte von Salvador da Bahia (Brasilien) und Havanna (Kuba). Passauer Kontaktstudium Erdkunde 9:65–78
Subanu LP (2008) Governing urban development in dualistic societies: a case study of the urban region of Yogyakarta, Indonesia. In: Kidokoro T, Subanu LP, Jessen J, Motte A, Seltzer EP (ed) Sustainable city regions: space, place and governance, vol 7. cSUR-UT Series: Library for Sustainable Urban Regeneration. Springer, pp 41–60
Sugiana K (2008) Problems and prospects for sustainable regeneration of traditional markets in Yogyakarta City. In: Kidokoro T, Subanu LP, Jessen J, Motte A, Seltzer EP (ed) Sustainable city regions: space, place and governance, vol 7. cSUR-UT Series: Library for Sustainable Urban Regeneration. Springer, pp 61–76
Sullivan J (1986) Kampung and state: the role of government in the development of urban community in Yogyakarta. Indonesia, Southeast Asia Program Publications at Cornell University 41:63–88
Sunaryo RG, Soewarno N, Ikaputra, Setiawan B (2011) The transformation of urban public space in Yogyakarta. A search for specific identity and character. Paper presented at the international symposium insular diversity: architecture—culture—identity in Indonesia, Vienna University of Technology, Vienna, Austria
Surendra KC, Khanal SK, Shrestha P, Lamsal B (2011) Current status of renewable energy in Nepal: opportunities and challenges. Renew Sustain Energy Rev 15:4107–4117. doi:10.1016/j.rser.2011.07.022
Suryanti E, Sumartono Hermawan (2014) Culture development planning in the special region of Yogyakarta (Management planning of cultural heritage in Kotagede District based on community empowerment conservation model). J Indonesian Tour Dev Stud 2(3):120–128
Tavare R (2014) Heritage Buildings Compensation in Jogja City. Jogja Today. http://jogjatoday.com/2014/11/25/heritage-buildings-compensation-in-jogja-city/. Accessed 15 April 2015

Teixeira FW (2012) Intervenções urbanas e a construção de uma imagem de cidade moderna no Recife de princípios do século XX. Revista Memória em Rede 2(7):1–18

Thapa RB, Murayama Y (2012) Scenario based urban growth allocation in Kathmandu Valley. Nepal. Landscape and Urban Planning 105:140–148

The Asia Foundation (2012) A guide to government in Nepal. Structures, Functions, and Practices. Himal Books, Kathmandu

The Consultative Group on Indonesia (2006) Preliminary damage and loss assessment. Yogyakarta and Central Java Natural Disaster BAPPENAS, the Provincial and Local Governments of D.I. Yogyakarta, the Provincial and Local Governments of Central Java, and International Partners, Jakarta

Timothy DJ (1999) Participatory planning: a view of tourism in Indonesia. Ann Tour Res 26 (2):371–391

Titz A (2012) Naturgefahren als Entwicklungshemmnis. Das Beispiel Nepal-Himalaya. Praxis Geographie 42(9):34–38

Tiwari SR (2015) Community participation in heritage affairs. In: Weise K (ed) Revisiting Kathmandu. Safeguarding Living Urban Heritage. International Symposium Kathmandu Valley 25–29 November 2013. UNESCO, UNESCO Office in Kathmandu, Kathmandu, pp 189–196

Tokya-Seid M (2003) Einleitung. In: Böhme H, Körte A, Tokya-Seid M (eds) WOHNEN - BAUEN - PLANEN. Erneuerung historischer Kernstädte in Südostasien und Europa im historisch-architektonischen Vergleich, Wissenschaftliche Buchgesellschaft, Darmstadt, pp 9–15

Truffi R (2014) A batalha pelo Cais José Estelita. Carta Capital, vol 2014. Recife

Tunas D, Peresthu A (2010) The self-help housing in Indonesia: the only option for the poor? Habitat International 34:315–322

UNDP (2014) Human Development Index and its components. United Nations Development Programme. https://data.undp.org/dataset/Table-1-Human-Development-Index-and-its-components/myer-egms? Accessed 15 Jan 2015

UNESCO (2012) Understanding World Heritage in Asia and the Pacific: the second cycle of periodic reporting 2010–2012. World Heritage Papers 35. United Nations Educational, Scientific and Cultural Organization, Paris

UNESCO (2014d) Lists of intangible cultural heritage and Register of best safeguarding practices. UNESCO. http://www.unesco.org/culture/ich/index.php?lg=en&%20pg=00011#tabs. Accessed 20 July 2014

UNESCO (2015b) Kathmandu Valley. UNESCO. http://whc.unesco.org/en/list/121. Accessed 20 Jan 2015

UNESCO (2015e) Tentative Lists. http://whc.unesco.org/en/tentativelists/state=id. Accessed 17 Sept 2015

UNESCO (2015g) UNESCO Database of National Cultural Heritage Laws. http://portal.unesco.org/culture/en/ev.php-URL_ID=33928&URL_DO=DO_TOPIC&URL_SECTION=201.html. Accessed 10 April 2015

UNESCO Office in Kathmandu (2015) Exhibition of art and research 'remembering the lost sculptures of Kathmandu'. UNESCO Office in Kathmandu. http://www.unesco.org/new/en/kathmandu/about-this-office/single-view/news/exhibition_of_art_and_research_remembering_the_lost_sculptures_of_kathmandu/#.VSTTz5MhEhR. Accessed 08 April 2015

UN-HABITAT (2008) Water Movements in Patan with Reference to Traditional Stone Spouts. UN-HABITAT Water for Asian Cities Programme Nepal, Kathmandu

UN-HABITAT (2010a) Nepal. Urban Housing Sector Profile. UN-HABITAT, Nairobi

UN-HABITAT (2010b) The State of Asian Cities 2010/11. UN-HABITAT, UN-HABITAT Foundation, UN-ESCAP, Fukuoka

United Nations (2014) World urbanisation prospects—highlights. United Nations, New York

United Nations/UNDP/UNESCO (2013) Creative economy report 2013, Special edn. Widening local development pathways, UNDP, New York

Van Oers R (2014) The Dutch urban heritage of Recife, Pernambuco, Brazil. J Architect Conserv 7 (1):56–71

Vasconcelos de Medeiros CR (2001) Expressões e Sentidos da Revitalização do Bairro do Recife: Limpeza. Universidade Federal de Pernambuco, Recife, Beleza e Política de Vida

Vickers A (2013) A history of modern Indonesia, 2nd edn. Cambridge University Press, New York

Vieira NM (2007) Gestão de sítios históricos: a transformação dos valores culturais e econômicos em programas de revitalização em áreas históricas. Editora Universitária UFPE, Recife

Villaça F (1998) Espaço intra-urbano no Brasil. Studio Nobel, São Paulo

von Droste B (2012) Empowerment of the convention. From theory to practical application (1981–2000). World Heritage, vol No. 64. UNESCO, Paris

Weise K (2012) Kathmandu valley world heritage property: conservation and community. Involving communities in world heritage conservation—concepts and actions in Asia. International conference in celebration of the 40th anniversary of the world heritage convention. UNESCO, Buyeo, Korea

Wolfram-Seifert U (1986) Die Entwicklung des Städtesystems in Indonesien: vergleichende Analyse der Kotamadya nach der Rang-Größen-Verteilung. Beiträge zur Stadtgeographie 76:33–101

World Bank (2011) Where there is a will there is a way: cash for work in early recovery post Merapi eruption. The Worldbank, Washington D.C.

World Bank (2014a) Data—Indonesia. World Bank. http://data.worldbank.org/country/indonesia. Accessed 08 Sept 2015

World Bank (2014b) Doing business 2015: going beyond efficiency. World Bank, Washington, DC. doi:10.1596/978-1-4648-0351-2

Yap KS (2012) The challenges of promoting productive, inclusive and sustainable urbanisation. In: Yap KS, Thuzzar M (eds) Urbanisation in Southeast Asia: issues and impacts. Institute of Southeast Asian Studies, Singapore, pp 10–77

Yunus HS (1991) The evolving urban planning: the case of the city of Yogyakarta. Indonesian J Geogr 21(61):1–14

Zahnd M (2005) Traditionelle Stadtquartiere in Semarang und Yogyakarta. Universität Stuttgart, Stuttgart, Indonesien. Möglichkeiten der Revitalisierung und innovativen Nutzung des historischen Städtebaus

Zancheti SM (2005) Development versus urban conservation in Recife a problem of governance and public management. City Time 1(3)(2):15–27

Zancheti SM, Gabriel J (2011) Salvador de Bahia, Brazil. In: Rojas E, Lanzafame F (eds) City development: experiences in the preservation of ten world heritage sites. Inter-American Development Bank, Washington, D.C., pp 87–132

# Chapter 6
# Discussion and Conclusions

**Abstract** Global heritage discourses indicate a growing recognition of heritage outside the Euro-American region over the past decades, emphasizing the importance of intangible heritage and its interlinkage with tangible assets. However, urban realities in the Global South are still somewhat different. Here often the understanding of heritage is still determined by a legislation rooted in outdated understandings, moreover not matching traditional and community-based belief systems which themselves have rather shaped today's heritage discourses. Manifold processes relate to urban heritage and the urban centre, making it a very difficult arena with different actors and particular interests, often dominated by economic interests while sacrificing those of marginalized or less powerful groups. Paradoxically it is exactly those groups that have contributed to constructing the urban tangible and intangible heritage. Attachment to the historic centre, its places, sites, uses and traditions itself as well as to urban intangible values nevertheless is still high among the urban population. To maintain historic urban cores is recommendable to see the built fabric, open spaces, underlying planning and meaning, as well as uses, functions and values associated to distinct sites as parts of a whole.

**Keywords** Urban future · Intangible heritage · Historic urban landscape · Heritage paradigms · Kathmandu · Recife · Yogyakarta

There are manifold processes related to urban heritage and the urban centre. As a consequence the approach used in this research is a quite broad one, with different scales and case studies. It is dealing not only with three cities, but with different aspects within each of them, namely

- Parts of the 'historic urban landscape' that compose the historic centre,
- Impacts that are influencing on its development,
- Urban regeneration over the past decades,
- Urban and heritage policies, and
- The perception of the historic centre and its tangible and intangible heritage.



S. Sandholz, *Urban Centres in Asia and Latin America*,
The Urban Book Series, DOI 10.1007/978-3-319-43735-4_6

All of them are key components to gain a deeper understanding on what is changing historic urban centres in the Global South and how much it matters to the inhabitants. Obviously a comparative case study of only three cities hardly allows for any universalizing conclusions on a global scale in terms of specific instructions or the content of legal documents. Nonetheless, it reveals parallels as well as differences that allow for interpretation and conclusions against global heritage discourses, as will be elaborated in the following chapter.

## 6.1 Suitability of Global Heritage Reference Frames on Local Scales

The case studies revealed various relations and linkages between urban cores, different actors and heritage paradigms on different scales, summarized in Fig. 6.1.

Global heritage discourses have been shaped by Euro-American understandings since the nineteenth century, as described in Sect. 3.3. Such understandings have trickled down to national and local scales worldwide, e.g. by introducing and perpetuating legal systems and heritage categories from colonial times or by following Western role models (cf. Sect. 3.9). Only since the 1990s the international heritage movement has shifted from a narrow understanding of 'cultural property' towards heritage as a 'public good' comprising of a *"representative value of the totality of creative expressions, practices and spaces that human communities recognize as part of their tradition and identity"* (Francioni 2012: 72). This normative development (Francioni 2012) was significantly induced by non-Euro-American actors which have tried to overcome Euro-American fabric-centred understandings by adding 'change' aspects (Falser 2010; Alberts and Hazen 2010; Schmitt 2009; Veldpaus et al. 2013) and the *"need to judge cultural heritage within its cultural context"* (von Droste 2012: 19, cf. Sect. 3.2 of Chap. 3).

While the global debate on heritage clearly indicates the growing recognition of heritage outside the Euro-American region over the past the decades the reality in these countries is somewhat different—and it is where things appear rather bizarre. The changing global heritage paradigms are nothing more than the manifestation of something that has always been there, namely different interpretations of what is heritage or 'of value' in different cultures and over time. Since the 1990s the exclusiveness of one way of interpretation only, rooted in Western understandings, has reversed on the global scale. But when taking a look on local scales, as done in this research, the following contradiction becomes apparent:

- In all three countries and likewise the case study cities tangible and intangible values are closely intertwined in the historic urban centres, rooted in local cultural contexts.
- Heritage legislation does not (yet fully) depict these changes, as it partly goes back to meanwhile outdated or even colonial understandings of what should be preserved and what not. Furthermore legislation on natural, cultural and

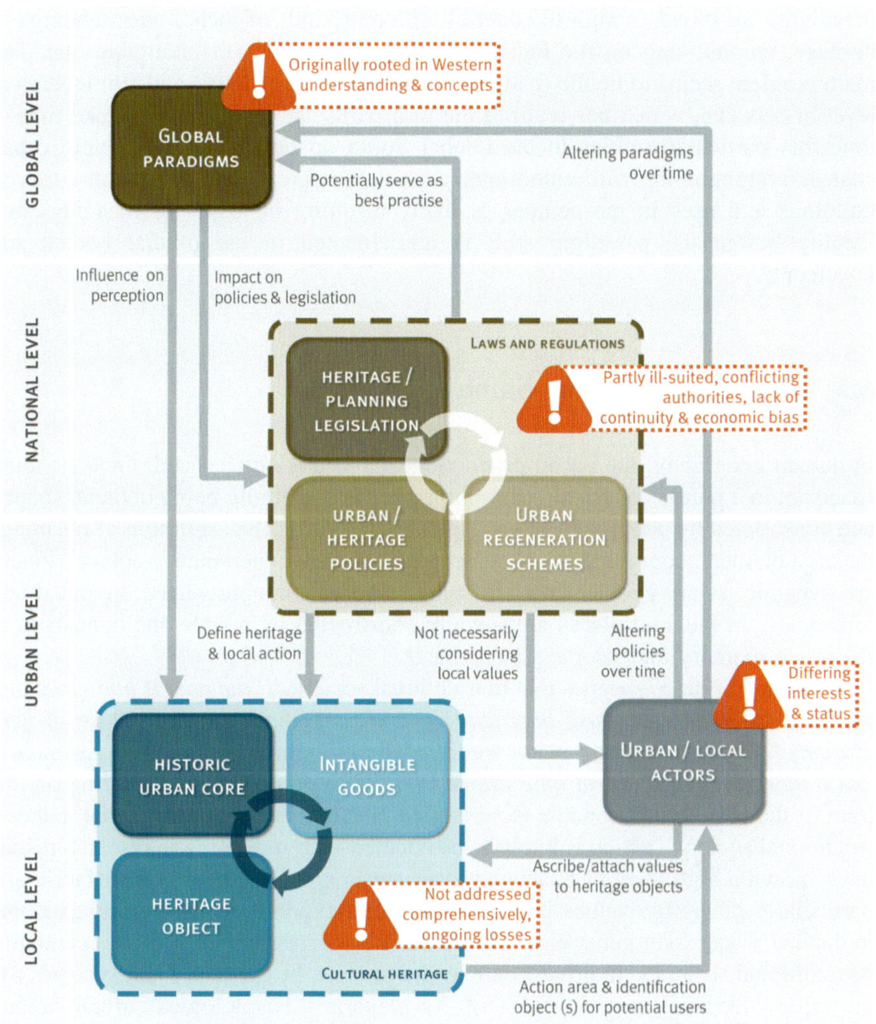


**Fig. 6.1** Linkages between spatial levels, paradigms, policies and urban heritage

intangible heritage—if there is any for intangible heritage—is often barely linked to each other and even under different authorities

- Contemporary urban regeneration projects are very often executed in popular PPP schemes. In this context the historic centre and associated intangible values are rather a benefit for new housing and tourism functions than a focal point.

Apparently, the phase where global concepts of heritage are competing with local understandings (Amen et al. 2011; Schmitt 2009, 2011) has been overcome in global discourses and paradigms within the past few years. Current discourses and paradigms are broad enough to cover all different kinds of local understandings of heritage, emphasizing on the linkages between tangible with intangible one. The main problem seems to be the (national and/or urban) legislative and administrative level in between, which has realized the turn partly, at best. This will take time—time that particularly cities in the Global South do not have. Here, rapid urban change, coming along with uninhibited economic interests and the ongoing loss of buildings and uses in the centres, is likely resulting in losses of meanings and identity before the paradigm shift is implemented in the official bodies and documents.

## 6.2 Heritage Is not the Same as Heritage

In human geography, the social production of space is emphasized, understanding space not in a purely 'physical' sense but rather as something being defined, shaped and constructed by people (Lefebvre 1991; Paasi 2000). The ascription of meanings makes a defined geographic location created by human experiences a 'place'. Places are dynamic, as are people; they are appreciated or non-appreciated, invented and reinvented. In the, end places are socially constructed by people and contribute to the urban memory and people's attachment.

Giddens (1990: 37) argues that in traditional societies, "*the past is honoured and symbols are valued because they contain and perpetuate the experience of generations.*" To him, tradition is not wholly static but reinvented by each generation, "*as it takes over its cultural inheritance from those preceding it.*" Building on this train of thought it is reasonable to scrutinize his approach by asking what makes a 'traditional society', which is likely to be equated with rurality. The research results have shown a high level of attachment to historic urban centres, to their fabric and particularly intangible values in terms of cultural expressions rooted and executed in distinct places over generations. From this perspective, urban societies can (still) be traditional societies. In urban contexts the past can be honoured and symbols can be valued likewise, e.g. in form of Yogyakarta's cosmological urban layout, Kathmandu's still worshipped temple areas or Recife's bridges as a symbol of the Dutch era. Urban traditions are continued and localized, like Kathmandu's processional ways in the occasion of certain festivals, Recife's carnival parade routes, or the importance of Yogyakarta's alun-aluns, although the uses have changed from more ritual and military to leisure ones.

Uses, fabric and the ascription of values can change over time. In parallel, the same building can be assessed positively in one place and negatively in the other. For instance buildings from Dutch colonial times are interpreted in a much more positive way in Recife than in Yogyakarta, because of what they signify in collective urban memory. Material and function hardly differ; it is the symbolic value that counts. In this context it is interesting to consider Pierre Nora's *lieux de mémoire* (cf. Sect. 2.1) which are made of three components: "*material, symbolic and functional.*" (Nora 1989: 18f). This understanding somewhat contradicts the traditional (Western) understanding of 'heritage' as a physical object or place with the duty to preserve its fabric.

To Norberg-Schulz (2013: 279) *"the structure of a place is not a fixed, eternal state. As a rule places change, sometimes rapidly. This does not mean, however, that the genius loci necessarily changes or gets lost."* He talks about the need to balance both, stability and change: *"To protect and to conserve the genius loci in fact means to concretize its essence in ever new historic contexts."* The genius loci of a place are composed of different aspects and their interplay, namely natural conditions, built form and symbolic meanings. Jive'n and Larkham (2003: 78) *"argue that it is the people—individuals and society—that integrate these features, through their value systems, to form a sense of place."*

Michel Rautenberg (1998, translated after Tweed and Sutherland 2007) suggests that cultural heritage can be treated either as heritage by designation, or heritage by appropriation. If heritage by designation is considered as such heritage declared by authorities, and heritage by appropriation are objects or spaces that are heritage because of being appreciated and declared by people, this research results suggest that both do exist in the historic urban cores of Kathmandu, Recife and Yogyakarta. Nevertheless they are not necessarily congruent, due to fundamental differences listed in Table 6.1.

One key reason for such noncongruency is the different cognition, with heritage by designation being something definable and static, and heritage by appropriation being dynamic and allowing for a broad ascription of values, depending on individual or social group perception.

**Table 6.1** Classification of heritage in historic urban cores, based on findings in Kathmandu, Yogyakarta and Recife (headings based on Rautenberg 1998)

| | Historic urban landscape by designation | Historic urban landscape by appropriation |
|---|---|---|
| Content | Physical places, buildings, sites, monuments, urban quarters, or intangible heritage such as artistic expressions | Physical places with distinct uses, e.g. open spaces built for certain functions that are still actively practised |
| Understanding | *'either-or'*—despite changing paradigms still restricted to one heritage category, either physical (cultural/natural) or intangible | *'and'*—amalgamation of physical and social spaces, intertwined physical (cultural/natural) and intangible objects |
| Genesis | Officially declared by authorities, by justifying its 'objective' value | Unofficially appreciated, used and/or practised by people, 'subjective' |
| Change aspect | Static, usually under legal protection or other regulation | Dynamic, continuously adapted to societal change |
| Obstacle | Likely to freeze a built object or artistic expression in time, as a consequence intangible heritage may lose its anchoring in society, built heritage could be musealized | Comparably more difficult to grasp, due to its subjective character there may be different meanings to different groups, can change or vanish over time |

Following Rautenberg's (1998) approach, historic urban landscapes (in other words the historic core areas) are both, heritage by designation and appropriation. They comprise of goods declared heritage by law, or as he says, by designation, and of heritage by appropriation, making it their own by means of using it. Having the case study results in mind, it is not an overstatement to say that a historic urban core needs both to sustain both. Historic urban centres are both, physical places and construed spaces, profane and sacral at the same time (in the meaning of Wöhler 2008), which make a *lieux de mémoire* (Nora 1989) when coinciding—a 'spatialization of local memories' in global socities (Werlen and Lippuner 2007; Werlen 2014).

## 6.3 Global Paradigms Against Local Realities

Section 3.2 has elaborated on the concept of 'authenticity', spearheaded by UNESCO in the global concept of World Heritage. 'Authenticity' was introduced in debates on cultural heritage to decide on a site's or building's fabric heritage value. However, it was mostly restricted to the pure fabric which has to be authentic, another word for original. As said earlier (cf. Sects. 3.2 and 3.9), this understanding for a long time excluded non-Western notions of heritage to which a perpetuated maintenance or change is inextricably linked to the understanding of heritage.

Such a broadened understanding of what is 'authentic' does neither ignore the importance of considering present understandings or uses of a certain heritage object nor does it bear the risk of protecting something 'fallen out of time'. This is particularly important for inhabited sites like historic cities or city centres. Initiatives like the 'historic urban landscape' approach tend to take account of the need for changing heritage paradigms (cf. Sects. 3.8 and 3.9).

By critically examining the outcomes of the three case studies presented in this book, one ends up wondering if the 'historic urban landscape' is nothing but the rediscovery of something that is still alive in people's daily life but rather lost in conservation and planning realities. Roy (2009: 820) argues that *"the centre of theory making must move to the Global South; that there has to be a recalibration of the geographies of authoritative knowledge"*. It seems that this has at least partly happened and is still happening in the case of heritage. The spirit of the place, the genius loci of the historic centres are still perceived in each of the three cities, as proven by empirical research (cf. Chap. 5).

In fact, in none of the case studies cultural, natural and intangible heritage is addressed together in the same legal framework, as proposed by different international charters over the past decades, particularly by the Historic Urban Landscape Approach (cf. Sect. 3.8). Rather, legal frameworks are stuck in out-of-date understandings, partly even induced by colonial governments which themselves have been replaced a long time ago, as, e.g. the case in Indonesia (cf. Sects. 5.1.6, 5.2.6 and 5.3.6). In the case of built heritage they still often base on classifiable indicators like the building age instead of considering added intangible values. In addition, legal frameworks for conservation and urban development differ and may involve a multitude of formal actors on different levels, as shown in each of the case studies. Depending on the country, even more local informal actors are significantly involved in inner-city processes, without being part of formal decision-making.

The slow-moving consideration of recent international paradigms shifts towards the recognition of intrinsically intertwined forms of heritage is surprising as it very often depicts local cultural understanding much more than the legal framework in place. Such holistic approach is badly needed in the rapidly changing urban landscapes in developing and emerging countries, however it may need too long to trickle down. It is the administrative system in place—thinking in categories and with a high level of persistence as well as inefficiency—which constitutes the main obstacle. As a consequence, the bizarre reality is that after having finally a broadened understanding of what is heritage in place, it is particularly those countries with per se comparable concepts which are struggling with its implementation.

Traditional architecture in Asia is linked to strong cosmological beliefs, reflecting the understanding of societal structure and/or universe; at the same time it is a manifestation of distinctive knowledge of craftsmanship skills, construction techniques and local wisdom (Kwanda 2010, cf. Sects. 5.1 and 5.2). In Indonesia and Nepal alike, the materialistic global approach of conservation is mismatching the local concepts of impermanence and continuous substitution or reassembling of building parts.

Maybe this is the biggest difference between Latin American and Asian cities (being aware that this is a generalisation which does not apply for each and every city)—many Asian cities comprise of core areas that were built in local styles and following the respective cultural and societal values, while Latin American cities are strongly influenced by Spanish or Portuguese planning, depicting the colonial society. As a consequence, there is nothing like a cosmological planning, there is no mystical values attached to the urban layout—except in the rare cases where pre-Hispanic relics are still perceived, like in Mexico City, built on the ruins of Tenochtitlan. In this context, the (re-)awakening of pre-Hispanic and/or societal groups' cultures oppressed during colonial times and their reappearance in the urban arena could be regarded as another indicator for a growing 're-value-ization' of urban spaces.

In each of the three cities the core area is composed of different sub-sections or quarters. Although all of them belong to the historic centre they are differing profoundly—in users, functions and condition. On the one hand this may complicate planning and urban regeneration, as each quarter has its distinct characteristics; on the other hand this can be seen as a chance for developing tailor-made regeneration schemes to avoid urban uniformity. The historic centre should not be seen divided into different thematic pieces. Rather, it is recommendable to see the built fabric, open spaces, underlying planning and meaning, as well as uses, functions and values associated to distinct sites as parts of a whole. Unfortunately, urban administrations are hardly prepared to do so, legal frameworks may even be hindering.

In Recife high-rise apartments are constructed in a historic surrounding which is the unique selling point of the investors, in Yogyakarta more tourist and conference hotels are popping up near the centre, disturbing the cosmological axis, and in Kathmandu the maintenance of the touristic hotspots seems to be more important to authorities than vernacular heritage. Tomorrow's city cores will look differently from today's. Historic urban cores are arenas where contemporary utilization, heritage (in its different facets) and market considerations meet. They are spaces of the past, used today and transformed for tomorrow, with the built fabric as the essential link.

Avoiding fundamental alterations of a historic city and the underlying values that at the same time is inhabited and where such values change over time is quite a challenge. In all three case study cities there was a general agreement that the historic centres should be protected as they are. At the same time the analyses revealed ongoing processes that could result in quite the contrary: Preferences to live in modern houses (in the cases of Recife and Kathmandu), a loss of historic building and/or their uses and also the outmigration of former user groups (again more distinct in Recife and Kathmandu than in Yogyakarta). In all cities shopping functions in the centre could be improved. Particularly in Recife the lack of housing functions was mentioned as one of the major problems to keep the area and its heritage alive.

## 6.4 Directions for Future Research

Obviously this study does not provide generally valid results as 'only' three cities are analyzed, although being from two continents, three culture regions (Latin America, South-East-Asia and South Asia) and three religious communities (Christian, Muslim and Hindu-Buddhist). Nevertheless obviously the African and Arab region is left aside. This would have exceeded the frame of this research work but could potentially be done in a follow-up research which would definitely contribute to underpin this research finding. Additionally, the inclusion of European or North-American cities would be interesting. In this book, the Euro-American background of heritage conservation is deduced from literature and backed by expert interviews in the case study cities. However, it was desisted to do my own empirical research on this aspect due to time constraints and the existence of a sufficient body of literature on the topic.

The comparative analysis of the case studies findings has revealed some impressive similarities in terms of the values attached to the historic centre and intangible values found in the respective cities. Furthermore, analogies were found in the impacts factors influencing on the historic centre. However, there are differences in terms of the impacts' intensity as well as their interlinkages. Therefore, it would be recommendable to add data from more cities. Choosing university students as peer group has the comparable advantage that this choice allows for a comparison among different cities. Nevertheless it would be interesting to broaden the research approach by including other societal groups, e.g. by including people who live or work in the centre areas.

Comparative approaches have the potential to reveal similarities and differences, which serves for an improved understanding on which processes and understandings can be generalized, and what kind of policies are needed for an improved dealing with urban heritage in historic centres. At the end of the day, such historic urban centres or landscapes do exist all around the world, admittedly created in different cultures and under very different economic, political and social conditions, but still facing the same global challenges. Learning on how different cities in different societies cope with these challenges (which certainly vary in their intensity) can contribute to the development of improved strategies for urban sustainability. Previous studies as e.g. by Böhme et al. (2003) who have analyzed and compared different South-East-Asian cities against the background of European ones, Costin (1993) who analyzed and compared legal and policy issues of cultural heritage in different Asia-Pacific states, or Gough (2012) who compared young people's experiences of urban life in three cities of different continents, reveal such additional values. Another example is the comparison of experiences gained in ten urban world heritage sites (IDB 2011). Roy (2009, 2011b) emphasizes on the experiences that particularly Latin American and Asian cities have gained e.g. in dealing with informality and the potential for European cities to benefit from such a pool of knowledge.

In the context of urban heritage and identity a comparable analysis is especially useful, as these topics are often tackled in a very local context, to do justice to the specific character of a site, city or society. While this is definitely true to e.g. find the right conservation techniques or locally suitable conservation measures, the comparative analysis can offer additional benefits in terms of a broader context. This research has revealed crucial similarities among the case studies that can contribute to an enhanced understanding on how to sustain historic urban centres. In addition, more case studies in the Global South will contribute to create a 'critical mass' of knowledge, further broadening the primordially euro-centric view on heritage. Here, postcolonial approaches can be suitable to further analyze the gaps between local understandings of heritage and legal documents to gain more knowledge on colonial or in general Western influences.

Dealing with heritage and identity is nothing uncommon in geography. Tourism geography is working a lot with the creation and invention of destinations, which are branded by means of being different from other spots. This research has tried to analyze historic urban cores in the Global South from a different angle, which combines many of the aspects tourism and new cultural geography are operating with, like the construction of urban identities, which has much in common with discourses on touristic place branding. Urban geography and development geography are dealing with cities in the Global South. Here, in recent years it is mostly aspects of urban poverty, fragmentation and exclusion that are dealt with. However, broadening this scope by recognizing the urban centre as an arena where such processes take place, intertwined with the question which actor groups are determining the centre's appearance and its uses, could be a rewarding subject. Even Latin American debates on the Right to the City can be broadened by asking if this does not intrinsically comprise the right to participate in the construction of urban identities, by simply recognizing the multitude of social practices from different actor groups that make and shape a city.

In addition, fostering transdisciplinary approaches can be very beneficial to tap the many facets of historic urban centres. In particular linkages between heritage studies and geography could be very beneficial, as heritage studies are traditionally dealing with processes of assigning heritage (by designation), while different areas of social geography work on appropriation of spaces, what potentially covers heritage spaces as well. In this respect this research is an appeal to investigate further on historic cities in the Global South, within and among disciplinary boundaries.

## References

Alberts HC, Hazen HD (2010) Maintaining authenticity and integrity at cultural world heritage sites. Geogr Rev 100(1):56–73

Amen M, Toly N, McCarney P, Segbers K (2011) Sighting or slighting cities in international relations. In: Amen M, Toly NJ, McCarney PL, Segbers K (eds) Cities and global governance: new sites for international relations. Ashgate, Farnham, pp 13–32

Böhme H, Körte A, Tokya-Seid M (Eds) (2003). WOHNEN - BAUEN - PLANEN. Erneuerung historischer Kernstädte in Südostasien und Europa im historisch-architektonischen Vergleich. Darmstadt, Wissenschaftliche Buchgesellschaft

Costin CL (1993). Legal and policy issues in the protection of cultural heritage in south asia and the pacific. Cultural heritage in asia and the pacific: Conservation and policy. In: Mac Lean MGH (ed) Proceedings of a symposium held in honolulu, Hawaii, September 8–13, 1991. Los Angeles, The getty conservation institute: 27–62

Falser MS (2010) From the Venice Charter (1964) to the Nara Conference (1994)—changing concepts of authenticity? In: Falser MS, Lipp W, Tomaszewski A (eds) Theory and practice in conservation and preservation—interactions between theory and practice. Proceedings of the international conference of the ICOMOS international scientific committee for the theory and philosophy of conservation and restoration, 23–27 April 2008 (Vienna). Edizioni Polistampa, Florenz, pp 115–132

Francioni F (2012) Forty years on: world heritage and international law. World heritage, vol 63. UNESCO, Paris

Giddens A (1990) The consequenses of modernity Polity Press. Cambridge University Press

Gough KV (2012). Reflections on conducting urban comparison. Urban Geography 33(6): 866–878

IDB (2011) City development. experiences in the preservation of ten world heritage sites. Inter-American Development Bank, Washington D.C

Jive'n G, Larkham PJ (2003) Sense of place, authenticity and character: a commentary. J Urban Des 8(1):67–81. doi:10.1080/1357480032000064773

Kwanda T (2010) Tradition of conservation: redefining authenticity in Javanese architectural conservation, 22–26 June 2010. In: Paper presented at the Heritage 2010—international conference on heritage and sustainable development, Evora, Portugal

Lefebvre H (1991) The production of space. Blackwell, Oxford

Nora P (1989) Between memory and history: les lieux de mémoire. Representations 26(Special Issue: Memory and Counter-Memory):7–24
Norberg-Schulz C (2013) The phenomenon of place. In: Larice M, Macdonald E (eds) The urban design reader (Routledge urban reader series), 2nd edn. Routledge, Milton Park, pp 272–284
Paasi A (2000) Territorial identities as social constructs. Hagar. Int Soc Sci Rev 1(2):91–113
Rautenberg M (1998) L'émergence patrimoniale de l'ethnologie: entre mémoire et politiques publiques. In: Poulot D (ed) Patrimoine et modernité. L'Harmatan, Paris, pp 279–291
Roy A (2009) The 21st-century metropolis: new geographies of theory. Reg Stud 43(6):819–830. doi:10.1080/00343400701809665
Roy A (2011) Postcolonial urbanism: speed, hysteria, mass dreams. In: Roy A, Ong A (eds) Worlding cities: Asian experiments and the art of being global. Studies in urban and social change. Wiley-Blackwell, Chichester, pp 305–335
Schmitt TM (2009) Global cultural governance. Decision-making concerning world heritage between politics and science. Erdkunde 63(2):103–121
Schmitt TM (2011) Cultural governance. Zur Kulturgeographie des UNESCO-Welterberegimes. Erdkundliches Wissen., vol 152. Franz Steiner, Stuttgart
Tweed C, Sutherland M (2007) Built cultural heritage and sustainable urban development. Landscape and Urban Planning 83:62–69
Veldpaus L, Pereira Roders AR, Colenbrander BJF (2013) Urban heritage: putting the past into the future. The Historic Environment 4(1):3–18
von Droste B (2012) Empowerment of the convention. From theory to practical application (1981–2000). World Heritage, vol 64. UNESCO, Paris
Werlen B (2014) Globalisierung. In: Lossau J, Freytag T, Lippuner R (eds) Schlüsselbegriffe der Kultur- und Sozialgeographie. Eugen Ulmer, Stuttgart, pp 54–66
Werlen B, Lippuner R (2007) Regionale Kulturen und globalisierte Lebensstile. Geographische Rundschau 59(7–8):22–27
Wöhler K (2008) Heritagefication: Zur Vergegenwärtigung des Kulturerbes. In: Luger K, Wöhler K (eds) Welterbe und Tourismus: Schützen und Nützen aus einer Perspektive der Nachhaltigkeit. StudienVerlag, Innsbruck, pp 43–58

# Chapter 7
# 'Our Town'!—Final Considerations

**Abstract** Urban environments are changing rapidly, particularly in the Global South. Whether intended or not, such change is impacting strongly on what is left from the historic urban centres. Attachment to the place itself as well as to urban intangible values nevertheless is still high among the urban population but receiving comparably little attention in formal planning and urban development processes. Cultural, natural and intangible heritage is hardly addressed together in the same legal framework, although they are hard to separate in people's perception. Historic urban centres, their places, sites, uses and traditions contribute to place attachment, which itself is a major component of urban identity. Particularly intangible heritage is a crucial point and increasingly recognized in global discourses, but yet not often considered on urban scales. Particularly urban regeneration projects are often biasing intangible values. Nevertheless, a major part of urban key actors in the case study cities of Kathmandu, Yogyakarta and Recife recognize the important role that urban heritage can play in sustainable urban development.

**Keywords** Conclusion · Urban heritage · Historic centre

Documents like the Nara Declaration show the growing self-consciousness of southern countries that appeared on stage and triggered a shift in heritage discourses, towards the recognition of intangible values. It is not surprising that such a shift in paradigms emerged in a non-Western context, where recognition of aboriginal concerns and beliefs has grown over the past 20 years. Examples like the T-Shirts stating "I did not climb the Ayer's Rock" sold to tourists and the hill's renaming back to Uluru show a fundamental shift in recognizing native tribes' or simply different value systems. The point is: Is such a shift easier in 'natural' settings than in 'urban' ones? Should there also be T-Shirts with "I did not build a high-rise along Malioboro Road"? And would people buy it? Recognizing the cosmological planning of Yogyakarta or the Mandala layout of Kathmandu Valley's late royal cities on a global scale is an achievement. But it has to be mainstreamed into urban development plans. Here, the awareness of key actors is essential.



S. Sandholz, *Urban Centres in Asia and Latin America*,
The Urban Book Series, DOI 10.1007/978-3-319-43735-4_7

Each and every day historic city centres around the world are changing. In no case such change has to mean something bad per se. Change is necessary, without change there is no progress and living cultures turn into something static and synthetic. But change can also easily lead to unintended losses, in particular if pressures are strong and planning is one-dimensional. In other words, historic urban centres can and are easily getting lost—not only their fabric, but also their uses, functions and the values that different user groups attach to the place —the tangible and intangible heritage that forms the urban landscape.

Urban environments are changing rapidly, particularly in the Global South, impacting strongly on what is left from the historic urban centres. Attachment to the place itself as well as to urban intangible values nevertheless is still high among the urban population. The historic centre, its places, sites, uses and traditions contribute to place attachment. Summing it up, the main research findings are

**#1: The historic urban centre matters**

The historic urban centre is not just like any other quarter. As the initial nucleus of the city, it is the place where much of the urban history is manifested. Furthermore, it comprises of a high or even the highest concentration of cultural heritage than other parts of the city. Both facts are perceived and appreciated by urban inhabitants.

**#2: Intangible heritage matters even more but is comparably difficult to grasp**

Intangible heritage is a comparably new phenomenon in the global heritage debate. Within this debate, it was foremost indigenous communities associated with it. Urban intangible heritage, however, has gained in importance over the last years at least in the international debate. On the urban scale of the case study cities, intangible heritage is a concern of urban authorities partly at the best, with the exception of major religious or cultural festivities. Only in Brazil larger programs focusing on the support of intangible values and popular culture do exist, while in the other two cities particularly popular culture is much less taken care of.

**#3: Uses and user groups matter to keep urban heritage alive**

The maintenance of mixed uses within the urban core is of utmost relevance. Housing functions are of particular importance and need to be promoted—also and particularly in the case of historic buildings—to have lively centres and to avoid turning the centre into an open-air-museum for temporary visitors. At the same time cultural expressions like handicraft, traditional skills and artistic expressions that can be localized in the centre should be considered in any development plan for the built environment.

**#4: Local culture receives little attention in urban realities**

Urban growth, densification and changing lifestyles are menaces for historic city centres. Particularly in the Global South urban change is occurring at rapid speed, resulting not only in the loss or alteration of fabric all around the city and particularly in its core areas. Another result is the fundamental transformation of local cultural environments, jeopardizing values, beliefs, arts and crafts rooted in historic surroundings. Not all of these changes are intended, yet they receive comparably little attention in urban development.

**#5: Globalized discourses on up-to-date cities bias urban regeneration schemes**

The global 'homogenized' understanding on what constitutes a modern city, following actual discourses of modernisation, globalization and/or neoliberalism, is impacting on the approaches and goals chosen for urban regeneration. In this context historic urban cores are likely providing the backdrop for new constructions instead of being a focal point. With such approaches particularly intangible values that local user groups ascribe to certain places may get lost, whether intentional or accidental.

**#6: Local heritage policies do not keep pace with changing international paradigms**

Cultural, natural and intangible heritage is hardly addressed together in the same legal framework. The slow-moving consideration of recent international paradigm shifts towards the recognition of intrinsically intertwined forms of heritage is surprising as it very often depicts local cultural understanding much more than the legal framework in place. It is the administrative system in place—thinking in categories and with a high level of persistence as well as inefficiency—which constitutes the main obstacle. As a consequence, the bizarre reality is that after having finally a broadened understanding of what is heritage in place, it is particularly those countries with communities still rooted in such understandings which are struggling with its implementation.

**#7: Awareness of urban key actors on the interlinkages of heritage types is essential**

Urban actors are not always aware of the side-effects of regeneration schemes, particularly when it comes to the impacts on intangible values. Losses of intangible assets may even be intended in such projects that emphasize economic upgrading, particularly economy-driven PPP schemes. Nevertheless, a major part of urban key actors in Kathmandu, Yogyakarta and Recife recognize the significance that urban heritage can play in sustainable urban development—including potential future key decision makers, as documented in the questionnaire results, which indicate the value attributed to the historic centre and urban heritage in general.

To cut a long story short and to answer the initial question of this research: Yes, it is still 'our town'. In other words, people from Kathmandu, Yogyakarta and Recife care about their historic centres and take pride in it, even or maybe precisely because of the rapidly changing urban environments.

# Chapter 8
# Postscript

**Abstract** In April and May 2015, Kathmandu has been hit by devastating earthquakes with severe impacts, among others on Kathmandu and its heritage. Much of the historic fabric in the city centre is lost, not only in the World Heritage Sites like the Durbar Squares but also small shrines, statues and community temples in other quarters. The loss of buildings and temples has also impacted severely on the intangible urban heritage which is often directly linked with the damaged heritage sites. Kathmandu is now at a crossroads, between restoring its urban past—tangible and intangible one—and converting the city or parts of it into something new. While it is intended that destroyed parts of the World Heritage Sites will be reconstructed over time this is quite unlikely for other sites that may simply vanish. Here the earthquake may be rather seen as chance to modernize the city that surely had to face many challenges already before the earthquake, but without paying much attention to its past. Hopefully the growing understanding about the importance of intangible values and assets, of cultural heritage as expression of the society's past, will be considered in the reconstruction process and not be left aside.

**Keywords** Kathmandu · Nepal · Urban heritage · Earthquake · Reconstruction

> In twenty years Kathmandu will be flat like a pancake, resembled because of an earthquake (K09).

Sadly, this is not the full end of the story. As mentioned in the previous chapters, urban heritage is at risk for various reasons, among them natural disasters. As pointed out before, each of the three case study cities is prone to different natural hazards. Particularly Kathmandu is facing a very high risk level. Unfortunately, on April 25, and May 12, 2015, Nepal was hit by two devastating earthquakes with magnitudes of 7.8 and 7.3 Richter scale.

These earthquakes and their aftershocks have led to more than 8,600 casualties and immense damages of buildings and infrastructure all around the country, among them about 2,900 structures with a cultural and religious heritage value (National Planning Commission 2015; UN OCHA 2015). Kathmandu itself was among the most affected areas. Overall 97,000 buildings in Kathmandu were



S. Sandholz, *Urban Centres in Asia and Latin America*,
The Urban Book Series, DOI 10.1007/978-3-319-43735-4_8

**Fig. 8.1** Kathmandu Durbar Square before and after the earthquake (*top* taken April 2016, *source* Rupesh Shrestha, *bottom* taken April 2013)

damaged irreparably (Bilham 2015) and most of the damage occurred in the city centre (Chiaro et al. 2015). Among others, various temples and shrines on Kathmandu Durbar Square were completely damaged; Hanuman Dhoka Palace was damaged seriously (cf. Fig. 8.1). Other landmarks have suffered as well, e.g. Dharahara Tower that collapsed entirely, the same is true for Kastamandap, Kathmandu's name-giving landmark building at Durbar Square (Fahad Hossain et al. 2015; Goda et al. 2016) (Fig. 8.2).

Much of the historic fabric in the city centre is lost. The same applies to other quarters, where small shrines, statues and community temples are buried under the

**Fig. 8.2** Kastamandap, at southern end of Kathmandu Durbar Square before and after the earthquake (*top* taken April 2016, *source* Rupesh Shrestha, *bottom* taken April 2013)

debris of collapsed buildings. Unfortunately the urban heritage of Kathmandu has been wiped out to a large part. As a consequence, the Kathmandu case study as presented in this research is somehow 'historic' as it depicts the city in a way that no longer exists.

However, the very essence of the research presented is still valid: both, tangible and intangible heritage do matter for urban inhabitants. It contributes to the formation of place attachment and an urban identity; as it gives a sense of feeling at home and offers people physical places their traditions, religious beliefs and habits are linked to. Despite the disaster and its impact on Kathmandu the field study results are still true—shockingly they are partly outdated already as many of the buildings I referred to are destroyed partly or even completely. Nevertheless how people relate to their past, to tangible heritage and intangible values, has not changed and will hopefully be considered in the reconstruction process. Decisions on how the reconstruction will take place, what and how to reconstruct are inseparable from decisions on the urban tangible and intangible heritage. It is unfortunately not very likely that all challenges that have been described in the Kathmandu chapter, particularly weaknesses in law enforcement and concerted action by the different stakeholders involved, will be overcome.

Besides all very necessary efforts to repair basic infrastructure and provide housing facilities, it is also very essential to put culture on the agenda. Not only as a source of income—tourism in Nepal is among the main sources of income generation (Muzzini and Aparicio 2013)—but rather as a basic right. This research has shown how much their heritage means to the people of Kathmandu, Yogyakarta and Recife. Although much has gone, the Kathmandu chapter still shows how much inhabitants appreciate their core and other heritage areas.

The earthquakes have impacted comparably more on traditional structures than on the popular reinforced concrete buildings. The initial layout of Newari houses is however seismic stable, but their vulnerability augmented due to modifications and alterations over time (Romão et al. 2015). Studies on the need and potentials of retrofitting traditional buildings have already been available before (cf. Shakya et al. 2014 on retrofitting of Pagoda temples) but were often not realized in reality. Retrofitting guidelines are ill-suited for the special requirements of historic building, governing the heritage sites is very complex. Rapid losses of the social cohesion system like the Guthis and ongoing vertical division of privately owned houses is constantly increasing the risk levels (Acharya and Pradhananga 2015; Maskey 2015; Watson 2016).

The loss of structures has also impacted severely on the intangible urban heritage which is often directly linked with the damaged heritage sites. Deities of destroyed temples and shrines were either shifted or are left on the sites with temporary constructions to maintain daily offering rituals and homage. In other cases worshipping has become more complicated, e.g. in the case of Chariot processions during Machhindranath and Indra Jatra festivals, where important locations are severely damaged. Nevertheless the communities have expressed their wish to continue the festival (Bahadur Dimal 2015).

"*For communities and individuals alike, culture and cultural heritage are undeniable sources of pride, identity, purpose and resilience*" (Bandarin et al. 2011: 20). There is not much to add to this statement, except the wish that this statement will be considered in the reconstruction process the years to come. Efforts to reconstruct a less disaster-prone built environment at the expense of cultural considerations like social norms and sensitivities will not be sustainable as they are very much integral to long-term reconstruction and enhanced risk preparedness (International Federation of Red Cross and Red Crescent Societies 2014; Rutherford 2015). As US author Elizabeth Enslin has stated in an article on the earthquake, Nepal's cultural past and its temples are part of the national identity and a source for the people's strength (Blake 2015).

While Swayambunath pagoda itself survived the earthquakes, 27 buildings around the stupa collapsed. As a consequence, 30 families involved in daily worshipping and religious rituals around Swayambunath had to live in tents. French archaeologist and art historian David Andolfatto, working as UNESCO consultant during the Post Disaster Needs Assessment states "*These people are intangible heritage that keeps the tangible heritage alive*" (Huët 2015). Kathmandu's heritage goes beyond the truly impressive Durbar Squares, which are among the key priorities for reconstruction, while other temples and shrines may be of even higher importance for the urban population, though less popular as tourist destination (Hutt 2015).

All three Durbar Squares have been reopened on June 16 2015, only few weeks after the disasters, hoping for tourists to come. In parallel, an urban regeneration strategy for the three core areas is planned to be elaborated by Kathmandu Metropolitan City. The plan does not foresee changes to the outer and inner outlook of the buildings, blocks (chowks) and monuments as they reflect the culture, tradition, architecture and settlements. However, it is planned to achieve the regeneration through land pooling, posing questions on the exact layout or reconstruction plans. Uttar Kumar Regmi, Chief of the Department of Physical Development and Construction at Kathmandu Metropolitan City stated that "*Each cluster will have 50–60 retrofitted houses equipped with solar lights and rainwater harvesting infrastructure. KMC will design some new houses suiting the local culture, traditions and architecture*" (Bogaty 2015). This in fact does sound like transformation and reminds of the plan that experts (K06 and K12) mentioned during the interviews. On the other hand the Post Disaster Needs Assessment (2015: 16) that was elaborated by hundreds of experts during summer 2015 emphasizes strongly on the need to consider the country's tangible and particularly the intangible heritage—"*historic value of a monument and the religious and cultural activity associated with it.*" (2015: xii)—in the reconstruction process:

> Indigenous and ethnic communities, too, have a strong and unique cultural heritage, which is an important part of their identity. Recovery interventions, particularly to do with housing and relocation should preserve rather than undermine these aspects of Nepal's proud cultural heritage.

Kathmandu is now at a crossroads, between restoring its urban past—tangible and intangible one—and converting the city or parts of it into something new. After the last big earthquake stock was rebuilt or repaired, restoring the urban environment. It remains to be seen, however, what will happen in this case—and what the city will look like in a few years. Realities after the earthquake as well as the research findings, particularly the ones on drivers of urban change and policies suggest a more pessimistic scenario. It is likely that destroyed parts of the World Heritage Sites will be reconstructed over time; however, this is quite unlikely for other areas. Here many site owners will take the opportunity to construct more storeys to maximize benefits. In case the historic fabric vanished, there is no legal framework regulating that reconstruction has to be done in a comparable way. Therefore it is most likely that—with the exception of the World Heritage Sites and their buffer zones—Kathmandu agglomeration will become a concrete jungle, once the debris is removed and building materials are available again. In the near future authorities there will have to decide how to proceed, what and how to rebuild.

Hopefully the growing understanding about the importance of intangible values and assets, of cultural heritage as expression of a society's past, will be considered in this process and not be left aside for economic reasons or the absence of planning. This way Kathmandu could become a role model for the Asian continent, where so many heritage sites are prone to geohazards (Pavlova et al. 2015) and beyond.

Interviewee K03 has put it in his own words, only a few days after the earthquake: "*Of course, we will restore this, it has to be that way…you cannot let your ancestry die just because the earth moved a few meter!*"

## References

Acharya KP, Pradhananga S (2015) Review of the integrated management plan of Kathmandu Valley World Heritage property. In: Weise K (ed) Revisiting Kathmandu. Safeguarding living urban heritage. International symposium Kathmandu Valley 25–29 November 2013. UNESCO, UNESCO Office in Kathmandu, Kathmandu, pp 127–132

Bahadur Dimal S (2015) Restoring ICH in Nepal since the earthquakes of 2015. Intangible Cult Herit Courier Asia Pacific 25:8–9

Bandarin F, Hosagrahar J, Sailer Albernaz F (2011) Why development needs culture. J Cult Herit Manag Sustain Dev 1(1):15–25. doi:10.1108/20441261111129906

Bilham R (2015) Seismology: raising Kathmandu. Nat Geosci 8(8):582–584. doi:10.1038/ngeo2498

Blake J (2015) How Nepal's history may be its greatest strength after the earthquake. CNN. http://edition.cnn.com/2015/04/26/world/nepal-snapshot/index.html. Accessed 28 April 2015

Bogaty T (2015) KMC mulling over urban regeneration targeting core areas of city. The Himalayan Times

Chiaro G, Kiyota T, Pokhrel RM, Goda K, Katagiri T, Sharma K (2015) Reconnaissance report on geotechnical and structural damage caused by the 2015 Gorkha Earthquake, Nepal. Soils Found 55(5):1030–1043. doi:http://dx.doi.org/10.1016/j.sandf.2015.09.006

Fahad Hossain ASM, Adhikari TL, Ansary MA, Bari QH (2015) Characteristics and consequence of Nepal earthquake 2015: a review. Geotech Eng J SEAGS AGSSEA 46(4):114–120

Goda K, Kiyota T, Pokhrel RM, Chiaro G, Katagiri T, Sharma K, Wilkinson S (2016) The 2015 Gorkha Nepal earthquake: insights from earthquake damage survey. Front in Built Environ 1 (8):1–15. doi:10.3389/fbuil.2015.00008

Huët S (2015) Saving Swayambhu. Kunda Dixit. http://www.nepalitimes.com/blogs/thebrief/2015/05/30/saving-swayambhu-earthquake-cultural-sites/

Hutt M (2015) Heritage, continuity, and Nostalgia American Anthropological Association. http://www.culanth.org/fieldsights/737-heritage-continuity-and-nostalgia Accessed 24 April 2016

International Federation of Red Cross and Red Crescent Societies (2014) World Disasters report 2014—focus on culture and risk. International Federation of Red Cross and Red Crescent Societies, Geneva. ISBN 978-92-9139-214-8

Maskey PN (2015) Disaster risk of culture heritage sites of the Kathmandu Valley. In: Weise K (ed) Revisiting Kathmandu. Safeguarding Living Urban Heritage. International symposium Kathmandu Valley 25–29 November 2013. UNESCO, UNESCO Office in Kathmandu, Kathmandu, pp 283–290

Muzzini E, Aparicio G (2013) Urban growth and spatial transition in Nepal: an initial assessment. Directions in Development. World Bank, Washington, DC. doi:10.1596/978-0-8213-9659-9

National Planning Commission (ed) (2015) Nepal earthquake 2015 post disaster needs assessment, vol A: Key Findings. Government of Nepal, Kathmandu

Pavlova I, Makarigakis A, Depret T, Jomelli V (2015) Global overview of the geological hazard exposure and disaster risk awareness at world heritage sites. J Cult Herit. doi:http://dx.doi.org/10.1016/j.culher.2015.11.001

Romão X, Paupério E, Menon A (2015) Traditional construction in high seismic zones: a losing battle? The case of the 2015 Nepal earthquake. In: Correia MR, Lourenco PB, Varum H (eds) Seismic Retrofitting: learning from vernacular architecture. CRC Press, Leiden, pp 93–100

Rutherford N (2015) How to turn awareness into action on disaster preparedness. http://www.euractiv.com/sections/sustainable-dev/how-turn-awareness-action-disaster-preparedness-314413. Accessed 11 May 2015

Shakya M, Varum H, Vicente R, Costa A (2014) Seismic vulnerability and loss assessment of the Nepalese Pagoda temples. Bull Earthq Eng 13(7):2197–2223. doi:10.1007/s10518-014-9699-5

Un OCHA (2015) Nepal flash appeal revision: Nepal earthquake April–September 2015. UN Office for the Coordination of Humanitarian Affairs, UN Resident and Humanitarian Coordinator for Nepal, Kathmandu

Watson I (2016) Resilience and disaster risk reduction: reclassifying diversity and national identity in post-earthquake Nepal. Third World Q 1–22. doi:10.1080/01436597.2016.1159913

# Appendix A
# International Heritage Documents and Charters



S. Sandholz, *Urban Centres in Asia and Latin America*,
The Urban Book Series, DOI 10.1007/978-3-319-43735-4

**Table A.1** Overview of documents, chartas and recommendations of supranational importance on cultural and intangible heritage and main goals/purposes (not including documents focusing on museums, archaeological sites, moveable properties, illicit trade, underwater heritage, artifacts, tourism, cultural routes), *Sources* Ahmad (2006), Ahmad (2006), Birabi (2007), Falser (2010), Bandarin (2011), Ruland (2011), Veldpaus et al. (2013), The Getty Conservation Institute (2015)

| Document Name | Year | Author | Main goals/purposes |
|---|---|---|---|
| Athens Charter for the Restoration of Historic Monuments | 1931 | First International Congress of Architects and Technicians of Historic Monuments | Promoting idea of a common world heritage; the importance of the setting of monuments; and defining the principle of integration of new materials |
| Athens Charter | 1933 | IV International Congress for Modern Architecture | Introducing principle of "the functional city" in urban development. includes urban ensembles in the definition of the built heritage and emphasizes the spiritual, cultural and economic value of the architectural heritage |
| Hague Convention: Convention for the Protection of Cultural Property in the Event of Armed Conflict | 1954 | UNESCO | Safeguard heritage during war, natural calamities and of recent, in terrorist attacks |
| European Cultural Convention | 1954 | Council of Europe | Safeguard European common cultural heritage, for wider community study and promote European civilization |
| Recommendation Concerning the Safeguarding of the Beauty and Character of Landscapes and Sites | 1962 | UNESCO | Recognizes the important aesthetic, cultural, spiritual and scientific value of landscapes and the threat of poorly regulated development, including also urban landscapes, recommends education for awareness raising |
| Venice Charter: International Charter for the Conservation and Restoration of Monuments and Sites | 1964 | ICOMOS, Second International Congress of Architects and Technicians of Historic Buildings | Sets internationally accepted principles of conservation based on the concept of authenticity and the importance of maintaining the historical and physical context of a site or building, states that monuments are to be conserved not only as works of art but also as historical evidence |

(continued)

**Table A.1** (continued)

| Document Name | Year | Author | Main goals/purposes |
|---|---|---|---|
| Recommendation on the Means of Prohibiting and Preventing the Illicit Import, Export and Transfer of Ownership of Cultural Property | 1964 | UNESCO | Proposes to set up national inventories of cultural property and recommends establishing national services to protect against illegal export of cultural property and developing a certification process to authorize exports |
| Norms of Quito: Final Report of the Meeting on the Preservation and Utilization of Monuments and Sites of Artistic and Historical Value | 1967 | Organization of American States (OAS) | Recognizes the abundance of monuments and sites that make up the cultural heritage of the Americas, on correlation of heritage, urban planning, economic Development and conservation in a sustainable context to promote economic development |
| Recommendation Concerning the Preservation of Cultural Property Endangered by Public or Private Works | 1968 | UNESCO | Protection of all important monuments and sites, not just those identified as major monuments, recommends preparation and maintenance of cultural property inventories, and suggests legislative and financial measures as well as procedures for setting and implementing priorities of action |
| Convention on the Means of Prohibiting and Preventing the Illicit Import, Export and Transfer of Ownership of Cultural Property | 1970 | UNESCO | Encourages cooperation among nations to prevent the illicit movement of cultural property across borders |
| Convention Concerning the Protection of the World Cultural and Natural Heritage | 1972 | UNESCO | Promotion of international perspective of conservation, comprising of both cultural and natural heritage. |
| Recommendation Concerning the Protection, at National Level, of the Cultural and Natural Heritage | 1972 | UNESCO | Calls for States to formulate, develop, and apply policies for the protection, conservation and presentation of their cultural and natural heritage, recommends international cooperation for safeguarding |

(continued)

**Table A.1** (continued)

| Document Name | Year | Author | Main goals/purposes |
|---|---|---|---|
| Resolutions of the Symposium on the Introduction of Contemporary Architecture into Ancient Groups of Buildings | 1972 | ICOMOS General Assembly | Gives principles for the harmonious introduction of contemporary architecture into groups of ancient buildings, encourages new use for revitalizing ancient buildings |
| European Charter of the Architectural Heritage | 1975 | Council of Europe | Aims to develop a common European policy for the protection of architectural heritage. The document defines the nature of the European architectural heritage, its importance to the European community and threats to the heritage |
| Declaration of Amsterdam | 1975 | Congress on the European Architectural Heritage, Council of Europe | Emphasizes role of planning, education, legal and administrative measures in protecting the region's architectural heritage, calls for integrative approaches and architectural conservation as integral part of urban and regional planning |
| Resolutions of the International Symposium on the Conservation of Smaller Historic Towns | 1975 | ICOMOS | Advocates for heritage-led urban regeneration initiatives, outlines features and structures of historic towns, proposes effective planning |
| Charter of Cultural Tourism | 1976 | ICOMOS | Urges built heritage to be properly preserved & plough back tourism money into heritage preservation |
| Recommendation Concerning the Safeguarding and Contemporary Role of Historic Areas | 1976 | UNESCO | Highlights multi-disciplinary collaboration in protecting and enhancing historic areas, recognizes the importance of the setting—buildings, spatial elements and surroundings make up historic areas |
| Convention on the Protection of the Archaeological, Historical, and Artistic Heritage of the American Nations, Convention of San Salvador | 1976 | OAS | Seeks to identify, register and protect the cultural heritage of the Americas in order to prevent illegal traffic in cultural property and to promote cooperation between the American states |

(continued)

**Table A.1** (continued)

| Document Name | Year | Author | Main goals/purposes |
|---|---|---|---|
| Burra Charter: Australia ICOMOS Charter for the Conservation of Places of Cultural Significance | 1979 (revised 1999) | ICOMOS Australia | Defines core heritage terminology. Justifies conservation planning and originality of the Heritage, introducing new terms of place, cultural significance and fabric particularly significant for the outlined standards for using cultural significance to manage and conserve cultural sites |
| The Florence Charter: Historic Gardens | 1982 | ICOMOS, IFLA, International Committee for Historic Gardens | Defines historic gardens as architectural compositions and recommends their preservation as living monuments, proposes maintenance strategies, adopted in 1982 as an addendum to the Venice Charter |
| Tlaxcala Declaration on the Revitalization of Small Settlements | 1982 | Third Inter-American Symposium on the Conservation of Building Heritage/ICOMOS Mexico | States that small settlements bear witness to our cultures and emphasizes the local and national governments' responsibility to help preserve smaller settlements and stem migration to urban centres, proposes interdisciplinary and community-based measures, taking into account local values and traditions |
| Appleton Charter for the Protection and Enhancement of the Built Environment | 1983 | ICOMOS Canada | Sound management of the built heritage and conservation as essential management elements, emphasizes management of the urban environment as an important aspect of conservation of cultural heritage |
| Convention for the Protection of the Architectural Heritage of Europe | 1985 | Council of Europe | Legislative measures and protections to safeguard the cultural heritage of the European states as well as conservation policies to protect the architectural heritage as component of town planning, awareness programs |
| Washington Charter: Charter on the Conservation of Historic Towns and Urban Areas | 1987 | ICOMOS General Assembly | Emphasizes conservation of historic towns as part of participatory urban and regional planning, establishes the principles and guidelines for the protection and conservation of historic towns |

(continued)

**Table A.1** (continued)

| Document Name | Year | Author | Main goals/purposes |
|---|---|---|---|
| First Brazilian Seminar About the Preservation and Revitalization of Historic Centers | 1987 | ICOMOS Brazil | Document sees urban historical sites are dynamic, socially created entities that are "part of a wider totality, comprising the natural and the built environment and the everyday living experience of their dwellers". Preservation of historic centres should therefore be central to urban planning, encourages inventories and social value over economic ones |
| Recommendation on the Safeguarding of Traditional Culture and Folklore | 1989 | UNESCO | Preservation of the intangible aspects of heritage: traditional culture and folklore, establishment of documentations and archives |
| Québec City Declaration | 1991 | OWHC | NKS management and conservation as a means to safeguard historic towns, calls for establishing a network of world heritage towns |
| Charter for the Conservation of Places of Cultural Heritage Value | 1992 | ICOMOS New Zealand | Establishes principles to owners, authorities, tradespersons and professionals in conserving places of cultural heritage value in New Zealand, acknowledging indigenous groups as responsible authorities beyond legal ownership |
| Fez Charter | 1993 | OWHC | Strategies for management of the urban heritage through intercity cooperation, recognizing the important role that cities play as cultural centres and exemplars of human achievements |
| Guidelines for Education and Training in the Conservation of Monuments, Ensembles and Site | 1993 | ICOMOS | Effort to develop standards and guidelines for education and training in the conservation of monuments, groups of buildings and sites |
| Nara Document on Authenticity | 1994 | ICOMOS | Builds on the Venice Charte, Emphasizes authenticity in conservation by recognizing cultural diversity, importance of considering the cultural and social values of all societies |

(continued)

**Table A.1** (continued)

| Document Name | Year | Author | Main goals/purposes |
|---|---|---|---|
| Charter for Cultural Tourism | 1995 | ICOMOS | Development of tourism towards management of resources for their viability. |
| Declaration of San Antonio | 1996 | ICOMOS Symposia | Links authenticity and cultural identity in the conservation and management of the cultural heritage of the Americas |
| Evora Appeal | 1997 | OWHC | Addresses the threats and benefits of cultural tourism in historic areas, proclaims importance of ensuring quality of life and respect for the cultural identity of residents |
| The Stockholm Declaration: Declaration of ICOMOS marking the 50th anniversary of the Universal Declaration of Human Rights | 1998 | ICOMOS Symposia | The right to cultural heritage is "integral part of human rights", right to have one's heritage respected as an expression of cultural identity |
| Charter on the Built Vernacular Heritage | 1999 | CIAV, ICOMOS | Outlines issues and principles for conservation of traditional and natural community housing, asks for adequate conservation practice and training |
| International Cultural Tourism Charter: Managing Tourism at Places of Heritage Significance | 1999 | ICOMOS | Promote and manage tourism in ways that respect and enhance the heritage and living cultures of the host communities and indigenous people |
| ASEAN Declaration on Cultural Heritage | 2000 | Foreign Ministers of the ASEAN Member Countries | Promoting the enhancement of cultural education, awareness and literacy for all kinds of heritage, tangible and intangible, cultural traditions integral part of intangible heritage |
| Convention for the Safeguarding of Intangible Cultural Heritage | 2003 | UNESCO | Builds on 1989 Recommendation on the Safeguarding of Traditional Culture and Folklore, establishes the necessary measures that States should take in the safeguarding of Intangible Cultural Heritage, defines its nature and forms |

(continued)

**Table A.1** (continued)

| Document Name | Year | Author | Main goals/purposes |
|---|---|---|---|
| Vienna Memorandum on World Heritage and Contemporary Architecture—Managing the Historic Urban Landscape | 2005 | UNESCO | Focuses on the impact of contemporary development on the overall urban landscape of heritage significance, gives guidelines for conservation management and urban development |
| Xi'an Declaration on the Conservation of the Setting of Heritage Structures, Sites and Areas | 2005 | ICOMOS | Acknowledges the contribution of setting to the significance of heritage monuments, sites and areas, stresses importance of both tangible and intangible values for heritage |
| ICOMOS Charter on the Interpretation and Presentation of Cultural Heritage Sites | 2007 | ICOMOS | Building on the Venice Charter, identifies heritage sites and the intangible elements associated with the site as a resource for learning from the past, calls to respect the site authenticity and stakeholder inclusion |
| Leipzig Charter on Sustainable European Cities | 2007 | Member States' Ministers responsible for Urban Development | First joint strategy paper on urban development within Europe, focusing on social balance, cultural diversity, urban heritage and environmental quality |
| The Valletta Principles for the Safeguarding and Management of Historic Cities, Towns and Urban Areas | 2011 | ICOMOS—International Committee on Historic Towns and Villages | Considers urban heritage as essential resource, proposes principles and strategies applicable to any historic town and urban area to safeguard their tangible and intangible values |
| Recommendation on the Historic Urban Landscape | 2011 | UNESCO | Promotes the integration of policies and practices of conservation of the built environment into the wider goals of urban development respecting inherited values and traditions of different cultural contexts |

# Appendix B
# List of Interview Partners

**Table A.2** List of interview partners in Kathmandu

| No. | Date | Institution |
|---|---|---|
| K01 | 17.07.2013 | Kathmandu Valley Preservation Trust (KVPT) |
| K02 | 19.07.2013 | Department of Urban Development and Building Construction, Ministry of Urban Development |
| K03 | 21.07.2013 | Institute of Engineering, Tribhuvan University |
| K04 | 21.07.2013 | Kathmandu Valley Development Authority (KVDA) |
| K05 | 22.07.2013 | Department of Archaeology, Ministry of Culture, Tourism and Civil Aviation |
| K06 | 25.07.2013 | National Society for Earthquake Technology, Nepal (NSET) |
| K07 | 29.07.2013 | Department of Architecture and Urban Planning, Institute of Engineering, Tribhuvan University |
| K08 | 20.07.2013 | Central Department of Geography, Tribhuvan University |
| K09 | 30.07.2013 | Institute of Engineering, Center for Disaster Studies, Tribhuvan University |
| K10 | 02.08.2013 | Shop owner, Thamel |
| K11 | 02.08.2013 | *'Nagarik Dainik' & 'Republica' journalist,* Newspapers |
| K12 | 21.07.2013 | Department of Urban Development and Building Construction, Ministry of Urban Development |

**Table A.3** List of interview partners in Yogyakarta

| No. | Date | Institution |
|---|---|---|
| Y01 | 02.01.2013 | *Pusat Pengendalian Operasional Badan Penanggulangan Bencana Daerah Provinsi DIY—BPBD* (Disaster Management branch of Yogyakarta Special Region) |
| Y02 | 04.02.2013 | Department of Public Management & Policy Studies, Universitas Gadjah Mada |
| Y03 | 04.02.2013 | *Kantor Balai Pelestarian Cagar Budaya (BPCB)* (Yogyakarta conservation authority) |
| Y04 | 04.02.2013 | Center for Disaster Studies, Universitas Gadjah Mada |

(continued)



S. Sandholz, *Urban Centres in Asia and Latin America*,
The Urban Book Series, DOI 10.1007/978-3-319-43735-4

**Table A.3** (continued)

| No. | Date | Institution |
|---|---|---|
| Y05 | 12.02.2013 | *Pusat Studi Pariwisata,* Universitas Gadjah Mada (Center for Tourism Studies) |
| Y06 | 14.02.2014 | Architecture and Planning Engineering Department, Universitas Gadjah Mada |
| Y07 | 14.02.2014 and 24.02.2014 | Jogja Heritage Society/ICOMOS Indonesia/Center for Heritage Conservation, Faculty of Engineering, Department of Architecture, Universitas Gadjah Mada |
| Y08a7b | 18.02.2013 | *Kementerian Pekerjaan Umum* (Ministry of Public Works) |
| Y09 | 20.02.2013 | Faculty of Geography, Universitas Gadjah Mada |
| Y10 | 22.02.2013 | Project manager, Misereor |

**Table A.4** List of interview partners in Recife

| No. | Date | Institution |
|---|---|---|
| R01 | 15.10.2013 | Direitoria de Preservação do Patrimônio Cultural—DPPC, Secretaria da Cultura |
| R02 | 15.10.2013 | Direitoria de Preservação dp Patrimônio Cultural—DPPC, Secretaria da Cultura |
| R03 | 15.10.2013 | Direitoria de Preservação dp Patrimônio Cultural—DPPC, Secretaria da Cultura |
| R04 | 16.10.2013 | Fundação do Patrimônio Histórico e Artístico de Pernambuco—FUNDARPE |
| R05 | 16.10.2013 | Fundação do Patrimônio Histórico e Artístico de Pernambuco—FUNDARPE |
| R06 | 17.10.2013 | Self-employed |
| R07 | 18.10.2013 | Departamento de Arquitetura e Urbanismo., Conservação Integrada, Universidade Federal de Pernambuco—UFPE Centro de Estudos Avançados da Conservação Integrada—CECI |
| R08 a/b | 18.10.2013 | Instituto da Cidade do Recife Engenheiro Pelópidas Silveira |
| R09 | 21.10.2013 | Fundação do Patrimônio Histórico e Artístico de Pernambuco—FUNDARPE |
| R10 | 21.10.2013 | Fundação do Patrimônio Histórico e Artístico de Pernambuco—FUNDARPE |
| R11 a/b | 23.10.2013 | AD Diper |
| R12 | 23.10.2013 | Centro de Formação, Pesquisa e Memória Cultural—Casa do Carnaval |
| R13 | 23.10.2013 | Diagonal—Transformação de Territórios (architecture and urban planning firm) |
| R14 | 23.10.2013 | Museu da Cidade do Recife, Forte das Cinco Pontas |
| R15 | 25.10.2013 | Laboratório da Paisagem, Departamento de Arquitetura e Urbanismo da UFPE |

# Appendix C
# Questionnaire

Note: The same questionnaires were used in Kathmandu, Yogyakarta and Recife, only questions 10 and 19 were adapted to the specific context by inserting pre-defined answers.



S. Sandholz, *Urban Centres in Asia and Latin America*,
The Urban Book Series, DOI 10.1007/978-3-319-43735-4

General questions – background

1. Nationality

| |
|---|
| |

2. Study program Name/Faculty

| |
|---|
| |

3. Sex

| m | f |
|---|---|
| | |

4. Age

| years |
|---|
| |

5. Type of study programme

| Msc | PhD |
|---|---|
| | |

6. How much time are you already staying in Kathmandu? *Please mark with an "x"*

| I grew up here | For more than 5 years | Between 2 and 5 years | Between 6 months and 2 years | Less than 6 months |
|---|---|---|---|---|
| | | | | |

7. If you didn't grow up in Kathmandu: why did you pick this place for studying? *Please mark with an "x"*

| | |
|---|---|
| Because of the university | |
| I didn't decide myself | |
| My family or friends are living there | |
| Because of the city itself | |
| Others (please indicate) | |

Urban history and place memory

**8. Imagine some friends from outside visit you in Kathmandu. Where in and around the town would you take them for sightseeing and leisure? Why?**

| | Place | Purpose/Reason |
|---|---|---|
| 1 | | |
| 2 | | |
| 3 | | |
| 4 | | |
| 5 | | |

9. **Are you interested in the history of Kathmandu? Please mark with an "x"**

| not at all | a bit | yes | very much |
|---|---|---|---|
| | | | |

**10. Which places/buildings do you consider important as place of remembrance of Kathmandu history? Please mark with an "x" – additionally you can give a reason in the right column**

| Place | | Very important | Important | Not that important | Not important | Reason |
|---|---|---|---|---|---|---|
| 1 | Kathmandu Durbar Square | | | | | |
| 2 | Kasthamandap | | | | | |
| 3 | Kumari Chowk | | | | | |
| 4 | Hanuman Dhoka | | | | | |
| 5 | Narayanhiti Palace | | | | | |
| 6 | Patan Durbar Square | | | | | |
| 7 | Pashupatinath | | | | | |
| 8 | Boudhanath | | | | | |
| 9 | Swayambhunath | | | | | |
| 10 | Thamel | | | | | |
| 11 | Singha Durbar | | | | | |

Ties to Kathmandu and the historic center

**11. How often do you go to the historic center of Kathmandu? Please mark with an "x"**

| I do not go there at all | Not more than once per month | Several times per month | several times per week | Every day | I am living there |
|---|---|---|---|---|---|
| | | | | | |

**12. How much time would you approximately need to get to the historic center from home (in minutes, please mark your "normal" means of transportation with an "x")**

| minutes |
|---|
| |

| Walking | Bicycle | Motorbike | Car | Bus |
|---|---|---|---|---|
| | | | | |

**13. For what purpose do you go to the historic center?**
***Please mark with an "x" (multiple answers possible)***

| | | not at all | Not more than once per month | Several times per month | several times per week | Every day |
|---|---|---|---|---|---|---|
| 1 | Shopping | | | | | |
| 2 | Sightseeing | | | | | |
| 3 | Cultural events | | | | | |
| 4 | Leisure activities | | | | | |
| 5 | Administrative aspects | | | | | |
| 6 | Others (please indicate) | | | | | |

Perception of the historic center

**14. What is your feeling towards the historic center?** ***In each line you find a pair of words. Please indicate to which of the two attributes you agree more and mark accordingly more to the left or the right side with an "x" (one "x" in each line)***

| | | | very | Some-how | Nei-ther/nor | Some-how | very | |
|---|---|---|---|---|---|---|---|---|
| 1 | The area is… | boring | | | | | | fascinating |
| 2 | The area is… | clean | | | | | | dirty |
| 3 | The area is… | heterogeneous | | | | | | homogenous |
| 4 | The area is… | beautiful | | | | | | ugly |
| 5 | The area is… | difficult to access | | | | | | easy to access |
| 6 | The area is… | vital | | | | | | abandoned |
| 7 | The area is... | uninhabited | | | | | | inhabited |
| 8 | The area is… | in good shape | | | | | | run-down |
| 9 | The area is… | not valuable/ un-necessary | | | | | | of value |
| 10 | The area is con-stantly… | deteriorating | | | | | | being upgraded |
| 11 | During DAYTIME the area is… | Dangerous/ violent | | | | | | safe |
| 12 | At NIGHTTIME the area is… | Dangerous/ violent | | | | | | safe |

*15.* **Which values do you attribute to the historic center of Kathmandu?** ***Please mark with an "x" on the scale to what extend you agree or disagree to each of the statements below.***

| | | Definitely do not agree | Do not agree | Partly agree | Agree | Definitely agree | Do not know |
|---|---|---|---|---|---|---|---|
| 1 | The historic center is the commercial center of the city. | | | | | | |
| 2 | The historic center is of higher cultural value than any other part. | | | | | | |
| 3 | The historic center is of higher spiritual value than any other part. | | | | | | |
| 4 | The historic center is the most beautiful part of the whole city. | | | | | | |
| 5 | The historic center represents our history more than any other part. | | | | | | |
| 6 | The historic center has a high recreational value for me. | | | | | | |
| 7 | The historic center or a part of it is of high spiritual/religious value for me. | | | | | | |
| 8 | The historic center must be preserved as it is. | | | | | | |
| 9 | I like to go to the historic center for shopping very much. | | | | | | |
| 10 | The historic center has no value for me at all. | | | | | | |
| 11 | I do not like the area and prefer more modern parts of the city. | | | | | | |
| 12 | Others (please specify) | | | | | | |

Processes and projects in the historic center

**16. Which processes do you perceive in the historic center of Kathmandu?** ***Please mark with an "x" on the scale to what degree each of the statements does reflects your feelings.***

| | | Definitely do not agree | Do not agree | Partly agree | Agree | Definitely agree | Do not know |
|---|---|---|---|---|---|---|---|
| 1 | The historic center is not used by locals, the only users are tourists. | | | | | | |
| 2 | Traffic is getting more and causing parking problems. | | | | | | |
| 3 | The historic center is outmoded; I rather prefer more modern areas. | | | | | | |
| 4 | Previous user groups (e.g. mobile traders) were forced out. | | | | | | |
| 5 | Gentrification processes are going on; former inhabitants cannot afford the rents any more. | | | | | | |
| 6 | Old buildings are replaced by modern ones, historic ensembles are getting lost. | | | | | | |
| 7 | Old buildings are preserved but their traditional uses are getting lost. | | | | | | |
| 8 | People prefer to live in more modern buildings instead of the old ones. | | | | | | |
| 10 | Current planning follows international trends and no more local traditions. | | | | | | |
| 11 | Others (please specify) | | | | | | |

**17. If you had the chance to suggest some project to upgrade the historic center, what and where would it be?**

| |
|---|
| |

Future urban development

**18. Many different paradigms are linked to "Sustainability" and "sustainable urban planning". In your opinion, which ones should Kathmandu follow to prepare for the future?**
***Please pick the five aspects that are most relevant in your opinion and mark them with an "x".***

| | Aspect | |
|---|---|---|
| 1 | Protection of local values and beliefs | |
| 2 | Protection of cultural and natural heritage | |
| 3 | Create a green city with parks and gardens | |
| 4 | Create an environmentally-friendly city | |
| 5 | Enhance disaster risk preparedness and resilience | |
| 6 | Adapt the city to climate change impacts | |
| 7 | Create social equity | |
| 8 | Establish stakeholder participation processes | |
| 9 | Build sustainable economic growth | |
| 10 | Reduce squatting and/or new slum formation | |
| 11 | Have a compact city and reduce urban sprawl | |
| 12 | Create a functioning environmentally-friendly mass transportation system | |
| 13 | Develop an energy-efficient city with alternative energy sources | |
| 14 | Create a safe city | |

Future urban development – intangible values

**19. Which intangible things or activities do you consider as typical for Kathmandu area, how important is it to preserve them?** ***Please mark with an "x" on the scale***

| | thing/activity | Very important | important | Not that important | Not important | Do not know |
|---|---|---|---|---|---|---|
| 1 | Traditional festivals | | | | | |
| 2 | Traditional Nepali/Newari food | | | | | |
| 3 | Traditional Thangka paintings | | | | | |
| 4 | Traditional pottery and metal art | | | | | |
| 5 | Traditional wood carving | | | | | |
| 6 | Traditional bazaars/markets | | | | | |
| 7 | Traditional music and dancing | | | | | |

## Place attachment

**20. What feelings do you have about places in Kathmandu?**
**Please mark with an "x" on the scale to what degree each of the statements reflects your feelings about the respective place.**

| | | My house | | | | | My neighborhood | | | | | My city district | | | | | Historic city center | | | | | City as a whole | | | | |
|---|---|---|---|---|---|---|---|---|---|---|---|---|---|---|---|---|---|---|---|---|---|---|---|---|---|---|
| | | 1 | 2 | 3 | 4 | 5 | 1 | 2 | 3 | 4 | 5 | 1 | 2 | 3 | 4 | 5 | 1 | 2 | 3 | 4 | 5 | 1 | 2 | 3 | 4 | 5 |
| 1 | I know the place very well | | | | | | | | | | | | | | | | | | | | | | | | | |
| 2 | I miss it when I am not here | | | | | | | | | | | | | | | | | | | | | | | | | |
| 3 | I don't like this place | | | | | | | | | | | | | | | | | | | | | | | | | |
| 4 | I am proud of this place | | | | | | | | | | | | | | | | | | | | | | | | | |
| 5 | I have no influence on its affairs | | | | | | | | | | | | | | | | | | | | | | | | | |
| 6 | I would like to be involved in what is going on here | | | | | | | | | | | | | | | | | | | | | | | | | |
| 7 | I am rooted here | | | | | | | | | | | | | | | | | | | | | | | | | |

(1) Definitely do not agree
(2) Do not agree
(3) Neither agree nor not agree
(4) Agree
(5) Definitely agree

**Thank you very much**

# Appendix D
# Expert Interview Guideline and Coding System

*Note* The same interview guideline was used in Kathmandu, Yogyakarta and Recife, below the example for Recife is given.

1. **Personal background**
    - Where educated? In what?
    - "role models" for urban renewal
2. **Unique features in planning in Brazil**
    - Most important things in urban planning in Brazil/Recife
    - What is planning tradition, what makes Brazil/Recife special?
    - How are these special features considered in the planning process?
3. **Priorities in urban planning in general**
    - What is most important to do in Recife?
    - Is there a master plan? Who is the Author? What is the vision for the city, which paradigm does it follow?
    - On which areas is planning focussed, why?
4. **Past and present projects in inner-city areas**
    - Thematic and geographic focus (which buildings/areas)
    - What were the main goals, actor groups, target groups, planning paradigms,
    - Is there a "role model" from another city/country?
    - Success stories? Failures/difficulties?
    - Are there changes over time, changes in paradigms?
5. **Place attachment/identity**:
    - What are objects, events, activities… of value—in the whole city and in the centre
    - Are there changes over time?
    - What is the Importance of "local identity"/values/ characteristics for the urban planning in Recife?



S. Sandholz, *Urban Centres in Asia and Latin America*,
The Urban Book Series, DOI 10.1007/978-3-319-43735-4

6. **Own perception of inner-city area**

- Own opinion towards past and ongoing projects
- Personal recommendations?

7. **What is your vision for Recife, what should the city look like in 20 years?**

**Table A.5** Expert interview coding system

| No. | Category | Definition | Characteristics |
|---|---|---|---|
| 1 | Urban planning on national level | Unique planning features in country, Goals and priorities in urban planning | • Description of urban planning<br>• Expressed uniqueness of urban/national planning<br>• Description of urban planning goals and priorities<br>• Mentioned laws, policies and actors |
| 2 | Personal perception on legislation, policies and planning | Own opinion towards planning, planning paradigms, projects | • Personal opinion on problems and best practise regarding legal documents<br>• Description of own work/opinion related to urban planning, paradigms, projects |
| 3 | Urban identity | Importance of 'identity' in planning | • Tangible/intangible values mentioned<br>• Implementation in planning, related actors |
| 4 | Past and present processes and regeneration projects in the historic centre | Past and present projects in inner-city area—location, role models | • Description of past and ongoing processes in the urban centre<br>• Description of impacts on the centre<br>• Actors in regeneration projects |
| 5 | Perception of the historic centre | Personal opinion on success/failure of urban conservation and regeneration projects | • Personal opinion on the historic centre and its features<br>• Personal opinion on problems and best practise regarding past and present regeneration and conservation |
| 6 | Tangible and intangible values ascribed to the historic centre | Valuable tangible and intangible urban assets/objects of value | • People's perception of the historic centre Tangible values described<br>• Intangible values described |
| 7 | Specific sites | Site example | • Specific sites mentioned<br>• Case study descriptions |
| 8 | Outlook | Recommendations, vision for future | • Own vision towards the urban future<br>• Recommendations how to realise the own vision |

# Appendix E
# Questionnaire Statistical Review



S. Sandholz, *Urban Centres in Asia and Latin America*,
The Urban Book Series, DOI 10.1007/978-3-319-43735-4

**Table A.6** Correlations between the respondents' sex, length of stay in the respective city, their interest in urban history and different values, feelings and processes related to the historic centres (Spearman correlation coefficient, only significant correlations are given)

| | sex | | | length of stay in town | | | interest in urban history | | |
|---|---|---|---|---|---|---|---|---|---|
| | Kathmandu | Yogyakarta | Recife | Kathmandu | Yogyakarta | Recife | Kathmandu | Yogyakarta | Recife |
| **Feelings towards the centre** | | | | | | | | | |
| boring - fascinating | | | | -,291* | | | | | ,279* |
| dirty - clean | | | | | | | | | |
| heterogenous - homogenous | | | | | | | | -,212* | |
| ugly - beautiful | | ,195* | ,261* | | | | | ,241* | |
| difficult - easy to access | | ,286** | | | -,198* | | | | |
| abandoned - vital | | | | -,341** | | | | | |
| uninhabinted - inhabited | | | | -,428** | | | | | |
| run-down - good shape | | | | | | | | | |
| not valuable - valuable | | | | -,413** | -,209* | -,231* | | ,235* | ,330** |
| deteriorating - being upgraded | | | | | | | ,243* | | |
| DAYTIME dangerous - safe | | | | -,250* | | | | | |
| NIGHTTIME dangerous - safe | | | | -,253* | | | | | |
| **Values attached: The historic centre...** | | | | | | | | | |
| ...is the commercial center of the city | ,316** | | | | | | | | |
| ...is of higher cultural value than any other part | | | | -,381** | | | ,280* | | |
| ...is the most beautiful part of the whole city | | | | | | | | | |
| ..represents our history more than any other part | | | | -,420** | | | | | |
| ...has a high recreational value for me | | | | -,342** | | | | | ,318** |
| ...or a part of it is of high spiritual/religious value for me | | | | | | | | ,319** | |
| ...must be preserved as it is | | | | -,324** | | | | | |
| ..is where I like to go to for shopping very much | | | ,282* | | | | | | |
| ...has no value for me at all | | | | ,416** | | | -,364** | | -,241* |
| I don't like area and prefer more modern parts of the city | | | -,438** | | | | | | -,296** |
| **Ongoing processes** | | | | | | | | | |
| The historic center is not used by locals, the only users are tourists | | | | | | | -,296** | | |
| Traffic is getting more and causing parking problems | | | -,251* | -,234* | -,256** | | | | |
| The historic center is outmoded; I rather prefer more modern areas | | | -,251* | ,276* | | | -,384** | -,269** | -,271* |
| Previous user groups (e.g. mobile traders) were forced out | | | | | | | | | |
| Gentrification processes are going on; former inhabitants cannot afford the rents any more | | | | | -,272** | | | | |
| Old buildings are replaced by modern ones, historic ensembles are getting lost | | | -,237* | | | | | | |
| Old buildings are preserved but their traditional uses are getting lost | | | | | | | | | |
| People prefer to live in more modern buildings instead of the old ones | | | | -,290* | | | | | |

| | |
|---|---|
| ** | p< 0,01 |
| * | p< 0.05 |

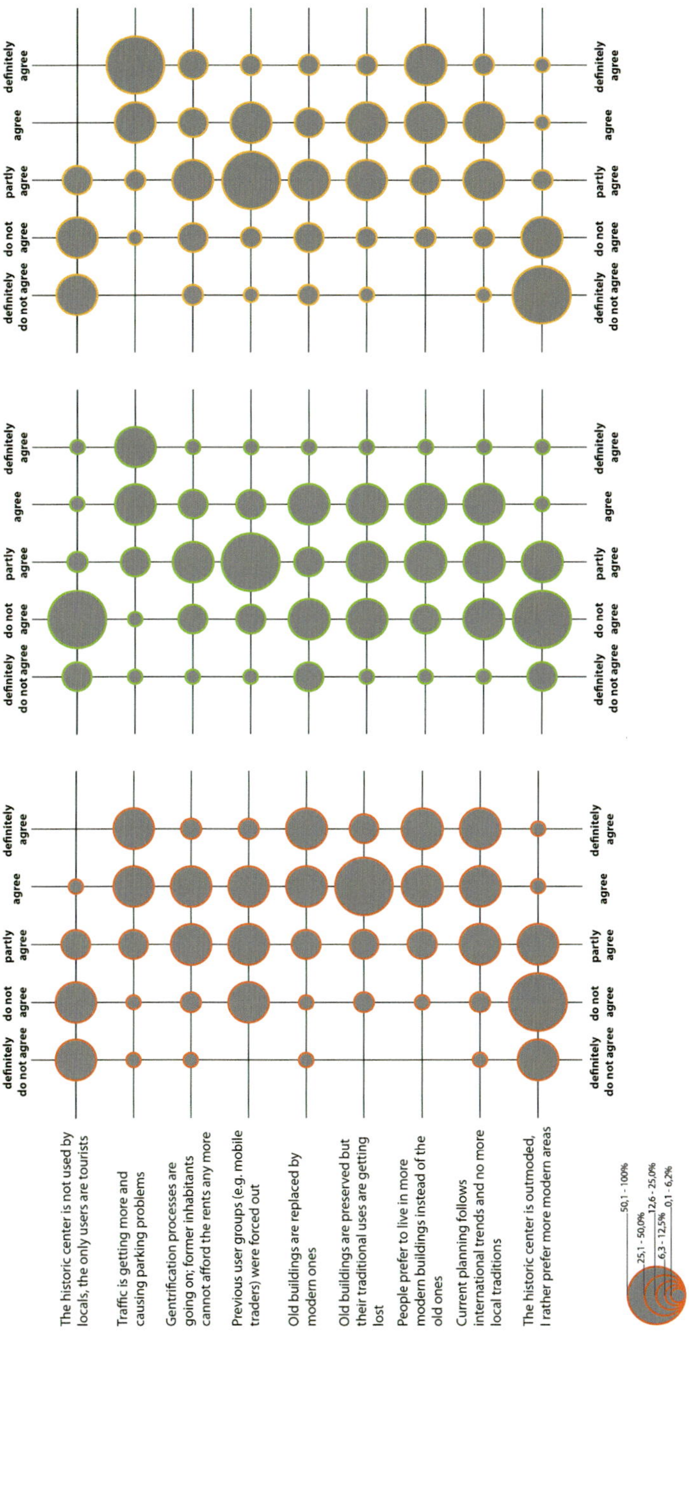


**Fig. A.1** Comparative analysis of values attributed to the three case study areas

**Table A.7** Correlations between different projects and processes in Recife's historic centre (Spearman $r_s$, matches with other cities framed)

| Recife | | | | | | | | | |
|---|---|---|---|---|---|---|---|---|---|
| **Correlations between different projects and processes** | The historic center is not used by locals, the only users are tourists | Traffic is getting more and causing parking problems | The historic center is outmoded; I rather prefer more modern areas | Previous user groups (e.g. mobile traders) were forced out | Gentrification processes are going on; former inhabitants cannot afford the rents any more | Old buildings are replaced by modern ones, historic ensembles are getting lost | Old buildings are preserved but their traditional uses are getting lost | People prefer to live in more modern buildings instead of the old ones | Current planning follows international trends and no more local traditions |
| The historic center is not used by locals, the only users are tourists | | -,083 | ,091 | -,026 | ,121 | ,046 | ,247* | ,134 | -,053 |
| Traffic is getting more and causing parking problems | | | ,049 | ,115 | -,059 | ,019 | -,016 | ,141 | ,017 |
| The historic center is outmoded; I rather prefer more modern areas | | | | ,096 | ,037 | ,147 | -,129 | ,122 | -,016 |
| Previous user groups (e.g. mobile traders) were forced out | | | | | ,349** | ,316** | ,008 | ,171 | ,382** |
| Gentrification processes are going on; former inhabitants cannot afford the rents any more | | | | | | ,394** | ,090 | ,151 | ,447** |
| Old buildings are replaced by modern ones, historic ensembles are getting lost | | | | | | | -,069 | ,173 | ,467** |
| Old buildings are preserved but their traditional uses are getting lost | | | | | | | | ,193 | ,082 |
| People prefer to live in more modern buildings instead of the old ones | | | | | | | | | ,194 |

*p< 0,05; **p< 0.01

**Table A.8** Correlations between different projects and processes in Yogyakarta' historic centre (Spearman $r_s$, matches with other cities framed)

| Yogyakarta | | | | | | | | |
|---|---|---|---|---|---|---|---|---|
| **Correlations between different projects and processes** | Traffic is getting more and causing parking problems | The historic center is outmoded; I rather prefer more modern areas | Previous user groups (e.g. mobile traders) were forced out | Gentrification processes are going on; former inhabitants cannot afford the rents any more | Old buildings are replaced by modern ones, historic ensembles are getting lost | Old buildings are preserved but their traditional uses are getting lost | People prefer to live in more modern buildings instead of the old ones | Current planning follows international trends and no more local traditions |
| The historic center is not used by locals, the only users are tourists | ,136 | ,095 | ,119 | ,167 | ,203* | ,123 | ,101 | ,144 |
| Traffic is getting more and causing parking problems | | -,108 | ,242** | ,109 | ,122 | ,176 | ,192* | ,071 |
| The historic center is outmoded; I rather prefer more modern areas | | | ,350** | ,284** | ,154 | ,091 | ,116 | ,058 |
| Previous user groups (e.g. mobile traders) were forced out | | | | ,352** | ,310** | ,115 | ,126 | ,148 |
| Gentrification processes are going on; former inhabitants cannot afford the rents any more | | | | | ,353** | ,309** | ,282** | ,139 |
| Old buildings are replaced by modern ones, historic ensembles are getting lost | | | | | | ,503** | ,494** | ,293** |
| Old buildings are preserved but their traditional uses are getting lost | | | | | | | ,567** | ,370** |
| People prefer to live in more modern buildings instead of the old ones | | | | | | | | ,470** |

*p< 0,05; **p< 0.01

**Table A.9** Correlations between different projects and processes in Kathmandu's historic centre (Spearman $r_s$, matches with other cities framed)

| Kathmandu | | | | | | | | |
|---|---|---|---|---|---|---|---|---|
| **Correlations between different projects and processes** | Traffic is getting more and causing parking problems | The historic center is outmoded; I rather prefer more modern areas | Previous user groups (e.g. mobile traders) were forced out | Gentrification processes are going on; former inhabitants cannot afford the rents any more | Old buildings are replaced by modern ones, historic ensembles are getting lost | Old buildings are preserved but their traditional uses are getting lost | People prefer to live in more modern buildings instead of the old ones | Current planning follows international trends and no more local traditions |
| The historic center is not used by locals, the only users are tourists | -,135 | ,510** | ,020 | -,141 | -,127 | -,082 | -,161 | ,025 |
| Traffic is getting more and causing parking problems | | -,061 | ,188 | ,337** | ,302** | ,121 | ,402** | ,176 |
| The historic center is outmoded; I rather prefer more modern areas | | | ,028 | -,068 | ,007 | ,088 | ,021 | ,019 |
| Previous user groups (e.g. mobile traders) were forced out | | | | ,195 | ,132 | ,077 | ,095 | ,145 |
| Gentrification processes are going on; former inhabitants cannot afford the rents any more | | | | | ,278* | ,321** | ,111 | ,339** |
| Old buildings are replaced by modern ones, historic ensembles are getting lost | | | | | | ,430** | ,070 | ,270* |
| Old buildings are preserved but their traditional uses are getting lost | | | | | | | ,282* | ,382** |
| People prefer to live in more modern buildings instead of the old ones | | | | | | | | ,376** |

*$p < 0,05$; **$p < 0.01$*

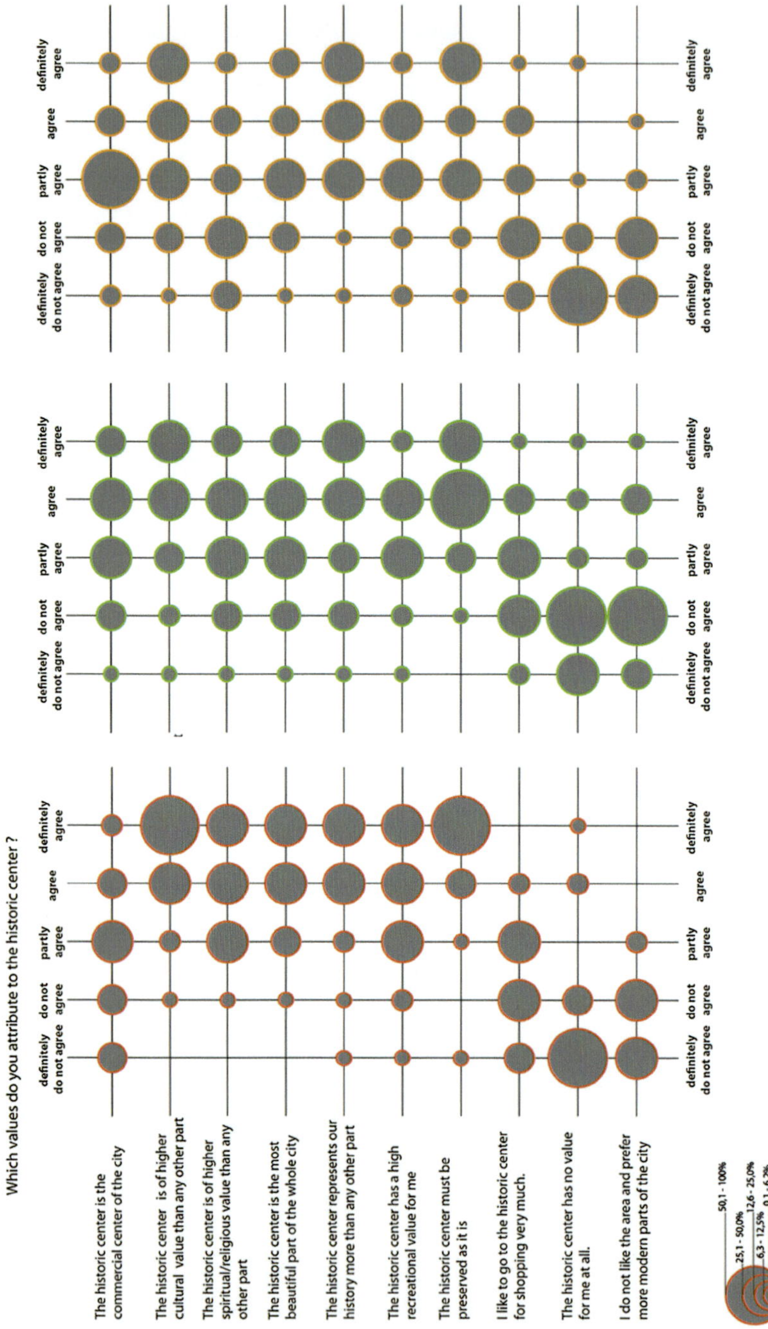


**Fig. A.2** History interest stated by the peer groups of the three case study cities

**Table A.10** Correlations between different values attributed to Yogyakarta's historic centre (Spearman $r_s$, matches with other cities framed)

| Yogyakarta | | | | | | | | | |
|---|---|---|---|---|---|---|---|---|---|
| **The historic centre...** | ...is of higher cultural value than any other part | ...is the most beautiful part of the whole city | ..represents our history more than any other part | ...has a high recreational value for me | ...or a part of it is of high spiritual/ religious value for me | ...must be preserved as it is | ...is where I like to go to for shopping very much | ...has no value for me at all | I do not like the area and prefer more modern parts of the city |
| ...is the commercial center of the city | ,185* | ,132 | ,034 | ,036 | -,045 | ,193* | ,195* | ,017 | ,056 |
| ...is of higher cultural value than any other part | | ,297** | ,423** | ,228* | ,031 | ,342** | -,119 | -,048 | -,014 |
| ...is the most beautiful part of the whole city | | | ,434** | ,390** | ,440** | ,265** | ,210* | ,195* | ,314** |
| ..represents our history more than any other part | | | | ,183* | ,070 | ,308** | ,112 | ,113 | ,130 |
| ...has a high recreational value for me | | | | | ,275** | ,258** | ,208* | -,061 | -,032 |
| ...or a part of it is of high spiritual/religious value for me | | | | | | ,094 | ,261** | ,127 | ,093 |
| ...must be preserved as it is | | | | | | | ,151 | -,107 | ,037 |
| ...is where I like to go to for shopping very much | | | | | | | | ,281** | ,288** |
| ...has no value for me at all | | | | | | | | | ,675** |

*p< 0,05; **p< 0.01

**Table A.11** Correlations between different values attributed to Recife's Historic centre (Spearman $r_s$, matches with other cities framed)

| Recife | | | | | | | | | |
|---|---|---|---|---|---|---|---|---|---|
| **The historic centre...** | ...is of higher cultural value than any other part | ...is the most beautiful part of the whole city | ..represents our history more than any other part | ...has a high recreational value for me | ...or a part of it is of high spiritual/ religious value for me | ...must be preserved as it is | ...is where I like to go to for shopping very much | ...has no value for me at all | I do not like the area and prefer more modern parts of the city |
| ...is the commercial center of the city | ,256* | ,086 | ,208 | ,004 | ,002 | ,089 | ,162 | -,046 | -,078 |
| ...is of higher cultural value than any other part | | ,349** | ,537** | ,137 | ,152 | ,228* | -,024 | ,070 | -,142 |
| ...is the most beautiful part of the whole city | | | ,500** | ,363** | ,031 | ,288** | ,184 | -,389** | -,254* |
| ..represents our history more than any other part | | | | ,105 | ,175 | ,209 | -,003 | -,228* | -,106 |
| ...has a high recreational value for me | | | | | ,260* | ,109 | ,299** | -,144 | -,354** |
| ...or a part of it is of high spiritual/religious value for me | | | | | | ,263* | ,240* | -,050 | -,243* |
| ...must be preserved as it is | | | | | | | ,331** | -,139 | -,055 |
| ...is where I like to go to for shopping very much | | | | | | | | -,192 | -,227* |
| ...has no value for me at all | | | | | | | | | ,368** |

*p< 0,05; **p< 0.01

**Table A.12** Correlations between different values attributed to Kathmandu's historic centre (Spearman $r_s$, matches with other cities framed)

| Kathmandu | | | | | | | | | |
|---|---|---|---|---|---|---|---|---|---|
| **The historic centre...** | ...is of higher cultural value than any other part | ...is the most beautiful part of the whole city | ..represents our history more than any other part | ...has a high recreational value for me | ...or a part of it is of high spiritual/ religious value for me | ...must be preserved as it is | ...is where I like to go to for shopping very much | ...has no value for me at all | I do not like the area and prefer more modern parts of the city |
| ...is the commercial center of the city | ,214 | -,033 | ,048 | ,126 | ,052 | -,013 | ,274* | ,005 | -,079 |
| ...is of higher cultural value than any other part | | ,317** | ,303** | ,337** | ,305** | ,466** | ,008 | -,163 | -,126 |
| ...is the most beautiful part of the whole city | | | ,427** | ,499** | ,436** | ,205 | ,063 | ,087 | -,035 |
| ..represents our history more than any other part | | | | ,415** | ,440** | ,383** | ,040 | -,007 | -,057 |
| ...has a high recreational value for me | | | | | ,585** | ,303** | ,215 | -,053 | -,208 |
| ...or a part of it is of high spiritual/religious value for me | | | | | | ,249* | ,272* | ,052 | -,019 |
| ...must be preserved as it is | | | | | | | -,051 | -,255* | ,000 |
| ...is where I like to go to for shopping very much | | | | | | | | ,119 | -,079 |
| ...has no value for me at all | | | | | | | | | ,457** |

*p< 0,05; **p< 0.01

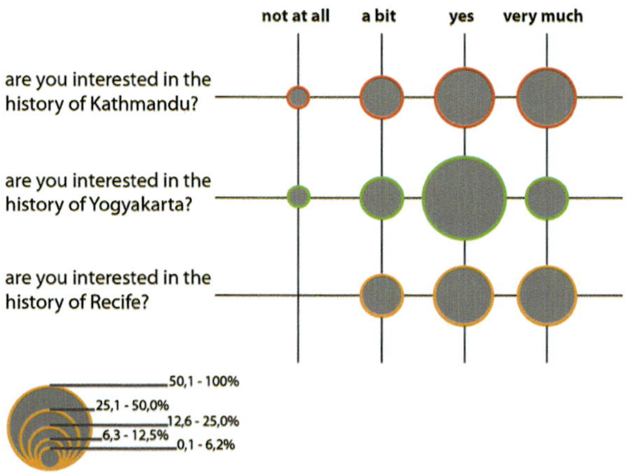


**Fig. A.3** History interest stated by the peer groups of the three case study cities

# Glossary

**Culture** "A patterned way of life shared by a group of people". *Source* Department of Economic and Social Affairs—Division for Social Policy and Development and Secretariat of the Permanent Forum on Indigenous Issues (2009: 52)

**Conservation/preservation** Maintaining a site in its original condition to the greatest extent possible, taking only measures that protect it from further damage, e.g. repairing a leaking roof. *Source* Alberts and Hazen (2010)

**Genius loci** 'Spirit of a place', impalpable but generally agreed upon unique character of a place a term already utilized in Roman culture to name the atmosphere and the uniqueness of a certain location, can be described as the typical nature or atmosphere of a place that leaves an impression. *Source* Brakman (2011), Lewicka (2008), Norberg-Schulz (1982)

**Place** A qualitative, 'total' phenomenon, which cannot be reduced to any of its properties, such as spatial relationships, includes perception and symbolic meaning. *Source* Freytag (2014), Norberg-Schulz (2013)

**Place attachment** An affective bond or link between people and specific places, forming a symbolic relationship, giving culturally shared emotional/affective meanings to a particular space or piece of land that provides the basis for the individual's or group understanding of and relation to the environment. *Source* Low (1992), Hidalgo and Hernández (2001)

**Place identity** "The process" of building and rebuilding meaning, in a space-time continuum, on the basis of emotive forces—the 'heart'—as formed and reformed by the flow of rational forces, the 'head'. *Source* Handal (2006: 51)

**Reconstruction** A new structure based on historic designs. *Source* Alberts and Hazen (2010)

**Restoration** Returning a structure to an earlier, often the original, state. In restoration work, keeping original elements in place, even when they are damaged, is preferable to replicating elements. *Source* Alberts and Hazen (2010)



S. Sandholz, *Urban Centres in Asia and Latin America*,
The Urban Book Series, DOI 10.1007/978-3-319-43735-4

**space** Geographically defined area, where objects & people can be located. *Source* Freytag (2014)

**Transformation** A process of change that involves the alteration of fundamental attributes of a system. *Source* Brown et al. (2013)

**Urban renewal/Urban regeneration** Action aiming at improving the physical, socialeconomic and ecological aspects of urban areas through various actions including redevelopment, rehabilitation, and heritage conservation, 'urban renewal' often used in similar meaning. *Source* Ercan (2011), Zheng et al. (2014)

# References


Dziomba M, Matuschewski A (2007) Grossprojekte in der Stadtentwicklung - Konfliktbereiche und Erfolgsfaktoren. disP 171(4):5–11

Freyre G (2007) Guia prático, histórico e sentimental da cidade do Recife. 5. ed. Ver., atual. e aum. edn. São Paulo, Global

Fundarpe (2015) Patrimônio Cultural - O que é. http://www.cultura.pe.gov.br/pagina/patrimonio-cultural/material/o-que-e/. Accessed 15 Sept 2015

Geddes P (1915) Cities in evolution: an introduction to the town planning movement and to the study of civics. Williams, London

Hengky SH (2015) Beneficial images: Batik handicraft tourism in Yogyakarta, Indonesia. Bus Econ Res 11–23. doi:10.5296/ber.v5i1.6760

Hilgert M (2015) Es sind auch unsere Wurzeln FAZ

His Majesty's Government of Nepal (1956) Ancient Monument Preservation Act, 2013 (1956). Kathmandu

ICOMOS (2004) The world heritage list: filling the gaps—an action plan for the future. ICOMOS, Paris

ICOMOS (2011) The valletta principles for the safeguarding and management of historic cities, towns and urban areas. ICOMOS. http://www.international.icomos.org/Paris2011/GA2011_CIVVIH_text_EN_FR_final_20120110.pdf. Accessed 28 Aug 2014

ICOMOS, Second International Congress of Architects and Technicians of Historic Buildings (1964) The venice charter: international charter for the conservation and restoration of monuments and sites ICOMOS. http://www.international.icomos.org/charters/venice_e.pdf. Accessed 28 Aug 2014

IDB (2011) City development. Experiences in the preservation of ten world heritage sites. Inter-American Development Bank, Washington D.C

IPHAN (2014) Livros do Tombo. Instituto do Patrimônio Histórico e Artístico Nacional. http://portal.iphan.gov.br/pagina/detalhes/608. Accessed 15 Sept 2015

Krieger A (2008) Dynamik und Management von Waterfront Development. In: „Waterfront Development" - Dokumentation der internationalen Städtekonferenz im Juni 2007 in Wien, Urania/TechGate Wien, 2008 2007. Stadtentwicklung Wien, pp 27–32

Leipzig Charter on Sustainable European Cities (2007) EU. http://ec.europa.eu/regional_policy/archive/themes/urban/leipzig_charter.pdf. Accessed 23 July 2014

Pavlova I, Makarigakis A, Depret T, Jomelli V (2015) Global overview of the geological hazard exposure and disaster risk awareness at world heritage sites. J Cult Herit. doi:10.1016/j.culher.2015.11.001

Rothfuß E, Dörfler T (2013) Raumbezogene Qualitative Sozialforschung. Konzeptionelle Überlegungen zwischen Geographie und Soziologie. In: Rothfuß E, Dörfler T (eds) Raumbezogene Qualitative Sozialforschung. Springer, Wiesbaden, pp 7–32


 
S. Sandholz, *Urban Centres in Asia and Latin America*,
The Urban Book Series, DOI 10.1007/978-3-319-43735-4

Roy A (2011) Postcolonial urbanism: speed, hysteria, mass dreams. In: Roy A, Ong A (eds) Worlding cities: Asian experiments and the art of being global. Studies in urban and social change. Wiley-Blackwell, Chichester, pp 305–335

The Economist (2013) The Bilbao effect. http://www.economist.com/news/special-report/21591708-if-you-build-it-will-they-come-bilbao-effect

The Getty Conservation Institute (2015) Cultural heritage policy documents. http://getty.edu/conservation/publications_resources/research_resources/charters.html. Accessed 01 Oct 2015

UNESCO (2006) Climate change and world heritage. UNESCO. Accessed 31 Aug 2014

UNESCO (2015c) Lists of intangible cultural heritage and register of best safeguarding practices. http://www.unesco.org/culture/ich/index.php?lg=en&pg=00011&multinational=3&type=00002&display1=inscriptionID&display=maps#tabs. Accessed 09 Oct 2015

UNESCO (2015d) Number of world heritage properties by region. http://whc.unesco.org/en/list/stat#d1. Accessed 09 Oct 2015

World Urbanisation Prospects: The 2014 Revision, CD-ROM Edition (2014) United Nations, Department of Economic and Social Affairs, Population Division. http://esa.un.org/unpd/wup/CD-ROM/Default.aspx. Accessed 05 Jan 2015

Weichhart P (2009b) Multilokalität - Konzepte, Theoriebezüge und Forschungsfragen. Informationen zur Raumentwicklung 1/2:1–14

World Bank (2014b) Doing business 2015: going beyond efficiency. World Bank, Washington, DC. doi:10.1596/978-1-4648-0351-2

World Commission on Environment and Development (1987) Our common future. Oxford University Press, Oxford

# Index




S. Sandholz, *Urban Centres in Asia and Latin America*,
The Urban Book Series, DOI 10.1007/978-3-319-43735-4